Fresh-Cut Fruits and Vegetables

Fresh-Cut Fruits and Vegetables

Editor

Andrick Cuevas

scitus academics

Fresh-Cut Fruits and Vegetables
Edited by **Andrick Cuevas**

ISBN: 978-1-68117-702-1
Library of Congress Control Number: 2016958741

SCITUS Academics LLC,
616, Corporate Way, Suite 2, 4766,
Valley Cottage, NY 10989

www.scitusacademics.com

Notice

Printed in United States of America on Acid Free Paper

Preface

Fruit and vegetable production and consumption globally have shown a marked upward trend over the past several years. Rising consumer demand in the world has come with greater awareness of food safety issues and increased need for convenience and quality. The fresh-cut produce sector has responded to these demands, and is currently at different stages of development across the globe. Assuring the safety and quality of fresh-cut produce necessitates the selection of high quality horticultural produce for processing, and the implementation of good practice during processing operations in order to maintain produce quality and assure safety of the final product. Today's society is characterized by an increasing health consciousness and growing interest in the role of food for maintaining and improving human well-being and consumer health. In addition to their nutritional and sensory properties, foods are currently recognized as active and protective agents. In the last years, their consumption has continued to grow rapidly linked to the increased public awareness of their health benefits. On the other hand, this new trend, has also hiked the chances of outbreaks of food poisoning and food infection related to consumption of fresh fruit and uncooked salad.

Fresh-Cut Fruits and Vegetables highlights on the use of emerging technologies to ensure microbiological stability and quality of fresh-cut fruits and vegetables. In particular, the aim of the content is to give an insight into the current knowledge on the technologies mostly suggested to ensure microbiological stability and quality of fresh-cut fruits and vegetables, in order to marketing and supply information about this subject which remains still a major challenge for the food industry. The book provides helpful guidelines to the industry for minimizing deterioration, keeping the overall quality, and lengthening the shelf life. Technology, physiology, quality, and safety of fresh-cut fruits and vegetables are emphasized.

It is well known that processing of fruit promotes a faster physiological deterioration, biochemical changes and microbial degradation of the products which may result in degradation of its color, texture and flavor, even when only slight processing operations are used. Prior to being packaged for consumption, minimally processed fruits are subjected to one or more mild unit operations, which include washing / sanitizing, peeling, cutting and/or slicing, dicing, shredding, etc. A special attention is necessary for mechanical operations considered very critical to delimit the shelf life of fresh-cut fruit commodities, causing the rupture of many cells and the liberation of intracellular products (such as oxidizing enzymes); at the same time, the surface of produce is exposed to air and to possible contamination with bacteria, yeasts and molds. It is important to underline that cutting increases the area of injured tissues favoring elevated respiration, promoting further rapid deterioration and microbial proliferation.

Table of Contents

Chapter 1

Environmental Friendly Sanitation to Improve Quality and Microbial Safety of Fresh-Cut Vegetables

Ji Gang Kim[1]

[1] *National Institute of Horticultural and Herbal Science, Rural Development Administration, Suwon, Korea*

1. INTRODUCTION

Fresh-cut products have a limited shelf-life due to rapid deterioration caused by microbial growth as well as physiological disorder. Cutting of fruits and vegetables increases microbial spoilage of fresh-cut produce through transfer of microflora on the outer surface to the interior tissue where microorganisms have access to nutrient-laden juice (Das & Kim, 2010a). Washing with sanitizer is an important step in reducing the microbial population and quality deterioration. The use of chemical compounds to extend postharvest life of fruit and vegetables has become lesser accepted by consumers since these compounds may be contaminant of the environment or harmful to human health. Therefore, the proper application of different sanitizing agents should be highly optimized to guarantee a minimal number of spoilage microorganisms.

Chlorine has been widely used in produce washes in order to inactivate microorganisms and ensure quality and safety. However, increasing public health concerns about the possible formation of chlorinated organic compounds and the emergence of new more tolerant pathogens, have raised doubts in relation to the use of chlorine by the fresh-cut industry (Kim, 2007). The use of chlorine as a sanitizing agent is prohibited in some countries due to the hazardous byproducts formed by chlorine with process water and other organic matters. Consequently, sanitization of fresh vegetables with chlorine in the industry renders a negative impact to the environment and human health as a whole. Recently, to avoid chlorine the use of non-chemical sanitizers in fresh-cut industries is becoming the more popular trend universally. Therefore, the industry is searching for alternative environment-friendly sanitizing methods to maintain the quality of fresh-cut produce at the best level.

Ozone is a strong antimicrobial agent with high reactivity, penetrability and spontaneous decomposition to a non-toxic product (Khadre et al., 2001; Kim et al., 2007a). Ozone is found in natural form in the atmosphere or it can be produced by generators. Ozone as an aqueous disinfectant was declared to be generally recognized as safe (GRAS) for food contact applications. Ozone's primary advantages include fast decomposition in water to oxygen, no residue, and improved microbial reduction efficacy against bacteria and fungal spores than hypochlorite. Ozone forms oxidated radicals in the presence of water that penetrate and act on cell membranes. The use of ozonated water has been applied to fresh-cut vegetables for sanitation purposes reducing microbial populations and extending the shelf-life of some of these products (Beltran et al., 2005; Hassenberg et al., 2007; Kim et al., 2007a). Ozone has been declared in many countries to have potential use for food processing including sanitation of fresh and fresh-cut vegetables. When compared to chlorine, ozone treated with optimum conditions has a greater effect against certain microorganisms and rapidly decomposes to oxygen, leaving no residues (Kim, 2007; Rico et al., 2007). Inactivation of microorganism by ozone is a complex process that attacks various cell membrane and wall constituents (e.g. unsaturated fats) and cell content constituents (e.g. enzymes and nucleic acids). The micro-organism is killed by cell envelope disruption or disintegration leading to leakage of the cell contents. Disruption or lysis is a faster inactivation mechanism than that of other disinfectants which require the disinfectant agent to permeate through the cell membrane in order to be effective (Kim et al., 1999). Bacteria are more sensitive than yeasts and fungi. Gram-positive bacteria are more sensitive to ozone than Gram-negative organisms and spores are more resistant than vegetative cells (Das & Kim, 2010b; Pascual, 2007).

Electrolyzed water (EW), the second most popular sanitizer in Korea (Kim, 2008) is considered as an environment-friendly sanitizer compared to chlorine. However, there are different opinions on the effect of EW among industrial users. EW generated by adding NaCl to pure water from non-diaphragm system is one of the major EW systems in Korea. This electrolytic process facilitates the conversion of chlorine oxidants (Cl_2, $HClO/\ OCl^-$) which are effective for inactivating a variety of microorganisms (Kim, 2007; Yang et al., 2003). The bactericidal effect of EW have been evaluated on several fresh-cut vegetables such as lettuce, carrots, spinach, and cucumber (Izumi, 1999; Nimitkeatkai & Kim, 2009). Acidic EW has a strong bactericidal effect against pathogens and spoilage microorganisms due to its low pH, high oxidation reduction potential (ORP) and the presence of residual chlorine (Kim, 2007). Using acidic EW resulted in moderate control of aerobic bacterial growth during storage of fresh-cut cilantro (Wang et al., 2004). Acidic EW was also tested for its efficacy in inactivating Salmonella on fresh-cut produce. However, the concentration of acid used can influence the organoleptic quality of vegetables, i.e., loss of texture (Kim, 2007; Zhang & Farber, 1996). Electrolyzed water at high pH (pH 6.8, 20 mg/L available chlorine) was tested as a disinfectant and the research found that it did not affect tissue pH, surface color, or general appearance of fresh-cut vegetables (Izumi, 1999).

Heat treatments such as hot water and hot air are non-chemical methods that have been used to control microorganism and senescence-related symptoms of fresh produce. Recently, mild heat treatment as physical technology to extend shelf-life of fresh and fresh-cut produce have become of interest. Heat treatment using hot water was used for fungal and insect control, but has been extended to improve the storage quality of fresh-cut produce. Hot water dips to control both decay and quality change of fresh and fresh-cut produce are often applied for 30 seconds to a few minutes at temperatures of 40-60 C (Kim, 2007; Kim et al., 2011). Heat treatments combined with other agents have also been used to prevent the microbial quality, browning, and maintaining texture in various vegetables (Das & Kim, 2010a). A combination of heat treatment followed by calcium dip has also been applied for the primary purpose of controlling postharvest pests and/or diseases and has been found to have very good results in maintaining or improving the texture of various horticultural products. Combined heat treatments with UV-C were applied to fresh-cut processed broccoli (*Brassica oleracea* L.) florets to investigate their effects on several quality and senescence parameters. Heat treatments and UV-C radiation have been utilized to extend storage quality of fresh-cut products (Lemoine et al., 2008).

Acidity is a commonly used factor to control the growth of microorganisms in foods. Organic acids such as citric acid and ascorbic acid have been applied for preserving physicochemical qualities (Rosen & Kader, 1989) and preventing microbial growth at levels that did not adversely affect taste and flavor (Yildiz, 1994). Therefore, organic acids could be a potential sanitizer for fresh-cut vegetables. However, different studies have shown that the inhibitory or bactericidal effect depends on the characteristics of the acid used to adjust the medium pH (Buchanan et al., 1993; Eswaranandam et al., 2004; Parish & Higgins, 1989). Application of organic acids as sanitizers at higher concentration can reduce the overall quality and produce off-flavor in leafy vegetables after few days of storage (Kim 2007; Chandra & Kim, 2011). The acid tolerance of fresh-cut vegetables varies among different microorganisms and products. Organic acid alone treatment is not successful sanitation in controlling pathogens and maintaining food quality during storage. Hurdle technology or combined technology, which involves simultaneous multiple preservation approaches, is generally better to control both microbial safety and food quality of fresh-cut produce.

Natural antimicrobial agents derived from fruit, herb, and shell have been investigated as preservatives. The interest in the possible use of natural alternatives to food additives to prevent bacterial and fungal growth has notably increased. Edible coatings containing natural antimicrobial agents are gaining importance as potential treatments to reduce the deleterious effects imposed by fresh-cut processing on fresh-cut fruits. However, application of natural edible coatings for fresh-cut vegetables has not received interest as much as fresh-cut fruits. Essential (volatile) plant oils occur in edible, medicinal and herbal plants which minimizes questions regarding their safe use in food products. Essential oils and their constituents have been widely used as flavouring agents in foods since the earliest recorded history and it is well established that many have wide spectra of antimicrobial action (Holley & Patel, 2005). No fresh-cut company used natural antimicrobial agent to wash or preserve fresh-cut produce in Korea due to less effects than chemical sanitizers and non-economic efficiency. However, these days natural agents are good candidate to replace tap-water washing because these agents have more potency against microorganisms. Natural agents can be used for washing microgreens and organic fresh-cut produce which are sold at a higher price. The growing demand for fresh and fresh-cut produce by consumers had led to the need for natural food preservation methods such as the use of natural antimicrobials and their combination with other hurdles, without adverse effects on the consumer or the food itself (Tiwari et al., 2009).

Although a wide range of different microbial agents are available for sanitizing fresh-cut produce, their efficacies vary and none are able to ensure elimination of pathogen completely without compromising sensory quality. In addition, recent studies have shown that chlorine lacks efficacy on pathogen reduction; the formation of the chlorine by-products are also deleterious to human health. Thus, there is much interest in developing a safer and more environmental friendly antimicrobial alternative to chlorine. Ozone, electrolyzed water, mild heat treatment, organic acid, and natural antimicrobial agents, or the combination of those sanitizing methods have been applied to various fresh-cut vegetables. It is generally accepted that an ideal sanitizing agent should have two important properties: a sufficient level of antimicrobial activity and a negligible effect on the sensory quality of the product.

2. OZONATED WATER WASHING

Ozonated water washing is getting more popular now-a -days due to its high biocidal efficacy, wide antimicrobial spectrum and environment friendly. Research has shown that treatment with ozone appears to have a beneficial effect in extending the storage life of fresh produce such as cucumber, apples, grapes, oranges, pears, raspberries and strawberries by reducing microbial populations and by oxidation of ethylene (Kim, 2007). The effect of ozonated waters with different concentrations and contact times on the quality attributes and microbial population of fresh-cut produce were studied. Two types of ozone generators were used to investigate the efficacy of microbial reduction and quality maintenance of fresh-cut lettuce, cilantro, carrot, broccoli, and paprika. One aqueous ozone solution was prepared by continuously circulating the water through an ozone generator and a stainless steel water tank. The circulating type ozone generator was equipped with a vortexer to facilitate dissolving of gaseous ozone in the water, and a de-gassing system to remove the undissolved ozone. The other ozone solution was prepared by flowing ozonated water into plastic bucket through an ozone generator. The flowing type ozone generator was equipped with a cylinder used as compression tank to facilitate high concentration of ozone. Both circulating type and flowing type ozone solutions were used immediately after the required ozone concentration were reached.

2.1. Sanitation With Low Ozone Concentration (≤1 Ppm)

Ozonated water using circulating type with low ozone concentration (less than 1 ppm) and insufficient contact time was not much effective compared to 100 ppm chlorine in reducing microbial population and maintaining quality of fresh-cut cilantro, iceberg lettuce, romaine lettuce, and baby leaves. Fresh-cut cilantro was washed in tap water, 100 ppm chlorine solution (pH 7), and 0.7 ppm ozonated water for 1 minute separately. The initial total aerobic plate count (APC) on the unwashed cilantro leaves was 6.45 log CFU/g. There was a significant decrease in APC between washed cilantro and unwashed sample after washing. However, no significant difference was found in microbial reduction of fresh-cut cilantro between tap water and ozonated water throughout 6 days storage at 5 C. The chlorine treatment maintained a low level of microbial count compared to other treatments. Fresh-cut romaine lettuce, spinach, microgreens, and baby leaves sanitized with 0.5-0.8 ppm of ozonated water had higher microbial population compared to samples washed in 50-100 ppm of chlorine. Ozonated water sanitation with low ozone concentration (less than 1 ppm) is not inadequate to be used in fresh-cut industry practically because the ozonated water was not effective in microbial decontamination and maintaining storage quality of fresh-cut vegetables (Kim, 2007).

Continuous flowing type ozonated water containing 1 ppm of ozone concentration was also used to sanitize fresh-cut iceberg lettuce if the sanitation could get much effect in reducing microbial population. The ozonated water was compared with tap water and 100 ppm chlorine solution (pH 6.5). During storage at 5 C, there was a significant increase in APC among all treatments. The highest numbers of APC and coliform plate count (CPC) were observed in tap water washing followed by ozone washing. Chlorine treatment had the most reduction on microbial population on fresh-cut iceberg lettuce throughout storage. In fresh-cut lettuce, cut edge browning commonly occurs during storage making unsuitable for consumers. Discoloration occurred in all fresh-cut iceberg lettuces on day 6. Samples washed in ozonated water had lower degree of cut edge browning index score than samples washed in tap water. However, chlorine solution showed the lowest degree of cut edge browning. Though ozonated water containing 1 ppm of ozone concentration was effective in delaying discoloration and reducing microbial population the effectiveness was lower compared to 100 ppm chlorine solution (Kim, 2007).

2.2. Sanitation With High Ozone Concentration (>1 Ppm)

2.2.1. Washing fresh-cut vegetables with 1.5 ppm of ozone

Flowing ozonated water washing was also used to sanitize 'Tah Tasai' Chinese cabbage baby leaves and fresh-cut romaine lettuce. Those fresh-cut products were washed in tap water, 100 ppm chlorine (pH 7.0), and continuous flow of 1.5 ppm ozonated water for 2 minutes separately. Samples treated with the ozonated water had lower APC compared to those washed in tap water. Ozonated water containing 1.5 ppm of ozone concentration reduced APC on fresh-cut baby leaves and romaine lettuce by 0.3 and 0.6 log CFU/g, respectively, on day 0. On the other hand, 100 ppm chlorine solution treatment reduced APC on fresh-cut baby leaves and romaine lettuce by 0.5 and 0.9 log CFU/g on day 0. Fresh-cut produce washed in ozonated water had lower microbial population than samples washed in tap water until middle period of 9 days-storage at 5 C. Ozonated water washing containing 1.5 ppm of ozone concentration and washed for 2 minutes was not sufficient to decontaminate microorganism of fresh-cut romaine lettuce and baby leaves as much as effectiveness of 100 ppm chlorine solution washing. Therefore, it has been required to find optimum ozonated water washing conditions for improving storage quality and microbial food safety of each fresh-cut product to apply to fresh-cut industry as a chlorine alternative.

2.2.2. Washing fresh-cut carrots with initial 2 ppm ozone

Higher ozone concentration with longer contact time has been applied to get much effect in reducing microbial population and maintaining quality of fresh-cut produce. Fresh-cut carrot shreds were washed in tap water for 1 minute, 50 ppm chlorinated water (pH 6.3) once or two times for each 1 minute, or initial 2 ppm ozonated water using circulating type at varying times (1, 5, and 20 minutes). The samples were then centrifuged to remove excess water, packaged in 50 µm PE film bags, and stored at 5 C. Different ozonated water washing time affected microbial growth, off-odor development, color, and overall quality of carrot shreds. A single chlorine wash and 20 minutes ozonated water wash treatments had lower APC and lactic acid bacteria compared to other washing treatments until 2 weeks storage (Fig. 1). The 20 minutes ozone treatment reduced APC on carrot shreds by1.4 and 1.1 log CFU/g on week 1 and week 2, respectively.

Ozonated water washing for 20 minutes maintained quality by inhibiting off-odor and high overall quality score due to less whiteness development (Fig. 2). The single chlorine water wash was effective and resulted in better quality compared to two time chlorine water wash. However, samples washed for 20 min in ozonated water had better quality with less off-odor and higher overall visual quality scores than samples washed in chlorine water washed once. The efficacy of optimum ozonated water washing on microbial reduction and quality of those fresh-cut produce was similar to chlorine or better than chlorine. Ozonated water containing initial 2 ppm ozone concentration with sufficient washing time would be effective in reducing microbial population and maintaining quality of fresh-cut carrot and could be an alternative method to maintain quality and shelf-life of fresh-cut carrot shreds (Kim et al., 2007a).

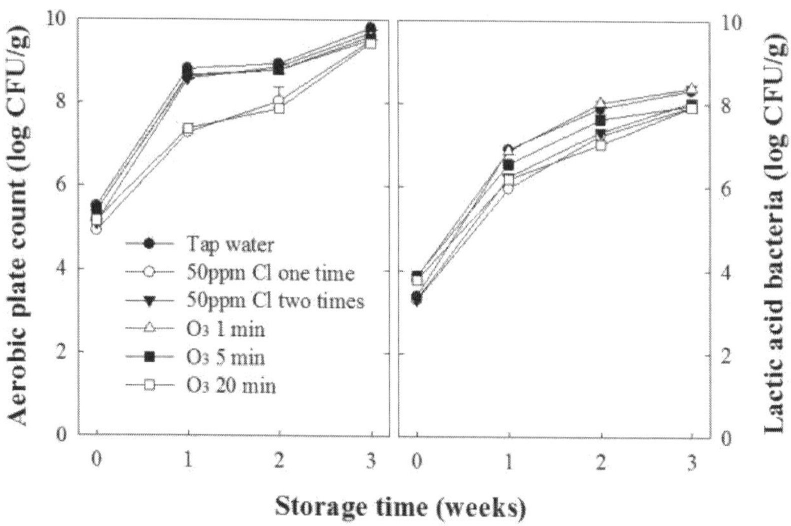

Figure 1. Aerobic plate count and lactic acid bacteria of fresh-cut carrot shreds washed in different sanitizers and stored at 5ºC for up to 3 weeks.

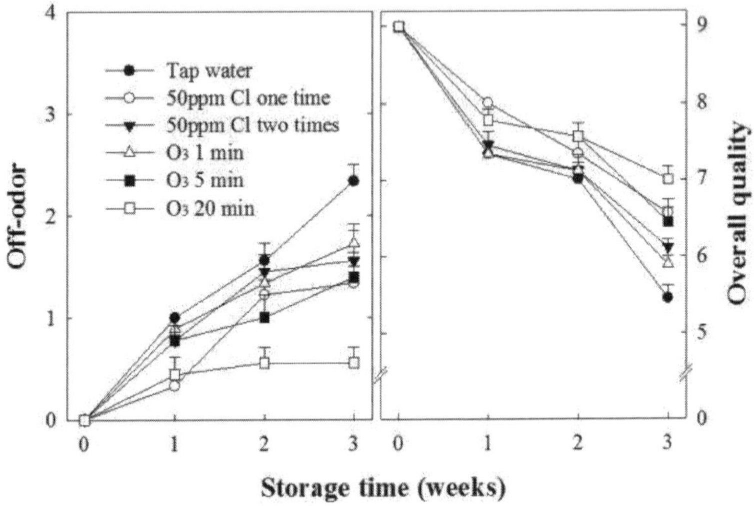

Figure 2. Off-odor development and Overall quality of fresh-cut carrot shreds washed in different sanitizers and stored at 5ºC for up to 3 weeks.

2.2.3. Washing fresh-cut broccoli with 2 ppm ozone

Ozonated water washing using flow type with different contact times on storage quality and microbial growth in fresh-cut broccoli was conducted to compare ozone with chlorine. Fresh-cut broccoli samples were washed each for 90 and 180 seconds in normal tap water, 100 ppm chlorinated water (pH 7), and 2 ppm ozonated water separately and respectively. Then, samples were packaged in 30 µm polyethylene bags and stored at 5ºC for 9 days. No significant differences were observed in gas composition and color among different sanitizers with contact times. No off-odor was detected during the 9 days storage. Sanitizers affected microbial population of fresh-cut broccoli. In the color characteristics no difference were marked in L* and a* value and hue angle of the samples among different washing solutions and contact times during the storage period. It was found that electrical conductivity increased with the longer contact time in all washing solutions compared to shorter contact time. Electrolyte leakage is generally considered as an indirect measure of plant cell membrane damage. Ozonated water washing for 180 seconds contact time initially showed highest electrical conductivity probably due to its highly oxidizing nature than chlorine and tap water washing, but at the end of the storage the value is low and nearly equal to the above washings (Fig.3, left). It may be due to quality maintenance without texture damage or decay.

Electrical conductivity is relatively high immediately after fresh-cut processing and decrease rapidly, then either decrease gradually or remain relatively stable until the samples have good quality in many fresh-cut produce (Kim et al., 2005a;Kim, 2007). This typical response pattern to processing and storage is similar to the result of fresh-cut broccoli sanitized in ozonated water and stored for 9 days at 5ºC. No color difference was found among the treatments during 9 days storage (Fig. 3, right).

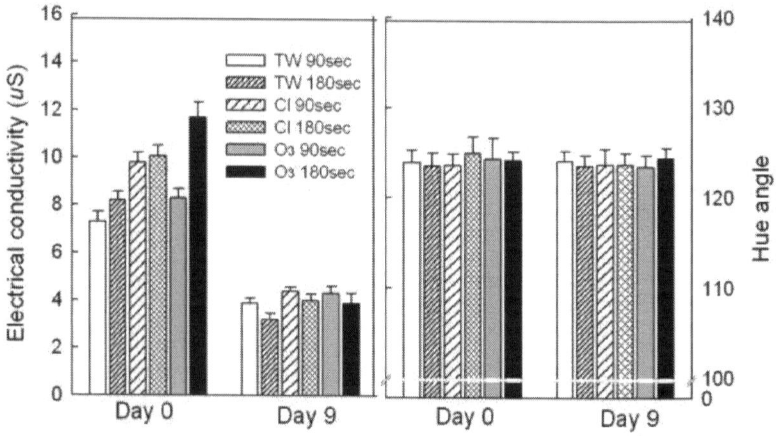

Figure 3. Electrical conductivity and hue angle of fresh-cut broccoli washed in different sanitizers and stored for 9 days at 5ºC. In figure, samples washed for 90 and 180 seconds in tap water (TW), 100 ppm chlorinated water (Cl), and 2 ppm ozonated water (O_3).

Among the sanitizers, ozonated water with 180 seconds maintained the lowest numbers of aerobic plate count throughout the storage days in comparison with others (Fig. 4). Ozonated water with 90 seconds was not much effective in reducing microbial population compared to chlorine. However, samples washed with ozonated water for 180 seconds showed the lowest coliform count. Absolutely no coliform were observed in ozonated water with 180 seconds washing treatment on day 0. The result reveals that longer contact time of ozone affects positively whereas other sanitizers don't affect on the microbial quality and safety aspects of fresh-cut broccoli. The difference of microbial population is may be due to the following causes; the surface wash off might be attached to the samples again during washing time, higher electrical conductivity observed during storage and the reactivity against pathogen may be less effective in comparison with 2 ppm of ozone. Ozone effectiveness against microorganisms depends not only on the amount applied, but also on the effectiveness of ozone delivery method, type of material, the target microorganisms, physiological state of the bacteria cells at the time of treatment (Das & Kim, 2010b).

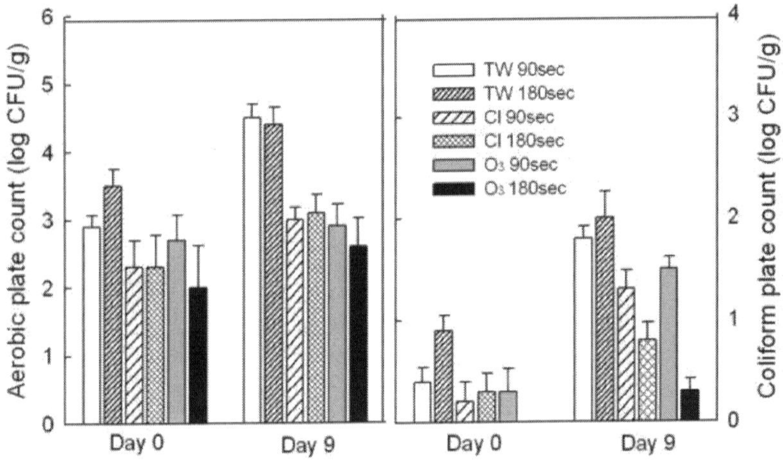

Figure 4. Aerobic plate count and coliform plate count of fresh-cut broccoli washed in different sanitizers and stored for 9 days at 5ºC. In figure, samples washed for 90 and 180 seconds in tap water (TW), 100 ppm chlorinated water (Cl), and 2 ppm ozonated water (O_3).

2.2.4. Washing fresh-cut vegetables with 3 to 4 ppm ozone

Ozonated water treatment containing 3 ppm of ozone with 3 minutes washing reduced APC and coliform/E. Coli count in both fresh-cut paprika and iceberg lettuce, similar to 100 ppm chlorine at day 0 (Table 1). Treatment with the ozonated water showed the lowest numbers of aerobic plate count and coliform/E. Coli count on day 6 in fresh-cut paprika. There was no difference in quality parameters such as color, off-odor, and visual quality among treatments. The highest numbers of aerobic and coliform/E. Coli were observed in tap water washing. Fresh-cut broccoli washed in 4ppm ozonated water for 5 minutes had lower microbial populations than samples washed in 100 ppm chlorine.

Table 1. Microbial population (log CFU/g) of fresh-cut paprika and iceberg lettuce washed in different sanitizers for 3 minutes, packaged, and stored for 6 days at 5ºC.

Sanitizer	Fresh-cut paprika				Fresh-cut iceberg lettuce			
	APC		Coliform/E. Coli		APC		Coliform/E. Coli	
	Day 0	Day 6	Day 0	Day 6	Day 0	Day 6	Day 0	Day 6
Tap water	2.3	5.3	1.5	1.2	3.6	5.3	0.8	0.9
100ppm Chlorine	2.0	4.3	0.5	1.0	3.1	4.7	0.2	0.7
3ppm Ozone	2.1	3.5	0.7	0.7	3.0	4.7	0.2	0.7

No quality deterioration or side effects of higher ozone concentration were found in fresh-cut broccoli. These results showed that ozonated water reduced microbial growth more effectively than 100 ppm chlorine solution. Therefore, ozonated water washing with optimum ozone concentration and sufficient contact time could be a favorite alternative sanitation to chlorine.

3. ELECTROLYZED WATER

Strong acidic EW (pH2.7) and weak acidic EW (pH 5-6.5), which were generated by electrolysis of NaCl solution and HCl, respectively have been used as disinfectant in fresh-cut industry. Weak alkaline EW (pH 7.5-8) from non-diaphragm EW generator, recently developed is getting popular among three types of EW. In general, strong acidic EW had stronger bactericidal effect compared to alkaline EW which has high pH levels. However, strong acidic EW can cause tissue damage in some fresh-cut produce, especially leafy vegetables during storage or distribution. Little information exists on the efficacy of weak alkaline or weak acidic EW on quality and microbial reduction in fresh-cut produce.

3.1. Washing Fresh-Cut Lettuce, Sesame Leaf, And Strawberry

The effect of strong acidic and weak alkaline EW containing 80ppm available chlorine concentration as well as general chlorine solution on storage quality and microbial growth of fresh-cut iceberg lettuce has been studied. The effectiveness of strong acidic EW on microbial reduction was greater than weak alkaline EW at initial storage. However, strong acidic EW affected quality deterioration due to texture damage after 6 days at 10ºC (Table 2). Weak alkaline EW reduced off-odor development and was as effective as chlorine in inhibiting total aerobic bacterial and coliform group on fresh-cut iceberg lettuces (cultivar; *U-lake*). Fresh-cut sesame leaves washed in weak alkaline and neutral EW had less total plate counts and better sensorial quality compared to samples washed in chlorine and strong acidic EW. The result reveals that, weak acidic EW affects positively whereas strong acidic EW was not effective in maintaining quality. Strong acidic EW may be used for sanitation of fruits which has firm texture like apple. Nimitkeatkai & Kim (2009) reported that apples washed in strong acidic EW for 5 minutes had less hue angle value throughout storage period and lower sensory evaluation score at the end of storage. However, apples washed in strong acidic EW for 2 min was effective in reducing microbial growth and maintaining sensorial quality of apples. Different result was found that microbial quality of strawberry (cultivar; *Maehyang*) washed in weak alkaline EW (pH 8.0, 60ppm HClO) was inferior compared to samples washed in chlorine (pH 6.5, 60ppm HClO). These results indicated that EW does not affect quality maintenance and microbial safety for all fresh and fresh-cut products. Though weak alkaline EW did not show the effectiveness in strawberry weak alkaline EW could be an effective alternative to chlorine for washing fresh-cut leafy vegetables.

3.2. Washing Fresh-Cut Broccoli

Weak alkaline EW was also used to sanitize fresh-cut broccoli. The fresh-cut samples were washed for 90 seconds in tap water, 80 ppm chlorinated water (pH 7), and EW (pH 7.2) containing 80 ppm free chlorine separately and respectively. Then, samples were packaged in 30 µm polyethylene bags and stored at 5ºC for 9 days. No significant differences were observed in gas composition and color among different sanitizers. No off-odor was detected during the storage. Samples washed with EW showed the lowest total aerobic bacterial population and coliform count. The result reveals that weak alkaline EW affects microbial population of fresh-cut broccoli positively.

Table 2. Gas composition, aerobic plate count, and quality of fresh-cut iceberg lettuce washed in different sanitizers and stored at 10ºC for 6 days

Treatment	Gas composition (%)		Total plate count (log CFU/g)	Off-odorz	Discolorationz
	O2	CO2			
Chlorine	0	25.8 b	7.0 a	3.3 b	0.5 b
Weak alkaline EW	0	31.2 a	7.3 a	3.8 a	1.8 a
Strong acidic EW	0	23.8 b	7.1 a	2.7 c	0.7 b

z 0 = none, 1 = slight, 2 = moderate, 3 = severe, 4 = strong

Table 3. Electrical conductivity and microbial population of fresh-cut broccoli washed in different sanitizers and stored at 5ºC for 9 days.

	Electrical conductivity		Aerobic plate count		Coliform plate count	
	Day 0	Day 9	Day 0	Day 9	Day 0	Day 9
Tap water	6.1	4.2	4.28	5.52	1.39	2.20
Chlorine	8.9	4.5	3.62	4.61	-	1.89
Electrolyzed water	8.2	4.7	3.53	4.24	0.22	1.02

3.3. Combined Ew Washing With Ma Packaging

Study on effect of combined EW washing with modified atmosphere (MA) packaging was carried out to investigate the influence of the combined treatment on quality maintenance and microbial food safety of fresh-cut iceberg lettuce. Fresh-cut iceberg lettuce were washed in alkaline EW (free chlorine 80 ppm), dried, and packaged with 35 µm P-Plus film to compare with conventional technology using combination of chlorine sanitation and vacuum packaging. Samples for control treatment were prepared following industrial practices; 100 ppm chlorine (100ppm, pH 7.5) wash and vacuum packaging with 80μm Ny/PE film. Combined EW and MA technology reduced off-odor development of packaged fresh-cut iceberg lettuce during storage (Table 4). The combined technology using EW washing and MA packaging was as effective as control using chlorine in inhibiting total aerobic bacterial counts on fresh-cut iceberg lettuces. The combined EW washing and MA packaging extended two more days of shelf-life of fresh-cut iceberg lettuce compared to control treatment.

Table 4. Gas compositi on, aerobic plate count, and quality of fresh-cut iceberg lettuce treated with hurdle technology and stored at 5ºC for 12 days.

Treatment	Gas composition		Aerobic plate count (log cfu/g)	Off-odor*	Dis-coloration*
	O2(%)	CO2(%)			
Chlorine + Vacuum pack.	0	19.0	6.4	2.3	0.6
EW + MA packaging	1.8	11.5	6.2	0.7	0.9

* 0 = none, 1 = slight, 2 = moderate, 3 = severe, 4 = strong

4. HEAT TREATMENT

Heat treatments is one of postharvest treatments that has been used to control postharvest decay and; or to improve the storage quality of fresh-cut produce. Heat treatments alone or combined with other agents have also been used to prevent the microbial quality, browning, and maintaining texture in various fresh-cut vegetables.

4.1. Heat Treatment For Fresh-Cut Winter Squash And Lotus Roots

The effectiveness of heat treatment for fresh-cut winter squash and lotus roots has been applied. Winter squash which had hard rinds can be treated with hot water at 60-65 C for 2-3 minutes to reduce microbial contamination before it is fresh-cut processed (Arvayo-Ortiz et al., 1994; Hawthorne, 1989). However, other vegetables which have soft texture are not recommended to treat at high temperature like winter squash. Winter squash (cultivar; *Bouzang*) harvested in summer season was immersed in 65 C hot water for 2 minutes before it is fresh-cut processed. The heat treatment reduced microbial population and maintained good quality with bright yellow or orange with fine, moist texture and high solids, and sugar contents. The changes in the quality of fresh-cut lotus roots treated with hot water were investigated. Lotus roots washed, peeled, and sliced with 1cm-thickness was dipped in water at 30, 55, and 80ºC for 45 seconds. Then, samples were air-dried at room temperature, packed in polyethylene films, and stored at 4ºC for 12 days. Generally, the weight loss of the lotus roots that were treated with hot water slightly increased. The application of the heat treatment delayed the browning of the lotus roots, especially the treatment with 55ºC hot water. The Hunter color 'L' and 'a' values of the lotus roots treated with 80ºC hot water significantly increased during their storage. The heat treatment effectively inhibited the growth of mesophilic microorganisms. Therefore, the organoleptic quality of the lotus roots that were treated with 55ºC hot water was the best among those temperatures (Chang et al., 2011).

4.2. Mild Heat Treatment For Peeled Potato

For peeled potato, the most popular method of retarding surface browning used in the Korean industry is vacuum packaging that induces high CO_2 and low O_2 levels. However, the presence of high $CO_2$and low O_2 concentrations may cause off-odor development due to anaerobic respiration. Hence, heat treatment methods that reduce browning and off-odor development were investigated. Potatoes (var. '*Jopung*') kept at 5 C after harvest were heat treated (30 C for 24 hours, 45 C for 3hours, or non-heated), peeled, and immersed in tap water at 5 C for 3hours.

Table 5. Effect of heat treatment before peeling on gas composition and quality of peeled potato fresh-cut processed and vacuum packaged, and stored at 5ºC for 5 days.

Treatment	Gas composition (%)		Color		Off-odor z
	O2	CO2	Lightness (L)	Redness(a)	
Control	0.67 b	34.7 a	69.8 b	-1.61 a	3.2 a
Heat treatment at 30°C for 24 hours	2.74 a	16.8 b	71.2 a	-2.47 c	2.5 b
Heat treatment at 45°C for 3 hours	0.56 b	32.4 ab	70.3 ab	-2.15 b	1.7 c
z 0 = none, 1 = slight, 2 = moderate, 3 = severe, 4 = strong					

Samples were then vacuum-packaged with 80µm Ny/PE film and stored at 10 C for up to 5 days. Mild heat treatment (30 C) was effective in reducing CO_2 concentrations and off-odor development in the package of samples throughout storage. The 30 C mild heat treatment also delayed browning of peeled potatoes and maintained the highest overall quality score. The mild heat treatment at 30ºC before peeling can be a practical method to delay browning and off-odor development of 'Jopung' potato (Kim et. al, 2009).

4.3. Combined Heat Treatment For Fresh-Cut Paprika

The combined effect of washing solutions and heat treatment were investigated as potential sanitizers for maintaining the quality and microbial safety of fresh-cut paprika. Fresh paprika shreds were washed in tap water and 1% calcium chloride combined with 19ºC (normal tap water) and 50ºC water temperature (heat treatment) for 2 minutes. Then, samples were packaged in 30 µm polypropylene bags and stored at 5ºC for 12 days. No significant differences were observed in color and gas composition of the package among treatments and no off-odor was detected until the end of 12 days storage. However, 50ºC water temperature with calcium chloride had lower microbial numbers upto the storage period in comparison with tap water (Fig. 5). The result reveals that 50 ºC water temperature with calcium chloride can be used as an washing solution to maintain the microbial quality in fresh-cut Paprika (Das & Kim, 2010a).

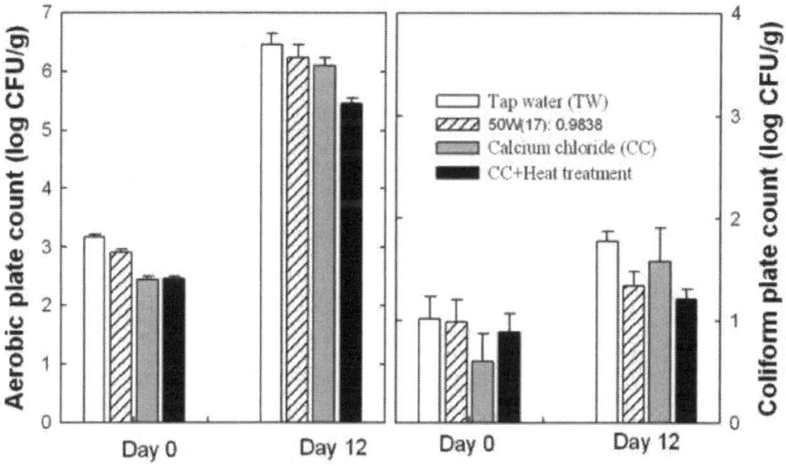

Figure 5. Aerobic plate count and coliform plate count of fresh-cut paprika during the storage period. In figure, samples washed for 90 seconds in tap water (TW) and calcium chloride solution at 5ºC.

Softening, textural changes are one of the main causes of quality losses in case of fresh-cut products. In general, fresh-cut vegetables that maintain firm and crunchy textures are highly desirable. Though there is no significant difference in gas composition and color among treatments calcium chloride and heat treatment tended to increase firmness of fresh-cut paprika during the beginning of the storage compared to tap water washing treatment. It is well known that calcium plays a major role in maintaining the quality of fruit and vegetables. Increasing the calcium content in the cell wall of fruit tissue can help delay softening of fresh-cut produce. The beneficial effects obtained with heat treatments have generally been explained in terms of pectin esterase activation. Calcium dips have been employed to improve firmness and extend the postharvest shelf-life of a wide range of fruit and vegetables. Similarly, certain commercial additives can maintain the quality of fresh-cut products (Encarna et al., 2008; Luna-Guzman & Barrett, 2000). The changes in electrical conductivity of fresh-cut paprika depend upon the type of washing solution and heat treatment (Fig. 6). Tap water washing treatments (both TW and TW + Heat treatment) showed lower electrolyte leakage compared to calcium chlorine washing during the entire storage period. A combination of heat treatment followed by calcium dip may need more research if it can be applied for the purpose of controlling postharvest pests and/or diseases and have very good results in maintaining or improving the texture of on various fresh-cut produce.

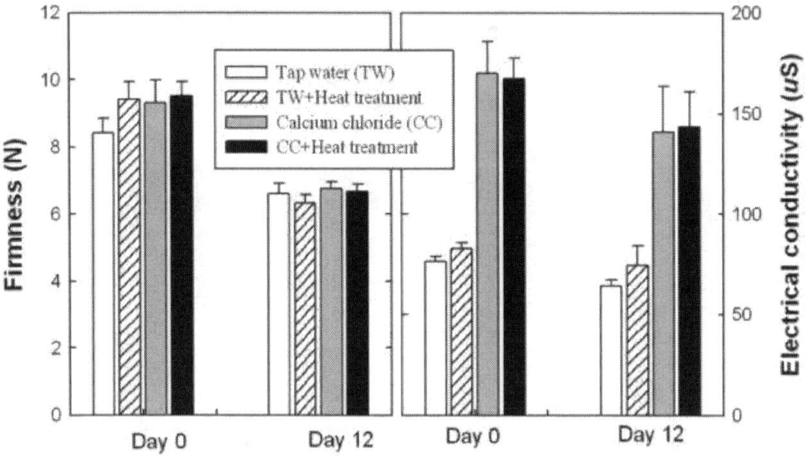

Figure 6. Firmness and electrical conductivity of fresh-cut paprika heat treated and stored at 5ºC for 12 days. In figure, samples washed for 90 seconds in tap water and calcium chloride solution at 5ºC.

5. ORGANIC ACID AND ACID COMPOUNDS

Organic acid and acid compound sanitizers have been used to sanitize fresh and fresh-cut produce. Organic acids are one of the important sanitizers that have been applied largely for preserving physicochemical qualities and for preventing microbial growth in many fresh-cut products. Organic acid with optimum condition did not adversely affect taste and flavor, but leaving no effect on environment. Citric acid can be used to extend the shelf life of fresh-cut produce by reducing the loss of eating quality and disease development. Ibrahim et al. (2009) reported that leaves of some selected vegetables decontaminated with 5% citric acid showed a considerable decrease in microbial count compared to water washing. However, the application of these acids at higher concentration may cause quality deterioration due to off-odor and texture damage in some fresh-cut leafy vegetables. Sequential treatment of citric acid and ethanol on the quality and microbial reduction of organic vegetables has also been examined. Hence, organic acid and the combined technology using have been carried out to find an alternative sanitizer to chlorine.

5.1. Washing Fresh-Cut Lettuce With Organic Acid And Combined Acid

Fresh iceberg lettuce leaves were sanitized separately with tap water, 100 μL L^{-1} chlorine, 0.2% citric acid, 50% ethanol, and the combination of citric acid solution and 50% ethanol spray. Samples were then dried with centrifugal dryer, packaged in 80μm Ny/PE films, and stored for 6 days at 5 C. The 50% ethanol solution dipping was the most effective treatment to reduce microbial population of fresh-cut iceberg lettuce (Fig. 7). However, fresh-cut iceberg lettuce sanitized in ethanol solution had severe injury with lowest visual quality score and highest electrical conductivity among treatments after 6 days storage (Fig. 8). The decline of overall visual quality in ethanol treated sample might be a consequence of tissue damage as reflected from the electrical conductivity data. Citric acid alone was not effective in reducing microbial population, similar to 100ppm chlorine solution treatment (Fig. 7). The combination of citric acid and ethanol spray reduced aerobic microbial population by 1.1 log CFU/g as compared to tap water. The combination with citric acid and ethanol spray also maintained good quality with high overall quality score at the end of 6 days storage (Fig. 8). Therefore, the combination of citric acid and ethanol spray could be an alternative to chlorine as an environment-friendly sanitizer for washing fresh-cut leafy vegetables (Kim et al., 2011).

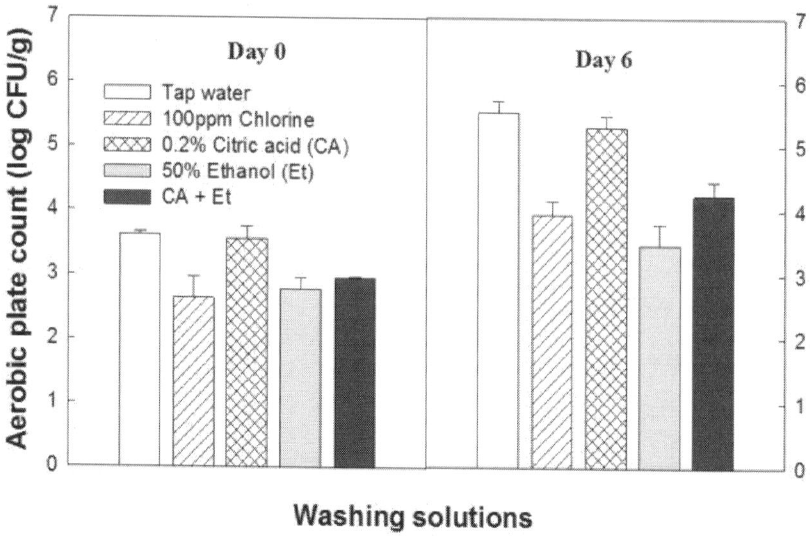

Figure 7. Aerobic plate count of fresh-cut iceberg lettuce treated with different sanitizers and stored at 5 ºC.

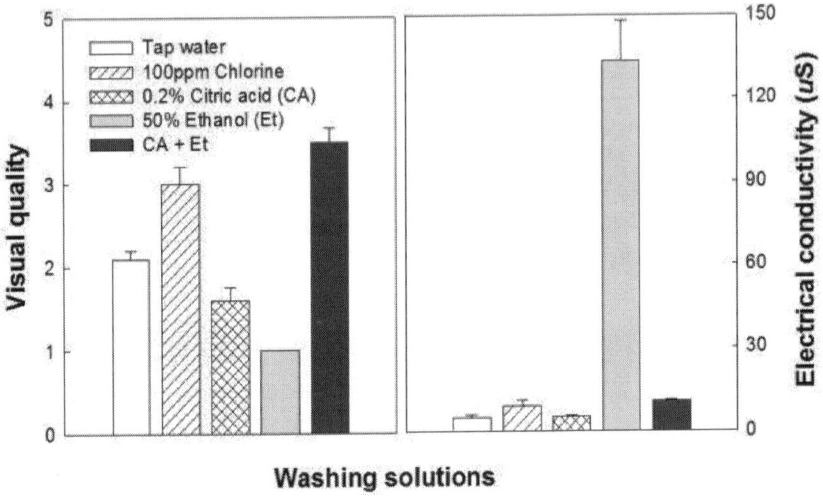

Figure 8. Visual quality and electrical conductivity of fresh-cut iceberg lettuce treated with different sanitizers and stored for 6 days at 5 ºC.

5.2. Washing Fresh-Cut Spinach, Baby Leaves, And Microgreens With The Combined Organic Acid

Fresh organic vegetables such as spinach, 'Tah Tasai' Chinese cabbage baby leaves, and microgreens were also sanitized separately with tap water, 100 ppm chlorine, 0.2% citric acid (CA), and the sequential treatment of 0.2% CA solution and 50% ethanol spray (CA+Et). In case of spinach, chlorine and CA+Et increased CO_2 partial pressures in the headspace of sample packages and generally had higher electrical conductivity compared to tap water. No significant differences were observed in color among different sanitizers during storage at 5 C in fresh-cut spinach samples. The chlorine and CA+Et treatments were effective in reducing microbial population of fresh-cut spinach. However, CA+Et treatment induced off-odor of microgreens resulting more aerobic plate count compared to chlorine treatment. In case of microgreens, samples treated with CA+Et did not have good quality score, worse than score of chlorine at the end of storage probably due to severe texture damage. In 'Tah Tasai' Chinese cabbage baby leaves, sanitizer chlorine treatment showed lower number of total aerobic count immediately after washing. However, citric acid in combination with ethanol spray treatment showed the lowest number until day 7. No significant difference was found in microbial number among the treatments at the end of 10 days storage. Wang et al. (2004) also reported that no significant difference in total aerobic plate count at the end of 14 days storage of cilantro leaves. The possible reason might be due to the baby leave samples became softer with the progress in storage which caused damage in texture for all samples. Citric acid and ethanol, on the other hand, both are used as anti-microbial agents leaving no effect to the environment. Their combined use was almost similarly effective as of chlorine possibly due to the dual sanitization effects on the sample used.

5.3. Acid Compound Sanitizers

5.3.1. Use of acidified sodium chlorite and peroxyacetic acid-based sanitizer

Acid compound sanitizers such as acidified sodium chlorite (ASC) and peroxyacetic acid-based sanitizer (PA) have been used for food safety of fresh and fresh-cut produce. Peroxyacetic acid is a strong oxidizing agent that has been used extensively to disinfect food processing equipment and has been approved by the U.S. Food and Drug Administration (FDA) as a disinfectant for fruits and vegetables (Gonzalez et al., 2004). Recent studies undertaken to determine the suitability of PA (Tsunami, Ecolab, USA) for washing fresh-cut vegetables showed it to be effective against *Listeria monocytogens*, *Salmonella* spp., and *E. coli* O157:H7 (Beuchat et al., 2004; Gonzalez et al., 2004). Acidified sodium chlorite has also been approved by the FDA for spray or dip application on various food products, including fresh and fresh-cut produce (Kim et al., 2007b). Studies have shown that ASC and PA have a strong antimicrobial efficacy against various human pathogens inoculated onto cantaloupes and asparagus (Kim 2007; Park & Beuchat, 1999). However, the effect of both ASC and PA on quality attributes such as texture and color of fresh-cut produce was not investigated with an in-depth study. It was found that carrot shreds washed in higher PA or ASC concentrations (200 and 500 ppm, respectively) lost firmness and looked melted during storage. There was a significant difference in firmness between 200 and 30 ppm PA concentration or 500 and 30 ppm ASC concentration. Therefore, high PA and ASC concentration may cause faster deterioration of carrot shreds because of inferior texture.

5.3.2. Washing fresh-cut carrot shreds

Fresh-cut shredded carrots were washed in tap water, 100 ppm chlorinated water, 30 ppm ASC, or 30 ppm PA. Samples were then packaged in 35 *u*m polyethylene film and stored for 14 days at 5 C. Sanitizers affected off-odor, skin whitening, and microbial population of carrot shreds. Chlorine and PA wash maintained a lower level of microbial count compared to water or ASC treatments. Though no significant difference was found in total aerobic count between chlorine and PA until day 11, samples treated with PA had a lower

APC than samples washed in chlorine at the end of storage. Chlorine caused the highest reduction (0.75 log), followed by PA (0.65 log) and ASC (0.56 log) at day 0. No significant difference in total aerobic count was found between ASC and water treatment after day 4. Therefore, 30 ppm ASC treatment was not effective in reducing microbial organism of fresh-cut carrot shreds during storage. The reason may be due to low ASC concentration (30 ppm). It was also found that 500 ppm ASC concentration was more effective in reducing microbial numbers than 50 ppm chlorine wash water. However, 500 ppm ASC concentration resulted in visual deterioration of the carrot shreds. Kim et al. (2006) reported that 30 ppm PA was effective in reducing microbial contamination as well as 100 ppm chlorine concentration. This would suggest that the residual effect of acetic acid released when PA is degraded causes reduced growth in microflora (Gonzalez et al. 2004).

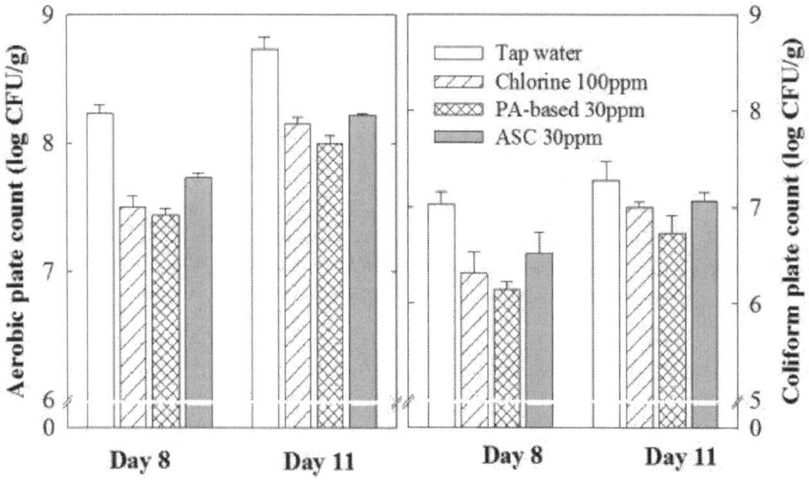

Figure 9. Aerobic plate count and coliform plate count of fresh-cut carrot shreds. In figure, samples washed for 2 minutes in tap water, chlorinated water, peroxyacetic acid-based sanitizer (PA), and acidified sodium chlorite (ASC).

Peroxyacetic acid-based sanitizer also maintained initial color values, inhibited off-odor development and skin whitening of samples, and achieved the highest overall quality score among those sanitizer treatments. Off-odor was detected in all samples on day 8 and day 11 (Fig. 10, left). Off-odor was lower in samples treated with PA than in any other treated samples. Fresh-cut produce is known to develop undesirable off-odors under low O_2 and elevated CO_2 atmospheres (Kim et al., 2005a). Carrot shreds treated with water, chlorine, or ASC reached score 2.0 (the limit of marketability) at the end of storage, whereas samples PA-treated had score 1.9 after 11 days of storage. In fresh-cut produce, patterns of off-odor development correlate with ethanol and acetaldehyde and there is also a strong relationship between package atmospheric conditions and off-odor development (Kim et al., 2005b). The degree of off-odor of fresh-cut carrot shreds may be influenced by fermentation due to anaerobic microorganism. Surface whitening which is one of major postharvest quality problems in carrot was significantly retarded by applying sanitizer PA. Samples treated with PA maintained inherent color and exhibited the lowest rate of increase in whitening scores (Fig. 10, right). Surface discoloration during storage is most detrimental to the quality of shredded carrots. The possible reasons for whiteness development on carrot surface are dehydration and lignification (Kim et al., 2006). Visual observation of carrot shreds treated with PA showed a moister surface compared to other sanitizer treatments. The PA-treated samples had the highest overall quality score, with relatively low levels of whitening. Therefore, sanitizer PA treatment significantly affected quality and shelf-life of fresh-cut shredded carrots. At present, chlorine is the most practical, efficient, and low cost disinfectant available. Due to concerns about the formation of by-products, however, a safer alternative is needed. Peroxyacetic acid-based sanitizer treatment resulted in comparable antimicrobial effectiveness and sensory quality of carrot shreds throughout storage.

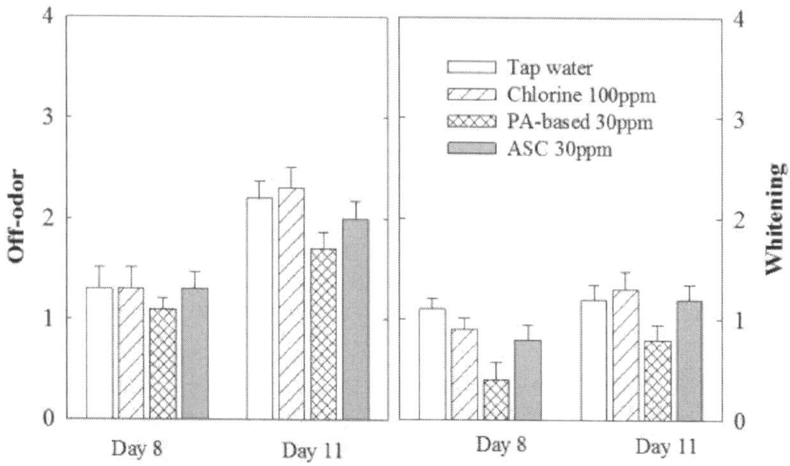

Figure 10. Off-odor development and surface whitening of fresh-cut carrot shreds after storage at 5ºC for 8 and 11 days. In figure, samples washed for 2 minutes in tap water, chlorinated water, peroxyacetic acid-based sanitizer (PA), and acidified sodium chlorite (ASC).

6. NATURAL ANTIMICROBIAL AGENTS

Natural compounds can serve as carriers for a wide range of food additives, including anti-browning agents, colorants, and antimicrobials that can extend product shelf-life and reduce the risk of pathogen growth on fresh-cut produce surface. In recent years there has been a considerable pressure by consumers to reduce or eliminate chemically synthesized additives in foods. Plants and plant products can represent a source of natural alternatives to improve the shelf-life and the safety of food. In fact, they are characterised by a wide range of volatile compounds, some of which are important flavour quality factors (Patrignani et al., 2008; Utama et al., 2002). A key role in the defence systems of fresh produce against decay microorganisms has been attributed to the presence of some of these volatile compounds (Patrignani et al., 2008). However, no plant volatiles have been used as natural antimicrobial agent for fresh-cut produce practically. Recently developed natural antimicrobial agents from marine resource product have been applied to fresh-cut vegetables.

6.1. Sanitation Fresh-Cut Lettuce With Calcinated Calcium

The heated scallop shell powder; calcinated calcium (CC) was investigated as potential sanitizers for maintaining storage quality and microbial safety of fresh-cut iceberg lettuce. Samples were washed in normal tap water, 50 ppm chlorinated water (pH 6.5), 1.5 g L^{-1} CC for 2 minutes separately. Samples were then packaged in 80 μm nylon/polyethylene bags and stored at 5ºC. The initial aerobic plate count of unwashed iceberg lettuce was 6.5 log CFU g^{-1}. The aerobic plate count on fresh-cut lettuce increased with storing time, reaching 6.05 to 7.05 log CFU g^{-1} on 12 days-storage. Washing in CC was effective in reducing aerobic plate count of fresh-cut lettuce samples by 0.4 to 1.0 log CFU g^{-1} as well as chlorine treatment throughout storage as compared to tap water (Fig. 11, left). Electrical conductivity of all samples decreased during the initial period of storage, remained stable thereafter or increased slightly at the end of storage (Fig. 11, right). Electrical conductivity of fresh-cut lettuces increased after 8 days. Electrical conductivity is generally considered as an indirect measure of plant cell membrane damage and deterioration of fresh-cut vegetables (Jiang et al., 2001; Kim et al., 2005b). Increased electrical conductivity after fresh-cut processing is a common phenomenon due to the leakage from cut ends of the samples or otherwise wounded tissues.

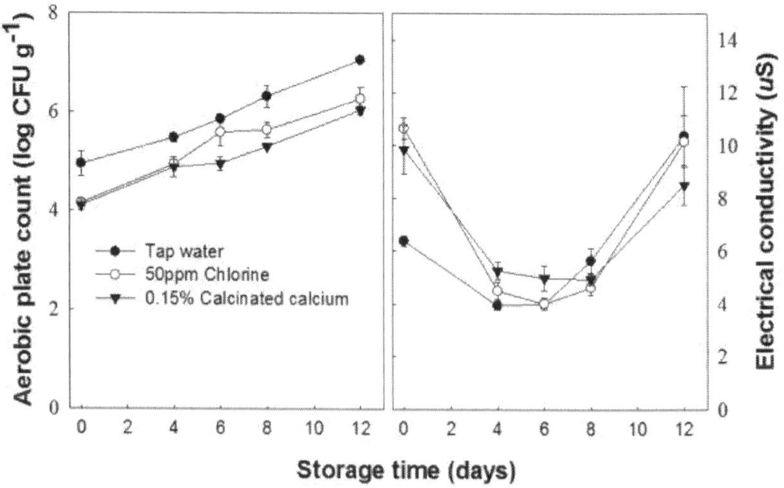

Figure 11. Aerobic plate count and electrical conductivity of fresh-cut iceberg lettuce sanitized with different washing solutions (tap water, chlorinated water, and calcinated calcium) and stored at 5ºC for 12days.

Samples treated with CC had good quality with low off-odor at the end of storage. Visual quality score of all fresh-cut iceberg lettuce samples was lower than score 3, which was considered the limit of marketability at the end of 12 days-storage (Fig. 12, left). The visual quality related to browning was probably induced by relatively high O_2 and low CO_2 concentration in sample packages. Off-odor was first detected in fresh-cut lettuce samples treated with TW after 6 days and increased relatively until the end of storage. Off-odor of fresh-cut lettuce samples washed in chlorine and CC was lower than tap water (Fig. 12, right). Samples sanitized with chlorine or CC reached score 1.3 which was lower score than the limit of marketability on the 12 days-storage. Fresh-cut produce is known to develop undesirable off-odors under low O_2 and elevated CO_2 atmospheres (Kim, 2007). In fresh-cut lettuce, patterns of off-odor development correlated with ethanol and acetaldehyde and there was also a strong relationship between package atmospheric conditions and off-odor development (Kim et al., 2005a;Kim et al 2005b). The degree of off-odor in the packaged fresh-cut lettuce was influenced by subsequently fermentation due to anaerobic microorganism.

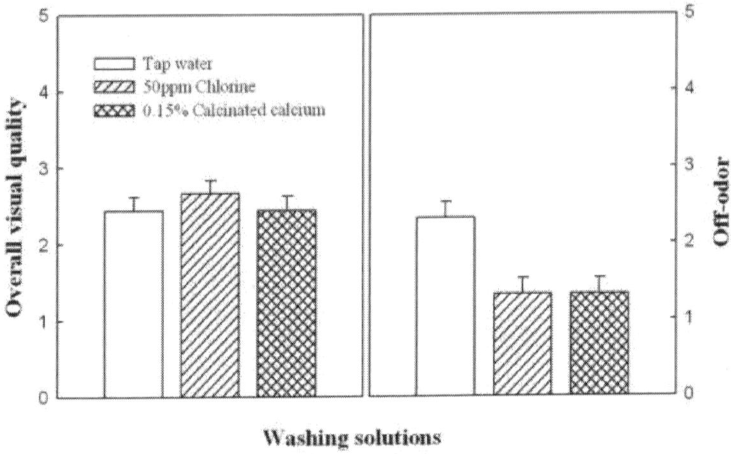

Figure 12. *Overall visual* quality and off-odor development of fresh-cut iceberg lettuce sanitized with different washing solutions (tap water, chlorinated water, and calcinated calcium) after 12 days storage at 5ºC.

6.2. Sanitation Fresh-Cut Bok Choi And Broccoli With Calcinated Calcium

Natural materials, calcinated calcium and fruit extract compound from Japanese apricot were used as sanitizer to maintain quality and reduce microbial population of bok choi with different maturities. Microgreen, baby leaf, and mature bok choies were washed in tap water, 50 ppm chlorine, and 500 fruit extract compound, for 2 min separately. Those samples were then packaged in 50 µm PE film and stored at 5 C for 6 days. One of natural compounds, the fruit extract compound was not effective significantly in reducing microbial population and quality such as off-odor. However, samples treated with CC had better quality with less off-odor until 4 to 6 days-storage in baby leaf and mature Bok choi. Calcinated calcium affected in reducing microbial population of microgreen, baby leaf, and mature Bok choi for 2, 4, and 6 days, respectively (Fig. 13). Bok choi micrggreens had highest microbial population, followed baby leaves, and mature samples in terms with maturity. Therefore, mature and baby leaf Bok choi samples can have 6 and 4 days of shelf-life with CC sanitation, respectively. Fresh-cut broccoli was also washed in CC at normal tap water temperature. Broccoli samples sanitized in CC solution had good quality with lower off-odor and microbial count at the end of 9 days-storage (Kim et al., 2010). To avoid chlorine which may lead to the formation of carcinogenic compounds, CC can be used as environmental friendly sanitizer and an alternative to chlorine washing for fresh-cut broccoli without affecting microbial and sensorial quality.

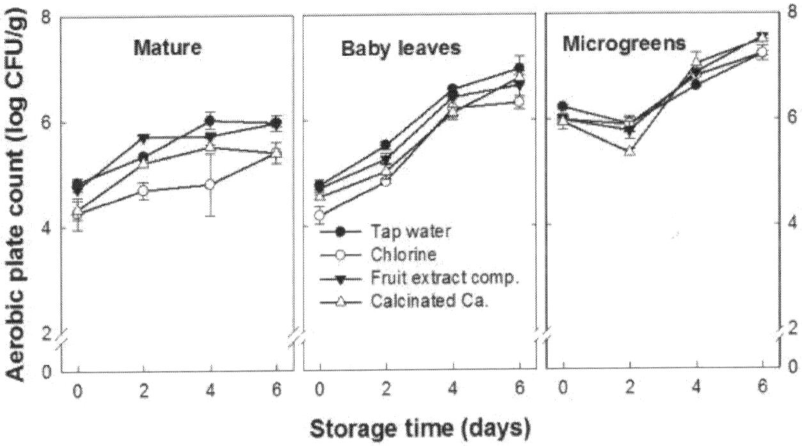

Figure 13. Aerobic plate count of mature, baby leaf, and microgreen fresh-cut bok choi samples sanitized with different washing solutions (tap water, chlorine, fruit extract compound, and calcinated calcium).

7. CONCLUSION

Use of chlorine to reduce microbial populations of fresh-cut vegetables has faced with challenges to find alternatives which are more environmental friendly and not harmful to human health. Wide ranges of different agents are available for sanitizing fresh-cut produce, their efficacies vary and none are able to ensure elimination of pathogen completely without compromising sensory quality. Therefore, application of different environment-friendly sanitizing agents has been conducted to investigate highly optimized condition to guarantee a minimal number of spoilage microorganisms in many fresh-cut vegetables. As alternatives to chlorine, ozone, electrolyzed water, mild heat treatment, organic acid with ethanol, and natural antimicrobial agents, or the combination of those sanitizing methods can be used for fresh-cut vegetables. But, selection of washing solution and use of optimum condition to meet each fresh-cut vegetable should be performed to get similar or better efficacy to chlorine. Ideal sanitizing agent should have effectiveness in two important properties: antimicrobial activity and sensory quality of the fresh-cut product.

For organic fresh-cut leafy vegetables which are facing challenges to find the means to extend shelf-life and to enhance microbial food safety, combined citric acid with ethanol spray or calcinated calcium alone solution can be used in fresh-cut industry. Those environment friendly agents are good candidate to replace tap water washing or organic acid solution because these agents have more potency against microorganisms. Heat treatment without chemical use can be used for washing of fresh-cut produce which have firm texture such as winter squash, potato, lotus roots, and paprika. To get much effectiveness in reducing microbial population of fresh-cut vegetables constant 1~4 ppm of ozonated water and week alkaline or weak acidic electrolyzed water can be used practically. However, the concentration of ozone and contact time is very important for microbial safety of fresh-cut vegetables such as broccoli, iceberg lettuce, carrot, etc. Calcinated calcium, a natural and an environment-friendly sanitizer can be an alternative to mild heat treatment for washing of fresh-cut vegetables without affecting sensorial quality.

REFERENCES

1. R. M. Arvayo-Ortiz, S. Garza-Ortega, E. M. Yahia, 1994 Postharvest response of Winter squash to hot-water treatment, temperature and length of storage. HortTechnology 4 253 255 .

2. D. Beltran, M. V. Selma, A. Marın, M. I. Gil, 2005 Ozonated water extends the shelf life of fresh-cut lettuce, Journal of Agricultural and Food Chemistry 53 5654 5663 .

3. L. R. Beuchat, B. B. Alder, M. M. Lang, 2004 Efficacy of chlorine and a peroxyacetic acid sanitizer in killing Listeria monocytogenes on on iceberg and romaine lettuce using simulated commercial processing conditions. Journal of Food Protection. 67 6 1238 1242 .

4. R. L. Buchanan, M. H. Golden, R. C. Whiting, 1993 Differentiation of the effects of pH and lactic or acetic acid concentration on the kinetics of Listeria monocytogenes inactivation. Journal of Food Protection, 56, 474e478.

5. M. S. Chang, J. G. Kim, G. H. Kim, 2010 Quality characteristics of fresh-cut lotus roots according to the temperature of the wash water. Korean Journal of Food Preservation. 18 3 288 293

6. D. Chandra, J. G. Kim, 2011 Effects of different sanitizers on the quality of'Tah Tasai' Chinese cabbage (Brassica campestris var. narinosa) baby leaves. Korean Journal of Food Preservation. In press

7. B. K. Das, J. G. Kim, 2010a Combined effect of heat treatment and washing solutions on the quality and microbial reduction of fresh-cut paprika. Horticulture, Environment, and Biotechnology. 51 4 257 261

8. B. K. Das, J. G. Kim, 2010b Microbial quality and safety of fresh-cut broccoli with different sanitizers and contact times. Journal of Microbiology and Biotechnology. 20 2 363 369

9. A. Encarna, H. E. Victor, A. Francisco, 2008 Effect of hot water treatment and various calcium salts on quality of fresh-cut'Amarillo' melon. Postharvest Biology and Technology. 47 397 406 .

10. S. Eswaranandam, N. S. Hettiarachchy, M. G. Johnson, 2004 Antimicrobial activity of citric, lactic, malic or tartaric acids and nisin-incorporated soy protein film against Listeria monocytogenes, Escherichia coli O157:H7 and Salmonella gaminara. Journal of Food Science, 69, 79e84.

11. R. J. Gonzalez, Y. Luo, S. Ruiz-Cruz, J. L. McEvoy, 2004 Efficacy of sanitizers to inactivate Escherichia coli O157:H7 on fresh-cut carrot shreds under simulated process water conditions. Journal of Food Protection. 67 11 2375 2380

12. K. Hassenberg, C. Idler, E. Molloy, M. Geyer, M. Plöchl, J. Barnes, 2007 Use of ozone in a lettuce-washing process: an industrial trial. Journal of the Science of Food and Agriculture. 87 (5), 914 919 .

13. B. T. Hawthorne, 1989 Effects of cultural practices on the inidence of storage rots in Cucurbita spp. New Zealand Journal of Crop Horticultural Science. 17 49 54 .

14. R. A. Holley, D. Patel, 2005 Improvement in shelf-life and safety of perishable foods by plant essential oils and smoke antimicrobials. Food Microbiology 22 2005 273 292

15. T. A. Ibrahim, B. S. Jude-ojei, E. O. Giwa, V. T. Adebote, 2009 Microbiological analysis and effect of selected antibacterial agents on microbial load of fluted pumpkin, cabbage and bitter leaves. Research Journal of Agriculture and Biological Sciences., 3 1143 1145

16. H. Izumi, 1999 Electrolyzed water as a disinfectant for fresh-cut vegetables. Journal of Food Science. 64 536 539 .

17. Y. Jiang, L. Pen, J. Li, 2004 Use of citric acid for shelf life and quality maintenance of fresh-cut Chinese water chestnut. Journal of Food Engineering. 63 325 328

18. M. A. Khadre, A. E. Yousef, J. G. Kim, 2001 Microbiological aspects of ozone applications in food: a review, Journal of Food Science 66 (9), 1242 1252.

19. J. G. Kim, A. E. Yousef, G. W. Chism, 1999 Applications of ozone for enhancing the microbiological safety and quality of foods: a review, Journal of Food Protection, 62 (9), 1071 87 .

20. J. G. Kim, 2007 Fresh-cut produce industry and quality management. Semyeong Press, Suwon, Republic of Korea. 978-8-99578-871-4

21. J. G. Kim, 2008 Quality maintenance and food safety of fresh-cut produce in Korea. The Asian and Australasian Journal of Plant Science Biotechnology. 2 1 6 .

22. J. G. Kim, S. T. Choi, D. H. Pae, 2009 Effect of heat treatment and dipping solution combination on the quality of peeled potato'Jopung'. Korean Journal of Horticultural Science Technology. 27(2):256 262 , 2009

23. J. G. Kim, Y. Luo, R. A. Saftner, K. C. Gross, 2005a Delayed modified atmosphere packaging of fresh-cut Romaine lettuce: Effects on quality maintenance and shelf-life. Journal of American Society for Horticultural Science. 130 1 116 123

24. J. G. Kim, Y. Luo, Y. Tao, R. A. Saftner, K. C. Gross, 2005b Effect of initial oxygen concentration and film oxygen transmission rate on the quality of fresh-cut Romaine lettuce, Journal of the Science of Food and Agriculture. 85 10 1622 1630

25. J. G. Kim, Y. Luo, C. H. Lim, 2007a Effect of ozonated water and chlorine water wash on the quality and microbial de-contamination of fresh-cut carrot shreds. Korean Journal of Food Preservation. 14 1 54 60

26. J. G. Kim, Y. Luo, Y. Tao, 2007b Effect of the sequential treatment of 1-MCP and acdified sodium chlorite on microbial growth and quality of fresh-cut cilantro. Postharvest Biology and Technology., 46 144 149

27. J. G. Kim, H. Nimitkeatkai, J. W. Choi, S. R. Cheong, 2011 Calcinated calcium and mild heat treatment on storage quality and microbial populations of fresh-cut iceberg lettuce, Horticulture, Environment, and Biotechnology. 52 4 408 412

28. J. G. Kim, H. Nimitkeatkai, B. K. Das, 2010 Effect of calcinated calcium washing solution and heat treatment on quality and microbial reduction of fresh-cut lettuce and broccoli. ACTA Horticulturae.875 237 242

29. J. G. Kim, K. F. Yaptenco, C. H. Lim, 2006 Effect of sanitizers on microbial growth and quality of fresh-cut carrot Shreds. Horticulture, Environment, and Biotechnology. 47 6 313 318

30. M. L. Lemoine, P. M. Civello, A. R. Chaves, G. A. Mart´ınez, 2008 Effect of combined treatment with hot air and UV-C on senescence and quality parameters of minimally processed broccoli (Brassica oleracea L. var. Italica). Postharvest Biology and Technology 48 15 21

31. I. Luna-Guzman, D. M. Barrett, 2000 Comparison of calcium chloride and calcium lactate effectiveness in maintaining shelf stability and quality of fresh-cut cantaloupes. Postharvest Biology and Technology. 19 61 72 .

32. H. Nimitkeatkai, J. G. Kim, 2009 Washing Efficiency of Acidic Electrolyzed Water on Microbial Reduction and Quality of'Fuji' Apples. Korean Journal of Horticultural Science and Technology. 27 2 250 255

33. M. E. Parish, D. P. Higgins, 1989 Survival of Listeria monocytogenes in low Ph model broth systems. Journal of Food Protection, 52, 144e147.

34. C. M. Park, L. R. Beuchat, 1999 Evaluation of sanitizers for killing Escherichia coli O157:H7, Salmonella and naturally occurring microorganisms on cantaloupes, honeydew melons, and asparagus. Dairy Food Environment and Sanitation.19 842 847 .

35. A. Pascual, I. Llorca, A. Canut, 2007 Use of ozone in food industries for reducing the environmental impact of cleaning and disinfection activities. Trends in Food Science & Technology. 18 (1): S29 S35.

36. F. Patrignani, L. Iucci, N. Belletti, F. Gardini, M. E. Guerzoni, R. Lanciotti, 2008 Effects of sub-lethal concentrations of hexanal and 2-(E)-hexenal on membrane fatty acid composition and volatile compounds of Listeria monocytogenes, Staphylococcus aureus, Salmonella enteritidis and Escherichia coli. International Journal of Food Microbiology. 123 1 8

37. D. Rico, A. B. Martín-Diana, J. M. Barat, C. Barry-Ryan, 2007 Extending and measuring the quality of fresh-cut fruit and vegetables: a review. Trends in Food Science & Technology. 18 7 373 386 .

38. J. Rosen, A. Kader, 1989 Postharvest physiology and quality maintenance of sliced pear and strawberry fruits. Journal of Food Science. 54 656 659

39. B. K. Tiwari, V. P. Valdramidis, C. P. O'Donnell, K. Muthukumarappan, P. Bourke, P. J. Cullen, 2009 Application of natural antimicrobials for food preservation. Journal of Agricultural and Food Chemistry, 57 5987 6000 .

40. I. M. S. Utama, R. B. H. Willis, S. Ben-Yehoshua, C. Kuek, 2002 In vitro efficacy of plant volatiles for inhibiting the growth of fruit and vegetable decay microorganisms. Journal of Agricultural and Food Chemistry 50 6371 6377 .

41. H. Wang, H. Feng, Y. Luo, 2004 Microbial reduction and storage quality of fresh-cut cilantro washed with acidic electrolyzed water and aqueous ozone. Food Research International. 37 949 956 .

42. H. Yang, B. L. Swem, Y. Li, 2003 The effect of pH on inactivation of pathogenic bacteria on fresh-cut lettuce by dipping treatment with electrolyzed water. Food Microbiology and Safety 68 1013 1017 .

43. F. Yildiz, 1994 Initial preparation, handling and distribution of minimally processed refrigerated fruits and vegetables. In: Minimally Processed Refrigerated Fruits and Vegetables, Wiley, R.C. (ed), Chapman & Hall, New York, USA, 15 49

44. S. Zhang, J. M. Farber, 1996 The effects of various disinfectants against Listeria monocytogenes on fresh-cut vegetables. Food Microbiology. 13 311 321 .

Chapter 2

Abiotic Stress in Harvested Fruits and Vegetables

Peter M.A. Toivonen[1] and D. Mark Hodges[2]

[1] Agriculture and Agri-Food Canada, Pacific Agri-Food Research Centre, Canada

[2] Atlantic Food & Horticulture Research Centre, Canada

1. INTRODUCTION

Harvested fruits and vegetables can be potentially exposed to numerous abiotic stresses during production, handling, storage and distribution (Hodges, 2003). Some of these stresses can be minor in nature, resulting in no quality loss or, in some cases, in quality improvement (Hodges et al., 2005) during distribution. However, when the abiotic stress is moderate or severe, quality losses almost always are incurred at market (Toivonen, 2003a& 2003b). As a consequence it is important to understand the nature and sources for abiotic stresses that affect fruits and vegetables. In addition, with improved understanding, options for better management or resistance become available (Toivonen, 2003b, Toivonen, 2005). One of the challenges facing fruit and vegetable production globally is that regional climate regimes are becoming more unpredictable from year to year. Hence understanding of effects of field abiotic stresses (e.g. drought, extreme temperatures, light and salinity) on postharvest stress susceptibility will become more important since postharvest stresses limit the storage and shelf life potential of fruits and vegetables (Toivonen, 2005). It is the intent of this chapter to first describe the nature of pre- and post-harvest abiotic stress events, delve into their importance for product quality and marketing and then explore the technologies available to begin managing the sensitivity of fruits and vegetables to stresses they encounter in the handling and distribution chain.

2. MAJOR PREHARVEST STRESSES WHICH INFLUENCE POSTHARVEST ABIOTIC STRESS RESPONSE

Abiotic stresses occurring during production can either be the primary cause (direct) for disorders that exhibit themselves during postharvest handling and storage practices or they can influence the susceptibility of a fruit or vegetable to postharvest conditions that cause abiotic stresses resulting in disorders (indirect) (Ferguson et al., 1999). It is important to characterize the relationship between preharvest abiotic stresses occurring during production and postharvest abiotic stresses that the fruit or vegetable is exposed after harvest and during storage and distribution, since the solution to these different problems will be best resolved by focusing on preharvest or postharvest abiotic stress amelioration, respectively. Moderate levels of preharvest stress can potentially work towards enhancing stress resistance of a fruit or vegetable through up-regulating genes and pathways which renders the tissues cross-tolerant to many stresses (Lesham & Kuiper, 1996, Bowler & Fluhr, 2000) which may occur subsequently in postharvest handling, storage and distribution.

2.1. Drought

The occurrence of drought conditions during production of fruit and vegetable crops is becoming more frequent with climate change patterns (Whitmore, 2000). While much work has been devoted to understanding of drought effects on production and productivity of these crops (Whitmore, 2000), there is limited published literature on the effects of pre-harvest water stress on responses to postharvest stresses and hence on subsequent quality and shelf life. However, the existing literature provides some insight which may lead to better understanding and perhaps also encourage future research.

Water stress during the production phase of some fruits and vegetables may affect their physiology and morphology in such a manner as to influence susceptibility to weight loss in storage. There have been both positive effects reported for field water deficits (stress) in tree fruits and root vegetables. In the case of peaches, it has been shown that lower levels of irrigation results in higher density of fruit surface trichomes and consequent lower weight losses in storage (Crisosto et al., 1994). In addition, two studies have shown that deficit irrigation of apples and pears could reduce water loss of these fruit in subsequent storage (Kilili et al., 1996, Lopez et al., 2011) and this was attributed to reduction in skin permeance of the deficit irrigated fruit (Kilili et al., 1996). Presumably, fruit grown under moderate water stresses imposed by deficit irrigation practices adapt by developing a less water permeable cuticle. In terms of understanding that water deficits can have negative effects on postharvest stress susceptibility, irrigation of apples has been shown to enhance apple size which was associated with lower to water losses during storage (Veličković, 1994). This observation highlights a main concern about using deficit irrigation, which is the reduced size of fruit from such treatments (Lopez et al., 2011). Size of fruit is important, since larger fruit have lower surface area to volume ratios, which confers lower relative water loss (Shibairo et al., 1997) Another negative affect associated with water deficits is the case of root vegetables, such as carrot, where preharvest water stress (watering to 25-75% of soil water field capacity) can weaken the cells, resulting in higher membrane leakage (i.e. cell damage) and consequently greater weight loss in storage (Shibairo et al., 1998b).

Timing of a water stress event can also be very important in determining response to postharvest abiotic stress response. One example is that 'Kensington' mango fruit (Mangifera indica L.) will be significantly more susceptible to postharvest chilling injury with exposure to water stress during the cell expansion phase of growth as opposed to being exposed to the stress during cell division or at a time near to harvest maturity (Léchaudel and Joas, 2007). Therefore it is critical to avoid water stress until the fruit has reached maximum size in order to minimize incidence of chilling-induced injury in storage. Water stress, particularly at the tuber forming stage, can also lead to a higher susceptibility of potatoes to postharvest development of black spot disorder (Hamouz et al., 2011). Black spot disorder is correlated primarily to susceptibility of cells in the potatoes to undergo decompartmentation in response to mechanical stress (i.e. bruising) (Stevens and Davelaar, 1997).

2.2. Plant Nutrition

There is limited literature regarding the effects of crop nutrition on the susceptibility of fruits and vegetables to postharvest abiotic stress. There is one review dealing with the effect of preharvest nutrition on postharvest physiology and disorders of fruits and vegetables (Sams and Conway, 2003), however most of the reviewed literature touches on nutrition effects on postharvest biotic stress effects (i.e. disease resistance).

Calcium nutrition during production has been well documented in regard to postharvest disorders of many fruits and some vegetables (Sams and Conway, 2003). Calcium is also been suggested as a putative signalling molecule involved in the development of cross-tolerance to abiotic stresses (Bowler and Fluhr, 2000). Therefore the role of preharvest calcium nutrition is postharvest stress resistance may be complex, and dependent on whether the fruit or vegetable is also exposed to environmental abiotic stresses.

Potassium nutrition has been shown to have a few important effects on postharvest abiotic stress susceptibility of vegetables. In carrots, deficiency in potassium is associated with greater weight loss (desiccation stress) in storage (Shibairo et al., 1998c). At levels below 1 mM potassium in the soil medium, weight loss was directly associated with increased membrane leakage (i.e. damaged cells) in the carrot tissues. Above 1 mM potassium, there were no significant differences in weight loss under standardized storage conditions (Shibairo et al., 1998c). Improved potassium nutrition has also been shown to reduce susceptibility of potatoes to internal bruising in response to mechanical stresses imposed during postharvest handling (Maier et al., 1986, McGarry et al., 1996).

Relatively high preharvest nitrogen is often associated with poor postharvest quality of many fruits and vegetables (Sams and Conway, 2003). In regards to affecting susceptibility to postharvest stress, applying higher than recommended levels of preharvest nitrogen for a specific crop have been linked to storage discoloration susceptibility in both cabbage and potato. In the case of cabbage, it appears that excessive nitrogen fertilization leads to high accumulations of zinc and aluminum and nitrate induced manganese deficiency (Berard et al., 1990). High nitrogen applications in the field resulted in increased incidence and severity of black midrib in cold storage, particularly for the susceptible cultivar, 'Safekeeper' (Berard, 1990). In the case of potatoes, black spot susceptibility (a consequence of bruising; Stevens and Davelaar, 1997) is influenced by nitrogen fertilization, particularly the balance of nitrogen applied in relation to levels of potassium applied (Horneburg and Wirsing, 1995). In contrast nitrogen deficiency or lower than recommended nitrogen application rates will most often results in increased vitamin C content in many fruits and vegetables (Lee & Kader, 2000). Vitamin C content has been tightly linked with storage life potential (Hodges et al., 2001), which is likely a consequence of the importance of this antioxidant nutrient in forestalling oxidative injury that leads to quality losses in storage (Noctor and Foyer, 1998).

2.3. Temperature Extremes

Susceptibility to high (heat injury inducing) or low (chill injury inducing) temperatures is known to be reduced by prior exposure of the sensitive fruit or vegetable to low ambient temperatures (Saltveit and Morris, 1990, Wang, 1990). However, if the preharvest temperature leads to chilling induced injury in the field, then susceptibility to postharvest chilling injury can be increased (Morris, 1954). Therefore, the level of the preharvest temperature extreme will be a determinant as to if the exposure will have positive or negative effects on postharvest stress sensitivity.

Extreme high temperatures can occur in the field and apple fruit exposed to direct sunlight can reach in excess of 40 C (Ferguson et al., 1999). High temperatures during the late season (leading up to harvest) can enhance susceptibility of apples to superficial scald which develops in storage (Bramlage and Weis, 1997). In contrast, the authors found that low temperatures in the preharvest interval could reduce susceptibility.

2.4. Salinity

Tomatoes grown under high salinity will produce smaller fruit with higher soluble solids (Mizrahi, 1982). Smaller fruit will have higher surface area to volume ratios, hence greater susceptibility to postharvest water loss (i.e. desiccation stress) (Shibairo et al., 1997). While there is no direct information in the literature to confirm that smaller tomato fruit from saline growing conditions would be subject to greater desiccation stress postharvest, firmness declines for tomatoes grown under 3 and 6 dS m^{-1} salinity levels were increased by 50 to 130%, respectively, at two weeks holding at 20 C compared with control fruit (Mizrahi et al., 1988).

2.5. Light

It would be considered logical to assume that effects of exposure to high light are difficult to dissociate from effects exposure to high temperatures. However, research has shown that low light (bagging of apples) in the preharvest interval reduced susceptibility of apples to developing superficial scald in cold storage (Barden and Bramlage, 1994) and, in contrast, high ambient temperatures resulted in increased susceptibility (Bramlage and Weis, 1997). Generally, only sun-exposed surfaces of susceptible cultivars of apples develop scald in storage (Ferguson et al., 1999).

In the case of ambient low light, when lettuce is grown under low light which is sub-optimal for photosynthetic activity, shelf life of fresh cut lettuce (i.e. lettuce subjected to mechanical stress) is much shorter than lettuce produced under optimal light conditions (Witkowska & Woltering, 2010). Tomato size is smaller when the crop is grown under ambient low light levels, such as in the early spring season in northern latitudes (Gruda, 2005) and since surface area to volume ratio is greater in smaller fruits, susceptibility to postharvest desiccation stress would increase (Shibairo et al., 1997). Low light also results in lower levels of ascorbate in many greenhouse-grown fruits and vegetables (Gruda, 2005), which would render them less fit to deal with postharvest stresses since ascorbate contents are general directly proportional to relative levels or stress tolerance (Noctor and Foyer, 1998).

3. POSTHARVEST STRESSES DURING HANDLING AND STORAGE

3.1. Temperature Extremes

Postharvest temperature abuse during distribution is an ongoing challenge for many products, particularly those being shipped by air or ocean container (East et al., 2008). Breaks in cool chain temperatures can result in acceleration of climacteric ripening and softening (i.e. reduction in shelf potential) for apples harvested and stored at the pre-climacteric stage of maturity (East et al., 2008). However, those authors also found that temperature breaks had minimal effects on apples harvested and stored at post-climacteric stage of maturity. Fruit harvested at post-climacteric stages of maturity are not generally of concern to industry since those fruit have shorter storage and shelf life potential (Toivonen and Beveridge, 2005).

Chilling injury susceptibility is a significant issue for many crops derived from subtropical and tropical growing regions. Generally fruits, fruit vegetables (fruits which are consumed as vegetables) and root and tuber crops are chilling sensitive (Kader, 2002, Toivonen, 2010). There is a significant literature pertaining to chilling sensitivity of crops and much effort has been devoted to better understanding of chilling injury mechanisms and approaches to ameliorating the disorder (Wang, 1993, Saltveit, 2003). While there are visible (surface pitting, internal browning) and textural (accelerated softening and development of mealiness) changes often associated with chilling injury (Wang, 1993, Saltveit, 2003), flavour generation capacity has been shown to be a sensitive early indicator of chilling stress effects (Maul et al., 2000).

Use of heat treatments has become popular for disinfestation and disinfection of fruits and vegetables (Lurie, 1998). Appropriately applied levels of heat treatment can also induce temperature tolerance to both high and low temperatures (Lurie, 1998). However, when excessively harsh heat treatments are applied, heat-induced damage can occur.

3.2. Low O_2 And High Co_2

Managing postharvest handling and storage atmospheres to avoid low O_2 or high CO_2 stress is a constant concern and the problem is more severe when handling products in modified atmosphere (MA) packages as opposed to controlled atmosphere (CA) systems since temperature is often not as easily controlled in the MA packages as the produce moves through a distribution chain (Toivonen, 2003a, Toivonen et al., 2009). As such, the importance atmospheric stress in postharvest systems deserves significant discussion.

While low O_2 levels are well-known to induce stress-induced changes in metabolism and resultant metabolite accumulations (Kanellis et al., 2009), acute low oxygen injury is not expressed until the tissue is re-aerated (Biemelt et al., 1998) and a consequent uncontrolled oxygen burst (consisting of hydrogen peroxide and other radicals) occurs, resulting in lipid peroxidation, protein denaturation and membrane injury (Blokhina et al., 2001, 2003). Different fruits and vegetables have varying thresholds for low O_2 stress, dependent on anatomy, temperature, physiological age, presence of supplemental gases (e.g. CO_2, CO, SO_2, C_2H_4) and duration of exposure (Kader and Saltveit, 2003, Kanellis et al., 2009). Threshold limits for low O_2 tolerance can range from approximately 10 kPa partial pressure (for early potatoes and asparagus) down to approximately 1 kPa partial pressure (for mushroom, broccoli, and chipping potatoes), assuming that the produce is cooled to its recommended temperature when it is placed into that atmosphere (Kader and Saltveit, 2003). When a fruit or vegetable is converted to fresh-cut format, it generally becomes tolerant to lower levels of oxygen.

One of the most notable effects of high CO_2 levels in postharvest handling is to competitively inhibit ethylene binding and action hence delaying ripening in climacteric fruits (Kanellis at al., 2009). High CO_2 will directly inhibit succinate dehydrogenase, thus impairing the functioning of the tricarboxylic acid cycle and aerobic respiration. There are numerous physiological disorders that can be attributed to high CO_2 stress, including black heart of potatoes (Davis, 1926), brown heart or core in apples and pears (Meheriuk et al., 1994), surface bronzing in apples (Meheriuk et al., 1994) and brown stain of lettuce (Kader and Saltveit, 2003). High CO_2 can also modulate chilling stress, ethylene induced disorders and susceptibility to pathogenic attack (Kader and Saltveit, 2003).

3.3. Mechanical Injury

There are two types of mechanical injury that can be incurred during harvest and handling of fruits and vegetables; 1) cuts or punctures, and 2) impacts leading to bruises. Cuts can lead to transitory increases in respiration, ethylene production, phenolics production and cell deterioration near the site of the injury (Toivonen and DeEll, 2002 , Toivonen et al., 2005). Several factors influence the severity and size of bruising sustained, including maturity, water potential, tissue or cellular orientation at the site of the injury, shape of the object imparting the bruising force, energy and angle of the impact, and temperature of the product (Miller, 2003).

Cut type injuries are most prevalent in fresh-cut fruit and vegetable products (Toivonen and DeEll, 2002). Severity of response to cutting is very much dependant on the tissue characteristics, maturity of the fruit or vegetable of interest, the coarseness or sharpness of the cutting implement used and the temperature at which the cutting is done (Toivonen and DeEll, 2002). Cut injuries occur during the harvest process of many fruits and vegetables and is more severe in machine-harvested as compared with hand-harvested product (Miller, 2003). Products such as tomatoes, pickling cucumber, snap beans, green peas, potatoes, garlic can sustain significant damage during machine harvest, whereas asparagus, Brussels sprouts, leafy greens and head lettuce are more tolerant (Miller, 2003). Cutting associated disorders are primarily visualized as cut-edge browning or blackening, but may also include yellowing in green tissues and whitening on carrots (Toivonen and DeEll, 2002).

Impact caused injuries are associated with loading of product for transport, events during transport (particularly when uneven or rough roads or lanes are encountered), during unloading and throughout packaging and processing lines (Miller, 2003). Impact bruising has been shown to induce significant levels of ethylene production in mature-green tomatoes, with the levels produced being concomitant with the number of impacts sustained (MacLeod et al., 1976). Both physically visible internal damage and slight declines in ascorbic acid content of the fruit were observed in response to the impact events imposed on the fruit. Impact bruising leads to development of internal black spots in potato tubers (Stevens and Vavelaar, 1997).

3.4. Desiccation

Loss of water leading to deterioration in fruit and vegetable tissues is a common issue for postharvest handling and distribution (Ben-Yehoshua and Rodov, 2003). In addition to wilting (Ben-Yehoshua and Rodov, 2003), water stress can lead to accelerated senescence or ripening which are expressed as softening of tissues, membrane deterioration and yellowing (Lurie et al., 1986, Burden et al., 1994). As mentioned previously, one of the main characteristics of a fruit or vegetable that defines susceptibility to water loss is surface area to volume ratio (Shibairo et al., 1997B). Burton (1982) described a range of surface area to volume ratios from as low as 0.2 cm^{-1} in winter cabbage or turnip to as high as 50-100 cm^{-1} in leafy vegetables, explaining differences in water loss characteristics in these very different types of vegetables. In beans, the density of hairs on the cuticle can modulate rates of water loss to some extent and damage to these hairs will lead to increased losses (Hoffman, 1967). In other fruits and vegetables, lenticel density or epicuticular wax thickness can modulate water loss (Ben-Yehoshua and Rodov, 2003).

The driving force for water loss is the vapour pressure deficit (vpd), which is the relationship that describes the difference in water activity of the fruit or vegetable and the water activity of the atmosphere surrounding it (Ben-

Yehoshua and Rodov, 2003). The greater the vapour pressure deficit, the greater the water loss. Three postharvest handling principles are important in minimizing water loss of any fruit or vegetable; 1) warm product loses water faster than cool product when placed into a cool room, hence the importance of rapid precooling before storage, 2) delays in cooling will lead to longer exposures to higher vapour pressure deficit conditions, hence timely cooling after harvest is of utmost importance, and 3) storing product at the coldest storage temperature and highest relative humidity possible will minimize water loss (Toivonen, 2010). Following these principles will result in the least desiccation stress for any fruit or vegetable.

4. EFFECTS OF ABIOTIC STRESSES ON PLANT METABOLISM

4.1. Metabolic Changes Induced By Stress

Heat stress induces metabolic changes associated with heat shock protein accumulations which are known to confer persistent levels of stress resistance in heat-exposed produce (Wang et al., 2004). Heat stress can also inhibit the production and accumulation of lycopene (Hall, 1964). The duration and temperature of exposure will determine if such an effect will occur, but tomato fruit exposed to 32 C continuously will not turn red, remaining yellow even at full ripeness.

Chilling stress results in the accumulation of lipid peroxidation products, superoxide anions and hydrogen peroxide and also to losses in flavor volatile production in sensitive fruits and vegetables (Wang, 1993, McDonald et al., 1996, Maul et al., 2000). Chillingstress affects ethylene metabolism (Wang & Adams, 1982, Obenland et al., 2008), leading to accelerated softening. It also modifies cell wall metabolism, with resultant up-regulation of cell wall dismantling enzymes such as pectin methylesterase and endopolygalacturonase (Nilo et al., 2010).

There are two distinct phases for response to exposure to anaerobic or anoxic atmospheres. The first phase involves the metabolic shift induced by impairment of the mitochondrial electron transport chain by limitation of the primary electron acceptor, molecular O_2 (Hanhijärvi and Fagerstedt, 1994). The resultant impairment has been well-characterized as declines in ATP levels, pyruvate dehydroxylase activity and cytoplasmic pH (Kader and Saltveit, 2003). These conditions lead to an increase in pyruvate decarboxylase, alcohol dehydrogenase and lactate dehydrogenase activities (Kader and Saltveit, 2003). Anaerobic respiration is induced and there is an accumulation in acetaldehyde, ethanol, ethyl acetate and/or lactate (Purvis, 1997). In addition, the impairment of the mitochondrial electron transport chain results in electron leakage and consequent generation of superoxide anions and hydrogen peroxide in the cells which can be decomposed by existing cellular antioxidant systems (Blokhina et al., 2001). Ascorbate and glutathione levels can also increase during an anoxic or hypoxic event (Biemelt et al., 1998). Any or all of these changes are indicators of oxygen stress in fruit and vegetable tissues and affect quality attributes. However, actual injury to the tissue does not occur until the second phase, where the fruit or vegetable is returned to higher O_2 atmospheres. On return to aerobic atmospheres, rates of oxygen radical production in the impaired electron transport chain becomes accelerated, leading to great accumulations of superoxide anion, hydrogen peroxide and hydroxyl radical that cannot be fully decomposed by existing antioxidant protection systems and consequent lipid peroxidation and protein denaturation ensues, leading to membrane damage, enzymatic browning and cell death (Blokhina et al., 2001, 2003).

Kubo et al. (1990) found that different fruits and vegetables responded in different ways to high CO_2. Those fruits and vegetables showing no change in O_2 uptake (respiration) in response to 60% CO_2exposure did not show any change in ethylene production. Those showing a decline in respiration generally also showed a decline in ethylene production, except banana (Kubo et al., 1990) and those which showed a significant increase in respiration also showed an increase in ethylene production (cucumber, eggplant, podded pea, spinach and iceberg lettuce). This suggests that these latter vegetables were more sensitive to CO_2 stress than the other fruits and vegetables tested. Mathooko et al. (1998) determined that CO_2 stress-induced ethylene biosynthesis was likely regulated at a post-transcriptional level and it involves the *de novo* synthesis of novel protein(s). They also identified that protein phosphorylation and dephosphorylation processes as being required in one or more of the steps leading to the induction of ACC synthase, which is the last enzyme in the ethylene biosynthesis pathway.

Initial response to wounding stress is characterized with a progressive accumulation of ACC synthase, ACC and ethylene production in tomatoes and this lasts for up to 2 hours, but if the cut tomatoes are held over longer durations, ethylene production declines (Kende and Boller, 1981). However, both ACC synthase and ACC continue to accumulate, indicating that there is a capacity to produce ethylene, but transformation of ACC to ethylene is blocked or inhibited (Kende and Boller, 1981). Wounding has also been shown to induce phenolic accumulation through up-regulation of phenylalanine ammonia lyase (PAL) (Ke and Saltveit, 1989). This up-regulation of PAL was associated with wound-induced ethylene production. Other metabolites increase in response to bruising or wounding stress, including isocoumarin in carrots (Lafuente et al., 1996), anthocyanins in midribs of red-pigmented lettuce (Ferreres et al., 1997), methanethiol, allyl isothiocyanates and dimethyl disulfide in cabbage (Yano et al., 1986, Chin and Lindsay, 1993), six-carbon aldehydes and alcohols in cut peppers (Wu and Liou, 1986) and suberin in tomatoes and bean pods (Dean and Kolattukudy, 1976). Some of the metabolic shifts are mediated by stress response messengers (e.g. phenolics, suberin and isocoumarin accumulation) and others are a direct consequence of cellular disruption that occurs during wounding or bruising (e.g. methanethiol, allyl isothiocyanate and dimethyl sulphide accumulations).

Desiccation stress has been shown to cause some metabolic changes in fruits and vegetables. In carrots, desiccation stress at extreme handling conditions (i.e. at 13 C) led to increases in osmotic potential, which are a function of free sugars in the roots (Shibairo et al., 2002). Increase in osmotic potential in response to water loss was most likely explained by increases in polysaccharide hydrolyzing enzymes activities in response to the stress. Hence, enzymes such as polygalacturonase and pectinesterase may increase in activity leading to loss of cell wall structure and concomitant increases in soluble sugars (Inari et al., 2002). This may explain at least a component of the loss of firmness that has been observed with carrots as they lose water (Shibairo et al., 2002). This hypothesis is borne out by results of work with cucumbers where water stress resulted in up-regulation of polygalacturonase activity, suggesting that water loss itself was not the only factor in causing softening of stressed fruit (Kubo et al., 2000). Another aspect of water stress is induction ethylene production (Kubo et al., 2000), which may explain why water stress leads to accelerated ripening in bananas (Burdon et al., 1994) and accelerated senescence in bell peppers (Lurie et al. 1986).

4.2. Effects On Quality

Postharvest abiotic stressors can lead to numerous quality problems in fruits in vegetables, including scald, core and flesh browning of fruits, sweetening, pitting, water-soaking appearance, abnormal ripening, russeting and tissue softening (Hodges et al., 2005). Stress can also result in the losses of nutrient constituents in the fruit or vegetable, with vitamin C loss being the most sensitive indicator of stress exposure (Noctor & Foyer 1998, Pignocchi & Foyer, 2003, Ioannidi et al., 2009).

4.3. Mechanisms For Abiotic Stress Response At The Biochemical And Molecular Levels

Response to all abiotic stresses can be acute and sub-acute in nature, where acute responses represent cases where cell death is a direct result and sub-acute responses represent cases where the stress leads to induction of adaptive changes in biochemical and gene expression (Toivonen, 2005). Many reactive oxygen species (ROS), particularly hydrogen peroxide, behave as signalling agents to trigger biochemical changes at the gene expression level (Jaspers & Kangasjärvi, 2010). In general, abiotic stressors will induce perturbations in the fruit or vegetable cellular homeostasis which will then result in the increased generation of ROS in the apoplast, mitochondria, peroxisomes, cytoplasm, chloroplasts and endoplasmic recticulum (Jaspers & Kangasjärvi, 2010). The ability of the cell to initially cope will depend largely on the endogenous free radical scavenging capacity (Mittler, 2002).

When free radical generation exceeds the endogenous scavenging capacity, the ROS interact with sensors, for which the full nature are not currently understood, that will initiate mitogen activated-protein kinase (MAPK) cascade reactions and also directly up-regulate transcription factors and calcium/calmodulin kinases (Mittler, 2002, Jaspers & Kangasjärvi, 2010). The MAPK cascade reaction will activate various transcription factors that enable *de novo* production of ROS, ROS scavenging systems, accumulation of heat shock proteins, and modulate NADPH supply in the cell (Mittler, 2002). Some of the MAPK cascade paths have been also shown to be linked specifically to ethylene production (Jaspers & Kangasjärvi, 2010), which is probably why ethylene production seems to be intrinsic to most stress responses. However, not all stressors produce identical response pathways, and so there is still a lot of work to be done in mapping of stress response networks (Jaspers & Kangasjärvi, 2010).

Chilling, bruising and cutting injuries all lead to increased activities of cell wall hydrolysing enzymes (polygalacturonase and pectinmethylesterase), which accounts for accelerated softening and abnormal ripening that occurs in response to those abiotic stresses (Van linden et al., 2008). Water stress will also lead to accelerated softening (Kubo et al., 2000), and that response has been associated with the induction of ethylene production in response to water stress. It quite likely that the effect of all these stresses on up-regulating the cell wall hydrolysing enzymes and consequently accelerating softening is an ethylene-mediated response.

Accumulation of heat shock proteins (HSPs) which are mediated via transcription factor activation downstream of the MAPK cascade (Mittler, 2002) have been reported to enhance persistent levels of stress resistance in affected tissues (Sabehat et al., 1996). HSPs are thought to be an important factor for protein folding, assembly, translocation and degradation under normal, stress free conditions (Wang et al., 2004). HSPs have also been linked to stabilization of proteins and membranes, and enabling protein refolding under stress conditions. As a consequence HSPs are thought to have a pivotal function of protecting plant tissues against stress by maintaining cellular homeostasis.

5. APPROACHES TO AMELIORATE ABIOTIC STRESS SENSITIVITY

5.1. Treatments To Enhance Stress Resistance

Numerous postharvest treatments have been evaluated for enhancing abiotic stress resistance of fruits and vegetables (Toivonen, 2003b). Generally, temperature modulation (including intermittent warming), extreme atmospheres (high O_2, CO_2 and low O_2), growth regulators, anti-transpirants, antioxidant dips, growth regulators, nitric oxide and ethanol have been tested (Toivonen, 2003b). Application of treatments in combination can often improve the stress resistance level to the fruit or vegetable (Toivonen, 2005& 2009).

One treatment that has not been widely studied to date is the concept of gradual cooling to enhance resistance to chilling injury. Gradual cooling (2 C per day) has been shown to reduce susceptibility of tomato to chilling injury, presumably by cooling slowly enough to allow the inherent stress resistance systems to develop before actual chilling temperatures were reached (Gálvez et al., 2010). The approach is perhaps more amenable to commercial practice than some others tested (i.e. intermittent warming).

While there has been significant effort placed on developing treatments to enhance postharvest abiotic stress resistance in fruits and vegetables, the efforts have only resulted in incremental success (Toivonen, 2004). This is largely due to the fact that stress response and resistance is a very complex matrix of processes and pathways, which are not fully understood at this point in time (Toivonen, 2005, Jaspers & Kangasjärvi, 2010). Hence, it is very difficult to design effective treatments to achieve the resistance required for any or all abiotic stresses that a fruit or vegetable may encounter during harvest, handling, storage and distribution.

Hot or warm water treatments have been shown to minimize cutting induced injuries in fresh-cut products (Lurie, 1998). Such treatments may also be used to control chilling injury via induction of heat shock proteins (Collins et al., 1995, Sabehat et al., 1996). Warm water treatment has also been shown to reduce sensitivity to irradiation in cut and packaged lettuce (Fan et al., 2003). Each product has a differing tolerance to temperature and so such treatments must be developed case by case for each vegetable or fruit of interest. The key is apply a sublethal temperature exposure, in order to enhance the adaptive response of the fruit or vegetable tissue (Saltveit, 2003).

Another approach is to apply conditioning treatments to enhance chill stress tolerance. This has been shown to provide enhanced chilling stress in several fruits and vegetables (Saltveit, 1991, Wang, 1993). It also has the advantage over heat treatments in that the risk of causing acute injury during treatment is very low.

Atmospheric treatments are quite often beneficial to controlling response to postharvest stresses. Modified or controlled atmospheres have been shown to help minimized chilling injury in a number of fruits and vegetables (Wang, 1993). However, it must be noted that in some cases, application of a controlled atmosphere led to detrimental effects on the fruit (e.g. asparagus, cucumber, limes, sweet bell peppers). Therefore, a case by case evaluation is required to determine if secondary stress is induced by the atmosphere treatments.

Growth regulator applications can also potentially enhance stress resistance, particularly for fruits and vegetables which are prone to show accelerated ripening or senescence in response to stress (Baldwin, 2003). Accelerated ripening or senescence is most often mediated by ethylene production in response to the stress. As consequence anti-ethylene products such as aminovinylglycine (AVG) and 1-methylcyclopropene (1-MCP) could enhance storage or shelf if ethylene production in response to stress was a main concern (Baldwin, 2003, Blankenship and Dole, 2003). Other growth hormones, such as methyl jasmonate (which promotes leaf senescence), can enhance chilling resistance in avocado, grapefruit, bell peppers and zucchini squash (Meir et al., 1996, Wang, 1994). Abscisic acid has been demonstrated to reduce chilling-induced injury in some crops (Wang, 1993). Other growth regulators have been suggested for use in preventing senescence in leafy vegetables (e.g. 2, 4-D), however their practical application is limited.

Quite often a single stress resistance enhancing treatment may not confer optimal resistance levels to all postharvest stresses encountered (Toivonen, 2003b). Therefore it may be productive to consider application of a combination of two or more stress tolerance enhancing treatments to achieve optimal levels of resistance (Toivonen, 2005& 2009). There are a several examples in the literature which show the added benefit of applying treatment combinations (Toivonen, 2009).

5.2. Germplasm Selection

There have been some reports on the selection of germplasm and cultivars from breeding programs that will have greater postharvest stress resistance and hence better storage capability. Hodges et al. (2001) were able to show that a cultivar which was more susceptible to yellowing was so because of differences in balance of antioxidant systems in the tissues, which resulted in higher accumulations of ROS, particularly hydrogen peroxide. They hypothesized that the higher levels of ROS were directly causal of the yellowing of the chlorophyll in the spinach leaves. A similar relationship was found for a broccoli cultivar resistant to yellowing when compared with a cultivar highly susceptible to yellowing during storage (Toivonen & Sweeney, 1997). Apple cultivars which have resistance to developing browning in response to cutting have greater levels of apoplastic antioxidant enzymes and lower levels of ROS in the apoplast after cutting than cultivars with lower levels of antioxidant enzymes (Toivonen et al., 2003). While measures of antioxidant enzymes and ROS levels can be instructive to understand the basis for difference in stress tolerance in selections of a breeding population, they are too labour intensive to consider for incorporation as selection tools into a breeding program.

In vitro selection is an approach where plant cells of a fruit or vegetable of interest are tissue-cultured and exposed to a stressor, taking surviving cells to regenerate new plants having a superior level of stress resistance (Rai et al., 2011). This approach has been highly successful to regenerate germplasm of many crop plants which can be regenerated via tissue culture techniques. *In vitro* selection is much less expensive approach than molecular engineering and laboratories can easily be set up almost anywhere in the world having access to basic utilities, using inexpensive technology (Rai et al., 2011). The fact that the approach relies on the cells to survive in response to the applied stress means that the complexity of the stress response network is incorporated into the plants regenerated in this way.

5.3. Use Of Molecular Probes For Marker - Assisted Breeding

There are a large number of genes and proteins associated with stress tolerance in plants and so the best approach to identifying stress tolerant lines is apply the stress of interest and perform quantitative trait loci (QTL) analyses (Foolad, 1999). This approach may be used with intact plants and/or harvested plant parts, generally using the plant part which is of interest in the breeding improvement strategy (Foolad, 1999). The approach requires analysis of adaptive changes in QTL expression as opposed to constitutive expression (Collins et al., 2008). A stress protocol, to which target fruit or vegetable will be exposed, must be established to provide clear differentiation between resistant and susceptible lines. However, as stated earlier, the stress response is complex and so success using QTLs will require interdisciplinary effort, integrating biochemistry, gene mapping and phenotyping activities to allow reliable interpretation and successful use of adaptive QTLs for selecting for stress resistance (Collins et al., 2008).

5.4. Molecular Engineering

Molecular engineering for stress resistance in fruits and vegetables is limited due two major factors; 1) the complexity of the stress response network (Toivonen, 2005, Jaspers & Kangasjärvi, 2010) means that modulating the stress resistance with single gene insertions is unlikely, and 2) methods to successfully transform many important fruit and vegetable crops have yet to be developed (Rai et al., 2011). One area where there has been some advance is insertion of anti-freeze genes to protect against low temperature injury (Kole & Hall, 2008). However, the future advances will require a better basic knowledge of the stress response network and control points at the molecular level. This approach may be a helpful adjunct to adaptive QTL analysis, since insertion of a putatively important gene may alter the adaptive QTL profile and provide a probe to better understand the functional changes induced by a stress (Collins et al., 2008).

5.5. Improved Postharvest Harvest And Handling Protocols

Most existing postharvest handling and storage procedures are generally not considered to be stress-inducing, except in cases where quality issues arise (Toivonen, 2009). However, there is increasing information in the research literature that suggests that there are significant stresses that could be modulated during the postharvest continuum. It is often difficult to avoid low temperature stress, since most produce is refrigerated as a necessary step to control spoilage and preserve food safety (Toivonen, 2009).

Simple modifications to postharvest handling systems can sometimes result in significant reduction in stress exposure and consequently result in storage and/or shelf life extension. One of the most successful strategies is the application of plastic film packaging or wraps to prevent desiccation, resulting in significant improvements and shelf life and quality of many fruits and vegetables (Ben-Yehoshua and Rodov, 2003). In many cases, modified atmosphere packaging is considered to largely control humidity around product and thus prevent moisture loss of fresh-cut and whole fruits and vegetables (Toivonen et al., 2009). Also, anti-transpiration coatings have been shown to be effective for maintaining quality through control of water loss (Baldwin, 2003). In regards to maintaining water content on the retail shelf, the application of misting systems can 'recharge' the vegetable and in so doing maintain quality over longer durations at less than ideal storage temperatures (Barth et al., 1990; Shibairo et al., 1998).

While rapid cooling is generally recommended to preserve quality, delayed or gradual cooling may be useful to chilling sensitive crops to allow them to acclimatize or adapt to storage and handling conditions. Delays in cooling of apples after harvest can be used to reduce the development of low temperature stress induced orders in 'Honeycrisp' apples (DeLong et al., 2004). Similarly, delays in cooling can reduce CO_2-induced internal browning in 'Braeburn' apples (Toivonen et al., 2003). Recently, slow cooling at a rate of 2 C per day from 12 C to 4 C of tomatoes has shown promise to reduce chilling injury when stored at the lower temperature (Gálvez et al., 2010).

6. CONCLUSIONS

Abiotic stresses are significant determinants of quality and nutritional value of fruits and vegetables during harvest, handling, storage and distribution to consumer. Crop management can have a significant influence on susceptibility to stress. In addition, climate change has created additional environmental variables which may influence postharvest stress susceptibility of fruits and vegetables. While breeding is underway for many crops to develop stress resistance that will allow them to adapt to climate change, it is not clear that breeding for stress resistance in the field will also extend stress resistance characteristics to the harvested portion. It is important understand the basis of molecular and biochemical response networks to various stresses encountered in the field and in the postharvest continuum to better evaluate the benefits that abiotic stress during production may yield for postharvest abiotic stresses.

The focus of effort should probably be on use of breeding or directed breeding to enhance stress tolerance in fruits and vegetables, as opposed to genetic engineering. This is because stress response networks are extremely complex and, as such, specific target transformations may be insufficient to confer significant increase in stress tolerance. An important aspect of the breeding and selection approach is that there must be stressors applied in reproducible ways to allow the breeder to identify expression of stress resistance since that characteristic is adaptive, rather than constitutive, in nature. The researcher must determine whether the adaptive stress response is best tested in the field, greenhouse or test tube.

In the context of postharvest handling treatments, there is some indication that such approaches may benefit in enhancing tolerance and thereby extending shelf and nutritional life of fruits and vegetables. However, again there is a limited amount of basic understanding to help guide the development of approaches to reliably confer useful levels of stress tolerance to stresses in general, and, more importantly, to specific stresses that a product is known to be subjected to during its normal distribution.

REFERENCES

1. E. A. Baldwin, 2003 Coating and Other Supplemental Treatments to Maintain Vegetable Quality. In: Postharvest Physiology and Pathology of Vegetables, 2nd Edition, J.A. Bartz & J. K. Brecht (Eds.), 413 456 , Marcel Dekker, Inc, 0-82470-687-0 York

2. C. L. Barden, W. J. Bramlage, 1994 Separating the Effects of Low Temperature, Ripening, and Light on Loss of Scald Susceptibility in Apples Before Harvest. Journal of the American Society for Horticultural Science, 119 1 February 1994), 54 58 , 0003-1062

3. M. M. Barth, A. K. Perry, S. J. Schmidt, B. P. Klein, 1990 Misting Effects on Ascorbic Acid Retention in Broccoli During Cabinet Display. Journal of Food Science, 55 4 July 1990), 1187 1188 , 1365-2621

4. S. Ben-Yehoshua, V. Rodov, 2003 Transpiration and Water Stress. In: Postharvest Physiology and Pathology of Vegetables, 2nd Edition, J.A. Bartz & J. K. Brecht (Eds.), 111 159 , Marcel Dekker, Inc, 0-82470-687-0 York

5. L. S. Berard, 1990 Effect of Nitrogen Fertilization on Stored Cabbage. I. Development of Physiological Disorders on Tolerant and Susceptible Cultivars. Journal of Horticultural Science, 65 3 May 1990), 289 296 , 1462-0316

6. L. S. Berard, M. Senecal, B. Vigier, 1990 Effects of Nitrogen Fertilization on Stored Cabbage. II. Mineral Composition in Midrib and Head Tissues of Two Cultivars. Journal of Horticultural Science, 65 4 July 1990), 289 296 , 1462-0316

7. S. Biemelt, U. Keetman, G. Albrecht, 1998 Re-Aeration Following Hypoxia or Anoxia Leads to Activation of Antioxidant Defense System in Roots of Wheat Seedlings. Plant Physiology, 116 2 February 1998), 651 658 , 0032-0889

8. S. M. Blankenship, J. M. Dole, 2003 1-Methylcyclopropene: A Review. Postharvest Biology and Technology, 28 1 April 2003), 1 25 , 0925-5214

9. O. B. Blokhina, T. V. Chirkova, K. V. Fagerstedt, 2001 Anoxic Stress Leads to Hydrogen Peroxide Formation in Plant Cells. Journal of Experimental Botany, 52 359 June 2001), 1179 1190 , 0022-0957

10. O. Blokhina, E. Virolainen, K. V. Fagerstedt, 2003 Antioxidants, Oxidative Injury and Oxygen Deprivation Stress: a Review. Annals of Botany, 91 2 January 2003), 179 194 , 0305-7364

11. C. Bowler, R. Fluhr, 2000 The Role of Calcium and Activated Oxygens as Signals for Controlling Cross-Tolerance. Trends in Plant Science, 5 6 June 2000), 241 246 , 1360-1385

12. W. J. Bramlage, S. A. Weis, 1997 Effects of Temperature, Light and Rainfall on Superficial Scald Susceptibility in Apples. HortScience, 32 5 August 1997), 808 811 , 0018-5345

13. J. N. Burdon, S. Dori, E. Lomaniec, R. Marinansky, E. Pesis, 1994 The Post-harvest Ripening of Water Stressed Banana Fruits. Journal of Horticultural Science, 69 5 September 1994), 799 804 , 1462-0316

14. W. G. Burton, 1982 Postharvest Physiology of Food Crops. Longman Group Ltd, 0-58246-038-7

15. H.w. Chin, R. C. Lindsay, 1993 Volatile Sulfur Compounds Formed in Disrupted Tissues of Different Cabbage Cultivars. Journal of Food Science, 58 4 July 1993), 835 830 , 841, 1750-3841

16. N. C. Collins, F. Tardieu, R. Tuberosa, (2008).Quantitative Trait Loci and Crop Performance under Abiotic Stress: Where Do We Stand? Plant Physiology, 147 147 2 June 2008), 469 486 , 0032-0889

17. G. G. Collins, X. L. Nie, M. E. Slatveit, 1995 Heat Shock Protein and Chilling Injury of Mung Bean Hypocotyls. Journal of Experimental Botany, 46 7 July 1995), 795 802 , 0022-0957

18. C. H. Crisosto, R. S. Johnson, J. G. Luza, G. M. Crisosto, 1994 Irrigation Regimes Affect Fruit Soluble Solids Concentration and Rate of Water Loss of'O'Henry' Peaches. HortScience, 29 10 October 1994), 1169 1171 , 0018-5345

19. W. B. Davis, 1926 Physiological Investigation of Black Heart of Potato Tuber. Botanical Gazette, 81 3 May 1926), 323 338 , 0006-8071

20. B. Dean, P. E. Kolattukudy, 1976 Synthesis of Suberin During Wound-Healing in Jade Leaves, Tomato Fruit, and Bean Pods. Plant Physiology, 58 3 September 1976), 411 416 , 0032-0889

21. P. J. Delaquis, S. Stewart, M. Cliff, P. M. Toivonen, A. L. Moyls, 2000 Sensory quality of Ready-to-Eat Lettuce Washed in Warm, Chlorinated Water. Journal of Food Quality, 23 6 December 2000), 553 563 , 1745-4557

22. J. M. De Long, R. K. Prange, P. A. Harrison, 2004 The Influence of Pre-Storage Delayed Cooling on Quality and Disorder Incidence

in'Honeycrisp' Apple Fruit. Postharvest Biology and Technology, 34 3 December 2004), 353 358 , 0925-5214 (Erratum Version)

23. A. R. East, D. J. Tanner, K. M. Maguire, A. J. Mawson, 2008 The Influence of Breaks in Storage Temperature on'Cripps Pink' (Pink Lady™) Apple Physiology and Quality. HortScience, 43 3 June 2008), 818 824 , 0018-5345

24. X. Fan, P. M. A. Toivonen, K. T. Rajkowski, K. B. Sokorai, 2003 Warm Water Treatment in Combination with Modified Atmosphere Packaging Reduces Undesirable Effects of Irradiation on the Quality of Fresh-Cut Iceberg Lettuce. Journal of Agricultural and Food Chemistry, 51 5 January 2003), 1231 1236 , 0021-8561

25. I. B. Ferguson, R. Volz, A. Woolf, 1999 Preharvest Factors Affecting Physiological Disorders of Fruit. Postharvest Biology and Technology, 15 3 March 1999), 255 262 , 0925-5214

26. F. Ferreres, M. I. Gil, M. Casteñer, F. A. Tomás-Barberán, 1997 Phenolic Metabolites in Red Pigmented Lettuce (Lactuca sativa). Changes with Minimal Processing and Cold Storage. Journal of Agricultural and Food Chemistry, 45 11 November 1997), 4249 4254 , 1520-5118

27. M. R. Foolad, 1999 Genetics of Salt Quantitative and Cold Analysis Tolerance in Tomato: and QTL Mapping. Plant Biotechnology, 16 1 March 1999), 55 64 , 0289-5770

28. L. R. Fukumoto, P. M. A. Toivonen, P. J. Delaquis, 2002 Effect of Wash Water Temperature and Chlorination on Phenolic Metabolism and Browning of Stored Iceberg Lettuce Photosynthetic and Vascular Tissues. Journal of Agricultural and Food Chemistry, 50 16 June 2002), 4503 4511 , 0021-8561

29. A. B. Gálvez, M. V. García, J. C. Corrales, A. C. López, J. A. L. Valenzuela, 2010 Effect of Gradual Cooling Storage on Chilling Injury and Phenylalanine Ammonia-Lyase Activity in Tomato Fruit. Journal of Food Biochemistry, 34 2 April 2010), 295 307 , 1745-4514

30. Y. Gong, P. M. A. Toivonen, O. L. Lau, P. A. Wiersma, 2000 Response of Anaerobic Metabolism to Delaying Controlled Atmosphere Storage in Braeburn Apples. HortScience, 35 3 June 2000), 411 0018-5345

31. N. Gruda, 2005 Impact of Environmental Factors on Product Quality of Greenhouse Vegetables for Fresh Consumption. Critical Reviews in Plant Sciences, 24 3 227 247 , 0735-2689

32. C. B. Hall, 1964 The Effect pf Short Periods pf High Temperature pn the Ripening pf Detached Tomato Fruits. Proceedings of the American Society for Horticultural Science, 84 501 506 .

33. K. Hamouz, D. Bečka, J. . Morava, 2011 Effect of Environmental Conditions on the Susceptibility to Mechanical Damage of Potatoes. In: Agris On-Line Papers in Economics and Informatics. 30.03.2011. March Available from http://www.agris.cz/vyhledavac/detail.php?id=116921&iSub=518&sHighLight=agria&PHPSESSID=997764de5509230f23dc06268b7b8344

34. A. M. Hanhijärvi, K. V. Fagerstedt, 1994 Comparison of Effect of Natural and Experimental Anoxia on Carbohydrate and Energy Metabolism in Iris pseudocorus rhizomes. Physiologia Plantarum, 90 3 March 1994), 437 444 , 1399-3054

35. D. M. Hodges, 2003 Postharvest Oxidative Stress in Horticultural Crops, Food Products Press, 1-56022-962-4 York, USA

36. D. M. Hodges, G. E. Lester, K. D. Munro, P. M. A. Toivonen, 2005 Oxidative stress: Importance for Postharvest Quality. HortScience, 39 5 August 2005), 924 929 , 0018-5345

37. D. M. Hodges, W. V. Wismer, C. F. Forney, 2001 Antioxidant Responses in Leaves of Two Cultivars of Spinach (Spinacia oleracea L.) Differing in Their Senescence Rates. Journal of the American Society for Horticultural Science, 126 5 September 2001), 611 617 , 0003-1062

38. J. C. Hoffman, 1967 Morphological Variations of Snap Bean Pods Associated with Weight Loss and Wilting. Proceedings of the American Society of Horticultural Science, 91 294 303

39. B. Horneburg, F. Wirsing, 1995 The Tendency of Potato Tubers to Black Spot. 2. Influence of Fertilizer Application and Irrigation on the Expression of Black Spot. Kartoffelbau, 46 8 August 1995), 318 320 , 0022-9156

40. T. Inari, R. Yamauhi, K. Kato, T. Takeuchi, 2002 Changes in Pectic Polysaccharides during the Ripening of Cherry Tomato Fruits. Food Science and Technology Research, 8 1 March 2002), 55 58 , 1344-6606

41. E. Ioannidi, M. S. Kalamaki, C. Engineer, I. Pateraki, D. Alexandrou, I. Mellidou, J. Giovannonni, A. K. Kanellis, 2009 Expression Profiling of Ascorbic Acid-Related Genes During Tomato Fruit Development and Ripening and in Response to Stress Conditions. Journal of Experimental Botany, 60 2 February 2009), 663 678 , 1460-2431

42. P. Jaspers, J. Kangasjärvi, 2010 Reactive Oxygen Species in Abiotic Stress Signalling. Physiologia Plantarum, 138 4 April, 2010), 405 413 , 0031-9317

43. A. A. Kader, 2002 Postharvest Biology and Technology: An Overview. In: Postharvest Technology of Horticultural Crops, A.A. Kader (Ed.), 39 47 , University of California Agriculture and Natural Resources Communications Services, 1-87990-651-1 CA

44. A. A. Kader, M. E. Saltveit, 2003 Atmosphere Modification. In: Postharvest Physiology and Pathology of Vegetables, 2nd Edition, J.A. Bartz & J. K. Brecht (Eds.), 229 246 , Marcel Dekker, Inc, 0-82470-687-0 York

45. A. K. Kanellis, P. Tonutti, P. Perata, 2009 Biochemical and Molecular Aspects of Modified and Controlled Atmospheres. In: Modified and Controlled Atmospheres for Storage, Transportation and Packaging of Horticultural Commodities, E. Yahia (Ed), 553 567 . CRC Press, 978-1-42006-957-0 Boca Raton, FL.

46. D. Ke, M. E. Saltveit, 1989 Wound-Induced Ethylene Production, Phenolic Production and Susceptibility to Russet Spotting in Iceberg Lettuce. Physiologia Plantarum, 76 3 July 1989), 412 418 , 1399-3054

47. H. Kende, T. Boller, 1981 Wound Ethylene and 1-Aminocyclopropane-1-Carboxylate Synthase in Ripening Tomato Fruit. Planta, 151 5 May 1981), 476 481 , 0032-0935

48. A. W. Kilili, M. H. Behboudian, T. M. Mills, 1996 Postharvest Performance of'Braeburn' Apples in Relation to Withholding of Irrigation at Different Stages of the Growing Season. Journal of Horticultural Science, 71 5 September 1996), 693 701 , 1462-0316

49. C. Kole, T. C. Hall, 2008 Compendium of Transgenic Crop Plants: Transgenic Temperate Fruit and Nuts. Blackwell Publishing Ltd., 978-1-40516-924-0 Hoboken, NJ, USA

50. Y. Kubo, A. Inaba, R. Nakamura, 1990 Respiration and C2H4 Production in Various Harvested Crops Held in CO2-Enriched Atmospheres. Journal of the American Society for Horticultural Science, 115 6 November 1990), 975 978 , ISSN0003-1062

51. Y. Kubo, Y. B. Xue, A. Nakatsuka, F. M. Mathooko, A. Inaba, R. Nakamura, 2000 Expression of a Water Stress-Induced Polygalacturonase Gene in Harvested Cucumber Fruit. Journal of the Japanese Society for Horticultural Science, 69 3 September 2000), 273 279 , 0013-7626

52. M. T. Lafuente, G. López-Gálvez, M. Cantwell, S. F. Yang, 1996 Factors Influencing Ethylene-Induced Isocoumarin Formation and Increased Respiration in Carrots. Journal of the American Society for Horticultural Science, 121 3 May 1996), 537 542 , 0003-1062

53. M. Léchaudel, J. Joas, 2007 An Overview of Preharvest Factors Influencing Mango Fruit Growth, Quality and Postharvest Behaviour. Brazilian Journal of Plant Physiology, 19 4 October-December 2007), 287 298 , 1677-0420

54. Y. Y. Lesham, P. J. C. Kuiper, 1996 Is There a GAS (General Adaptation Syndrome) Response to Various Types of Environmental Stress? Biologia Plantarum, 38 1 March 1996), 1 18 , 0006-3134

55. G. Lopez, C. Larrigaudière, J. Girona, M. H. Behboudian, J. Marsal, 2011 Fruit Thinning in'Conference' Pear Grown Under Deficit Irrigation: Implications for Fruit Quality at Harvest and After Cold Storage. Scientia Horticulaturae, 129 1 May 2011), 64 70 , 0304-4238

56. S. Lurie, B. Shapiro, S. Ben-Yehoshua, 1986 Effects of Water Stress and Degree of Ripeness on Rate of Senescence of Harvested Bell Pepper Fruit. Journal of the American Society for Horticultural Science, 111 6 December 1986), 880 885 , 0003-1062

57. S. Lurie, 1998 Postharvest Heat Treatments of Horticultural Crops: Commodity Responses. Postharvest Biology and Technology, 14 3 November 1998), 257 269 , 0925-5214

58. Leod. R. F. Mac, A. A. Kader, L. L. Morris, 1976 Stimulation of Ethylene and CO2 Production of Mature-Green Tomatoes by Impact Bruising. HortScience, 11 6 December 1976), 604 606 , 0018-5345

59. N. A. Maier, A. P. Dahlenburg, A. B. Frensharn, 1986 Potassium Nutrition of Irrigated Potatoes in South Australia. 3. Effect on Specific gravity, Size and Internal Bruising of Tubers. Australian Journal of Experimental Agriculture, 26 6 November/December 1986), 737 744 , 1836-0939

60. F. M. Mathooko, A. Inaba, R. Nakamura, 1998 Characterization of Carbon Dioxide Stress-Induced Ethylene Biosynthesis in Cucumber (Cucumis sativus L.) Fruit. Plant and Cell Physiology, 39 3 March 1998), 285 293 , 0032-0781

61. F. Maul, S. A. Sargent, C. A. Sims, E. A. Baldwin, M. O. Balaban, D. J. Huber, 2000 Recommended Storage Temperatures Affect Tomato Flavor and Aroma Quality. Journal of Food Science, 65 7 October 2000), 1228 1237 , 1750-3841

62. R. E. Mc Donald, T. G. Mc Collum, E. A. Baldwin, 1996 Prestorage Heat Treatments Influence Free Sterols and Flavor Volatiles of Tomatoes Stored at Chilling Temperature. Journal of the American Society for Horticultural Science, 12s 3 May 1996), 531 536 , 0003-1062

63. A. Mc Garry, C. C. Hole, R. L. K. Drew, N. Parsons, 1996 Internal Damage in Potato Tubers: A Critical Review. Postharvest Biology and Technology, 8 4 September 1996), 239 258 , 0925-5214

64. M. Meheriuk, R. K. Prange, P. D. Lidster, S. W. Porritt, 1994 Postharvest Disorders of Apples and Pears. Agriculture Canada, 0-66221-237-1 Canada

65. S. Meir, S. Philisoph-Hadas, S. Lurie, S. Droby, M. Akerman, G. Zauberman, B. Shapiro, E. Cohen, Y. Fuchs, 1996 Reduction of Chilling Injury in Stored Avocado, Grapefruit, and Bell Peppers by Methyl Jasmonate. Canadian Journal of Botany, 74 6 June, 1996), 870 874 , 1916-2790

66. A. R. Miller, 2003 Harvest and Handling Injury: Physiology, Biochemistry, and Detection, In: Postharvest Physiology and Pathology of Vegetables, 2nd Edition, J.A. Bartz & J. K. Brecht (Eds.), 177 208 , Marcel Dekker, Inc, 0-82470-687-0 York

67. R. Mittler, 2002 Oxidative Stress, Antioxidants and Stress Tolerance. TRENDS in Plant Science, 7 9 September 2002), 405 410 , 1360-1385

68. Y. Mizrahi, 1982 Effect of Salinity on Tomato Fruit Ripening. Plant Physiology, 69 4 April 1982), 966 970 , 0032-0889

69. Y. Mizrahi, E. Taleisnik, V. Kagan-Zur, Y. Zohar, R. Offenbach, E. Matan, R. Golan, 1988 A Saline Irrigation Regime for Improving Tomato Fruit Quality Without Reducing Yield. Journal of the American Society for Horticultural Science, 113 2 April 1988), 202 205 , 0003-1062

70. L. L. Morris, 1954 Field and Transit Chilling of Fall-Grown Tomatoes. In: Proceedings of the Conference on Transportation of Perishables, 101 105 , University of California, Davis, CA, USA.

71. R. Nilo, C. Saffie, K. Lilley, R. Baeza-Yates, V. Cambiazo, R. Campos-Vargas, M. González, L. A. Meisel, J. Retamales, H. Silva, A. Orellana, 2010 Proteomic Analysis of Peach Fruit Mesocarp Softening and Chilling Injury Using Difference Gel Electrophoresis (DIGE). BMC Genomics, 11 43 January 2010), ISSN1471-2164

72. G. Noctor, C. H. Foyer, 1998 Ascorbate and Glutathione: Keeping Active Oxygen Under Control. Annual Review of Plant Physiology and Plant Molecular Biology, 49 (June 1998), 249 279 , 0601-0249

73. D. M. Obenland, W. H. Vensel, W. J. Hurkman, 2008 Alterations in Protein Expression Associated with the Development of Mealiness in Peaches. Journal of Horticultural Science & Biotechnology, 83 1 January 2008), 85 93 , 1462-0316

74. C. Pignocchi, C. H. Foyer, 2003 Apoplastic Ascorbate Metabolism and Its Role in the Regulation of Cell Signalling. Current Opinion in Plant Biology, 6 4 August 2003), 379 389 , 1369-5266

75. A. C. Purvis, 1997 The Role of Adaptive Enzymes in Carbohydrate Oxidation by Stressed and Senescing Plant Tissues. HortScience, 32 7 December 1997), 1165 1168 , 0018-5345

76. M. K. Rai, R. K. Kalia, R. Singh, M. P. Gangola, A. K. Dhawana, 2011 Developing Stress Tolerant Plants Through In Vitro Selection- An Overview of the Recent Progress. Environmental and Experimental Botany, 71 2011 , 89 98 , ISSN

77. M. E. Saltveit, 1991 Prior Temperature Exposure Affects Subsequent Chilling Sensitivity. Physiologia Plantarum, 82 4 August 1991), 529 536 , 1399-3054

78. M. E. Saltveit, 2003 Temperature Extremes, In: Postharvest Physiology and Pathology of Vegetables, 2nd Edition, J.A. Bartz & J. K. Brecht (Eds.), 457 483 , Marcel Dekker, Inc, 0-82470-687-0 York

79. M. E. Saltveit, L. L. Morris, 1990 Overview of Chilling Injury of Horticultural Crops. In: Chilling Injury of Horticultural Crops, C.Y. Wang (Ed.), 3 15 , CRC Press, 0-84935-736-5 Raton, FL, USA

80. C. E. Sams, W. S. Conway, 2003 Preharvest Nutritional Factors Affecting Postharvest Physiology, In: Postharvest Physiology and Pathology of Vegetables, 2nd Edition, J.A. Bartz & J. K. Brecht (Eds.), 161 176 , Marcel Dekker, Inc, 0-82470-687-0 York

81. A. Sabehat, D. Weiss, S. Lurie, 1996 The Correlation Between Heat-Shock Protein Accumulation and Persistence and Chilling Tolerance in Tomato Fruit. Plant Physiology, 110 2 February 1996), 531 537 , 0032-0889

82. S. I. Shibairo, M. K. Upadhyaya, P. M. A. Toivonen, 1997 Postharvest Moisture Loss Characteristics of Carrot (Daucus carota L.) Cultivars

During Short-Term Storage. Scientia Horticulturae, 71 1-2 (November 1997), 1 12 , 0304-4238

83. S. I. Shibairo, M. K. Upadhyaya, P. M. A. Toivonen, 1998a Replacement of Postharvest Moisture Loss by Recharging and its Effect on Subsequent Moisture Loss During Short-Term Storage of Carrots. Journal of the American Society for Horticultural Science, 123 1 January 1998), 141 145 , 0003-1062

84. S. I. Shibairo, M. K. Upadhyaya, P. M. A. Toivonen, 1998b Influence of Preharvest Water Stress on Postharvest Moisture Loss of Carrots (Daucus carota L.). Journal of Horticultural Science and Biotechnology, 73 3 May 1998), 347 352 , 1462-0316

85. S. I. Shibairo, M. K. Upadhyaya, P. M. A. Toivonen, 1998c Potassium Nutrition and Postharvest Moisture Loss in Carrots (Daucus carota L.) Journal of Horticultural Science and Biotechnology, 73 6 November 1998), 862 866 , 1462-0316

86. S. I. Shibairo, M. K. Upadhyaya, P. M. A. Toivonen, 2002 Changes in Water Potential, Osmotic Potential, and Tissue Electrolyte Leakage During Mass Loss in Carrots Stored Under Different Conditions. Scientia Horticulturae, 95 1-2 (August 2002), 13 21 , 0304-4238

87. L. H. Stevens, E. Davelaar, 1997 Biochemical Potential of Potato Tubers To Synthesize Blackspot Pigments in Relation to Their Actual Blackspot Susceptibility. Journal of Agricultural and Food Chemistry, 45 11 November, 1997), 4221 4226 , 0021-8561

88. P. M. A. Toivonen, 2003a Effects of Storage Conditions and Postharvest Procedures on Oxidative Stress in Fruits and Vegetables. In: Postharvest Oxidative Stress in Horticultural Crops, D.M. Hodges (Ed.), 69 90 , Food Products Press, 1-56022-962-4 York

89. P. M. A. Toivonen, 2003b Postharvest Treatments to Control Oxidative Stress in Fruits and Vegetables. In: Postharvest Oxidative Stress in Horticultural Crops, D.M. Hodges, (Ed.), 225 246 . Food Products Press, 1-56022-962-4 York

90. P. M. A. Toivonen, 2005 Postharvest Storage Procedures and Oxidative Stress. HortScience, 39 5 August 2005), 938 942 , 0018-5345

91. P. M. A. Toivonen, 2009 Benefits of Combined Treatment Approaches to Maintaining Fruit and Vegetable Quality. Fresh Produce, 3 1 December 2009), 58 64 , 1749-4788

92. P. M. A. Toivonen, 2010 Postharvest Physiology of Vegetables, In: Handbook of Vegetables and Vegetable Processing, Y.H. Hui, N. Sinha, J. Ahmed, E.Ö. Evranuz and M. Siddiq (Eds.), 199 220 , Wiley-Blackwell Publishing, 978-0-81381-541-1 Ames, IA

93. P. M. A. Toivonen, T. H. J. Beveridge, 2005 Maturity, Ripening and Quality Relationships. In: Produce Degradation: Pathways and Prevention. O. Lamikanra, S.H. Imam & D.O. Ukuku (Eds.), 55 77 . CRC Press, 0-84931-902-1 Raton, FL

94. P. M. A. Toivonen, J. S. Brandenburg, Y. Luo, 2009 Modified Atmosphere Packaging for Fresh-cut Produce. In: Modified and Controlled Atmospheres for Storage, Transportation and Packaging of Horticultural Commodities, E. Yahia (Ed), 463 489 . CRC Press, 978-1-42006-957-0 Boca Raton, FL

95. P. M. A. Toivonen, J. R. De Ell, 2002 Physiology of Fresh-Cut Fruits and Vegetables, In: Fresh-Cut Fruits and Vegetables: Science, Technology, and Market, O. Lamikanra, (Ed.), 91 123 , CRC Press, 1-58716-030-7 Raton, FL

96. P. M. A. Toivonen, C. Hampson, S. Stan, 2005 Apoplastic Levels of Hydroxyl Radicals in Four Different Apple Cultivars are Associated with Severity of Cut-Edge Browning. Acta Horticulturae, 682 (June 2005), 1819 1824 , 0567-7572

97. P. M. A. Toivonen, C. Hampson, D. Neilsen, G. Neilsen, S. Stan, 2003 Relationship of Antioxidant Protection Systems to Browning or Flesh Browning-Related Disorders of Different Cultivars of Apples (Malus x domestica) and in Response to Phosphorus and Calcium Nutrition. HortScience, 38 3 June 2003), 798 0018-5345

98. P. M. A. Toivonen, M. Sweeney, 1997 Differences in Chlorophyll Loss at 13 C for Two Broccoli (Brassica oleracea L.) Cultivars Associated With Antioxidant Enzyme Activities. Journal of Agricultural and Food Chemistry, 46 1 January 1997), 20 24 , 1520-5118

99. V. Van linden, Sila. D. Ndaka, T. Duvetter, J. De Baerdemaeker, M. Hendrickx, 2008 Effect of Mechanical Impact-Bruising on Polygalacturonase and Pectinmethylesterase Activity and Pectic Cell Wall Components in Tomato Fruit. Postharvest Biology and Technology, 47 1 January 2008), 98 106 , 0925-5214

100. M. Veličković, 1994 The Influence of Irrigation of Fruit Weight Change of Some Major Cultivars During Refrigerated Storage. Review of

Research Work at the Faculty of Agriculture, Belgrade. 39 1 139 143 , 0522-8530

101. C. Y. Wang, 1993 Approaches to Reducing Chilling Injury of Fruits and Vegetables. Horticultural Reviews, 15 63 95 , 0163-7851

102. C. Y. Wang, 1994 Reduction of Chilling by Methyl Jasmonate. Acta Horticulturae, 368 901 907 , 0567-7572

103. C. Y. Wang, D. O. Adams, 1982 Chilling-Induced Ethylene Production in Cucumbers (Cucumis sativus L.). Plant Physiology, 69 2 February 1982), 424 427 , 0032-0889

104. W. Wang, B. Vinocur, O. Shoseyov, A. Altman, 2004 Role of Plant Heat-Shock Proteins and Molecular Chaperones in the Abiotic Stress Response. Trends in Plant Science, 9 5 May 2004), 244 252 , 1360-1385

105. I. Witkowska, E. J. Woltering, 2010 Pre-Harvest Light Intensity Affects Shelf-Life of Fresh-Cut Lettuce. Acta Horticulturae, 877 223 227 , 0567-7572

106. J. S. Whitmore, 2000 Drought Management on Farmland, Kluwer Academic Publishers, 0-79235-998-4 The Netherlands

107. C. M. Wu, S. E. Liou, 1986 Effect of Tissue Disruption on Volatile Constituents of Bell Peppers. Journal of Agricultural and Food Chemistry, 34 4 July 1986), 770 772 , 1520-5118

108. M. Yano, R. Saijo, Y. Ota, 1986 Inhibition of Browning and Ethylene Production in Shredded Cabbage by Isothiocyanates. Journal of the Japanese Society for Horticultural Science, 55 2 June 1986), 194 198 , 1882-3351

Chapter 3

Non-Conventional Tools to Preserve and Prolong the Quality of Minimally-Processed Fruits and Vegetables

Maria Rosaria Corbo *, Daniela Campaniello, Barbara Speranza, Antonio Bevilacqua and Milena Sinigaglia

Department of the Science of Agriculture, Food and Environment, University of Foggia, Via Napoli, 25, 71122 Foggia FG, Italy

ABSTRACT

The main topic of this paper is a focus on some non-conventional tools to preserve the microbiological and physico-chemical quality of fresh-cut fruits and vegetables. The quality of fresh-cut foods is the result of a complex equilibrium involving surface microbiota, storage temperature, gas in the headspace and the use of antimicrobials. This paper proposes a short overview of some non-conventional approaches able to preserve the quality of this kind of product, with a special focus on some new ways, as follows: (1) use of edible or antimicrobial-containing coatings (e.g., chitosan-based coatings) on fruits or vegetables; (2) alternative modified atmospheres (e.g., high O_2-modified atmosphere packaging (MAP)) or the use of essential oils in the headspace; (3) conditioning solutions with antimicrobials or natural compounds for fruit salad; and (4) biopreservation and use of a probiotic coating.

Keywords: fresh-cut; conditioning liquid; coatings; spoiling microorganisms; probiotics

1. INTRODUCTION

In recent years the attention towards a healthy diet is considerably growing; the consumer is paying more attention to the importance of the freshness and healthiness of foods. Consumption of fruits and vegetables is considered a successful way to contrast the current tendency toward obesity (especially in children). It is known that fruits and vegetables contain essential nutrients and are the major source of various antioxidants (vitamin C, vitamin E and β-carotene) that are necessary for the health, growth and development of the children. The World Health Organization (WHO), Food and Agriculture Organization (FAO), United States Department of Agriculture (USDA) and European Food Safety Authority (EFSA) recommended an increase of fruit and vegetable consumption to decrease the risk of cardiovascular diseases, cancer and ageing [1].

In addition, the consumer profile is changing as an effect of a stressed lifestyle: the number of working women and singles is gradually increasing; they are short on time and demand minimally-processed foods (also called ready-to-use) to save time on food preparation. To meet the consumer needs, the products should have an adequate shelf-life; it should be at least 4–7 days or even 21 days, depending on the products.

The International Fresh-cut Produce Association (IFPA) defines fresh-cut products as fruits or vegetables that have been trimmed and/or peeled and/or cut into a 100% usable product that is bagged or pre-packaged to offer consumers high nutrition, convenience and flavor, while still maintaining its freshness [2]. However, it is known that the processing of fruits and vegetables promotes physiological deterioration, biochemical changes and microbial degradation, beginning with raw materials, through processing methods and ending with packaging factors that affect the quality and shelf-life. These changes are due to tissue wounding, which includes browning, weight loss, accelerated respiration rate, off-flavor development, texture breakdown and increased susceptibility to microbial spoilage [3].

Thus, the search for new inexpensive and effective methods (such as chemical-based washing treatments, physical treatments, hurdle technology and packaging requirements) able to minimize these negative effects is of great interest to all of the stakeholders involved in the production and distribution of fresh fruits and vegetables.

The aim of this paper is to give insight into some non-conventional approaches able to ensure the microbiological stability and quality of minimally-processed fruits and vegetables with a special focus on some new ways currently available: (1) the use of edible or antimicrobial containing coatings (e.g., chitosan-based coatings) on fruits or vegetables; (2) alternative modified atmospheres (e.g., high O_2-modified atmosphere packaging (MAP)) or the use of essential oils in the headspace; (3) conditioning solutions with antimicrobials or natural compounds for fruit-salad; and (4) biopreservation and the use of probiotic coatings.

2. USE OF EDIBLE OR ANTIMICROBIAL-CONTAINING COATINGS (E.G., CHITOSAN-BASED COATINGS) ON FRUITS OR VEGETABLES

Edible coatings (ECs) and edible films belong to the modern food protection system; over the past few years, interest in the use of edible coatings for perishable foods has considerably increased due to their advantages and potential applications [4]. Edible coatings and edible films are terms that are often used interchangeably; however, a distinction is necessary.

An edible coating is defined as a thin layer of edible material applied to the surface of foods in addition to or as a substitution for natural protective coatings, able to form a barrier to moisture, oxygen and solute movement for the food [5,6,7,8,9,10].

On the other hand, edible film is defined as a thin layer of edible material formed on a product surface as a coating or placed (pre-formed) on or between food components [11].

Thus, an edible film is a thin skin, which has been pre-formed (for example, by casting a biopolymer solution separately from the food to form a film later applied to the food), whilst an edible coating is a suspension or an emulsion, which is applied directly to the food surface and later forms a film [12].

Edible coatings and films do not replace traditional packaging materials, but provide an additional factor to be applied for food preservation; ECs are consumed along with the food, thus the composition must conform to the regulations applied to the food product.

One of the advantages in using edible coatings and films is the reduction of water loss, considered one of the main factors in the deterioration of perishable foods. In fact, this thin layer protects fruits and/or vegetables against moisture loss, maintaining the texture and extending the shelf-life of the product, forming a protective barrier. On the other hand, when edible coatings are poor in water vapor barrier properties, a weight or moisture loss of the product could be recovered.

Numerous benefits result when edible coatings are applied, and these are summarized in Figure 1.

In addition, edible coatings may enhance sensory characteristics, can be consumed along with the food, provide additional nutrients and include quality-enhancing antimicrobials. Furthermore, they may reduce the cost and also the amount of traditional packaging used [13].

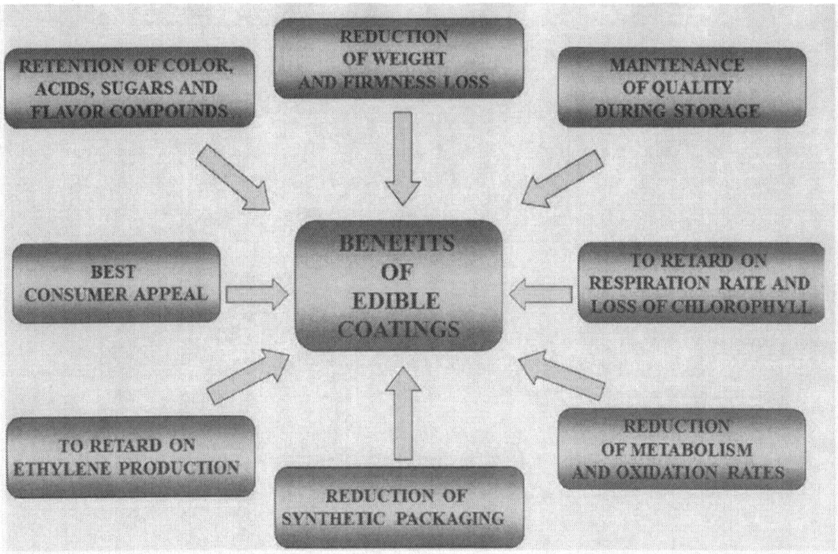

Figure 1. Main benefits of edible coatings.

2.1. Edible Coating Materials

Numerous varieties of fruits and vegetables are characterized by cuticle, a natural waxy layer on the surface, which, generally, has a low permeability to water vapor. To enhance the barrier properties of cuticle and/or substitute it when processing operations remove it, edible coatings could be applied.

Water and ethanol (or a blend of these compounds) are the solvents generally used for edible coating production.

Various coating materials could be added, but it is necessary that they possess some specific requirements (water solubility, hydrophobic of hydrophilic nature, easy formation of coatings, good sensorial properties) to obtain the resulting coating. In addition, an essential requirement is the plasticizing capacity to provide films with good flexibility properties; for this purpose, glycerol is generally used, followed by sorbitol, polyethylene glycol (PEG) and sugars [13].

Hydrocolloids have a good aptitude for forming continuous and cohesive matrices thanks to the hydrogen bonding of their polymeric chains.

Polysaccharides, proteins and lipids, alone or in combination, can also be used to produce edible films and coating. Lipids (together with waxes and fatty acids) do not have a suitable stand-alone filmmaking nature; they are usually opaque and relatively inflexible; the resulting films could also be quite fragile and unstable (rancidity) [9]. For this reason, lipids are generally incorporated into hydrocolloids-based films formulations [13].

The main polysaccharide, protein and lipid compounds used in edible coatings are reported in Table 1.

Polysaccharides include starch, dextrins, pullulan, cellulose and derivatives, alginate, carrageenan, gums, pectins and chitosan. Polysaccharides render transparent and homogeneous edible films; these films are oxygen, aroma and oil barriers (due to their tightly-packed, ordered hydrogen-bonded network structure and low solubility), but are not effective moisture barriers due to their hydrophilic nature. However, when applied in the form of high-moisture gelatinous coatings, they can retard the moisture loss of food [15].

Proteins can be obtained from animal sources, such as casein and whey protein (the main milk protein fractions: 80% and 20%, respectively), and from plant sources, such as zein, gluten and soy proteins [34].

Different proteins are able to form films and coatings; this ability depends on their molecular weight, conformations, electrical properties (charge *vs.* pH), flexibilities and thermal stabilities [18]. Nevertheless, proteins have been studied less extensively than polysaccharides.

Generally, protein-based films have good oxygen, carbon dioxide and lipid barrier properties and mechanical properties; on the other hand, the poor water vapor resistance limits their application; this can be attributed to the inherent hydrophilicity of proteins.

Table 1. Main polysaccharides, proteins and lipid compounds used for edible coatings.

Main Component	Description
	Starch contains amylose ((1→4)-α-D-glucopyranosyl) and amylopectin (amylose branched with side units of D-glucopyranosyl linked by α-1,6-glycosidic bonds). Amylose has a film-forming ability, rendering strong, isotropic, odorless, tasteless and colorless films. It is inexpensive, widely available and easy to handle. Due to its good oxygen barrier, starch is used for coating fruits and vegetables characterized by high respiration rates, thus suppressing respiration and retarding the oxidation of the coated products.
	Dextrins derive from starch and are characterized by a smaller molecular size. Coatings resulting from dextrins provide a better water vapor resistance than starch coatings [14].
	Pullulan derives from starch and is edible and biodegradable; relative films are transparent, elastic, odorless and tasteless [15]. Pullulan-based coatings have shown potential for preserving fresh strawberries and kiwifruits because of their barriers to moisture, O_2 and CO_2 [16]. Pullulan films themselves do not demonstrate antimicrobial activity [17].
Polysaccharides	**Cellulose** is the structural material of plant cell walls. It is composed of linear chains of (1→4)-β-D-glucopyranosyl units. Cellulose ethers (carboxymethylcellulose (CMC); methylcellulose (MC); hydroxypropyl cellulose (HPC); hydroxypropyl methylcellulose (HPMC)) are obtained by partial substitution of hydroxyl groups in cellulose by ether functions. Cellulose-derived films are tough, flexible, totally transparent, water soluble presence and are resistant to fats and oils [15,18], but are too expensive. Crosslinking treatments decrease the water solubility of cellulose ethers [19]. Cellulose derivative-based coatings were applied to some fruits and vegetables for providing barriers to oxygen, oil or moisture transfer [15]. CMC coatings help: (i) to retain the original firmness and crispness of apples, berries, peaches, celery, lettuce and carrots when used in a dry coating process; (ii) to preserve flavor components of some fresh fruits and vegetables; and (iii) to reduce oxygen uptake without increasing the carbon dioxide level in the internal environment of coated apples and pears by simulating a controlled atmosphere environment [15]. Li and Barth [20] observed that, after 3 weeks of storage, cellulose-based edible coating-treated carrots maintained a fresh appearance and higher carotene retention compared to controls.
	Alginates are extracted from brown seaweeds. They are sodium salts of alginic acid, which is a linear (1→4) linked polyuronic acid containing poly-β-D-manopyranosyluronic acid (M) blocks, poly-α-L-gulo pyranosyluronic acid (G) blocks and M–G blocks containing both polyuronic acids. Alginates possess good film-forming properties, producing uniform, transparent films with poor water resistance. Alginates react irreversibly with polyvalent metal cations (for example, calcium ions) to produce water-insoluble polymers. Alginate coatings are good oxygen barriers and retard lipid oxidation in fruits and vegetables. They show good ability to reduce weight loss and natural microflora counts in minimally-processed carrots [21]. Calcium alginate coatings reduce shrinkage, oxidative rancidity, moisture migration, oil absorption and sealing-in volatile flavors, improving appearance and color and reducing the weight loss of fresh mushrooms in comparison with uncoated ones [22].
Polysaccharides	**Carrageenan** is extracted from red seaweeds. It is a complex mixture of several water-soluble galactose polymers. There are 3 different carrageenans (lambda-, iota- and kappa-carrageenan), differing in sulfate ester content and the distribution of 3,6-anhydro-α-D-galactopyranosyl residues [13]. Carrageenan-based coatings were applied to fresh apples for reducing moisture loss, oxidation or disintegration of the apples [23,24]. In combination with ascorbic acid, carrageenan-based coatings resulted in positive sensory results and reduction of microbial levels on minimally-processed apple slices [24].
	Gums include exudate gums (arabic, tragacanth and karaya), seed gums (locust bean and guar) and microbial fermentation gums (xanthan and gellan). Xanthan gum provides uniform coatings. Gum arabic was used for coating pecan nut halves to eliminate a moist and oily appearance [25].
	Pectins are polymers mainly composed of (1→4) α-D-galactopyranosyluronic acid units naturally esterified with methanol. According to their content of methyl esters or the degree of esterification (DE), pectins are divided into high methoxyl (HM, DE > 50%) or low methoxyl (LM, DE < 50%). The DE has a decisive effect on pectin solubility and gelation properties. The literature on pectin-based coatings is poor. Maftoonazad *et al.* [26] evaluated the protective effect of a pectin-based edible emulsion coating on the activity of *Lasiodiplodia theobromae*, a plant pathogen, in avocados, finding that the pectin-based coating was effective at controlling the spread and severity of stem end rot in avocados.

	Chitosan is a natural polysaccharide prepared by the alkaline deacetylation of chitin β-1,4-*N*-acetylglucosamine (found in fungi, arthropods and marine invertebrate); commercially, it is produced from exoskeletons of crustacean, such as crab, shrimp and crawfish. Chitosan is used for its film-forming ability; it is environmentally friendly, due to its biodegradability, biocompatibility, antimicrobial activity, non-toxicity and versatile chemical and physical properties. When applied on fruit and vegetables, chitosan-based coatings have shown effectiveness in delaying ripening and decreasing respiration rates [27,28].
Proteins	**Casein** forms transparent, flavorless and flexible films. Casein-based emulsion films (emulsified with lipid-based materials) were more effective than pure caseinate films in controlling moisture loss of fruits and vegetables [29].
	Whey proteins produce transparent, flavorless and flexible films, similar to caseinate films. Whey protein-based films possess excellent oxygen barrier properties comparable to the synthetic polymer films [30]; they also are good grease barriers [31,32].
	Zein is the principal protein of corn. It is characterized by water vapor permeabilities. Applied on vegetable and fruits, zein-based coatings are able to retard the ripening of tomatoes, to maintain the color and firmness of broccoli florets, to provide an adhesive and stable coating, to reduce the growth of *Listeria monocytogenes* on cooked sweet corn and to maintain the gloss and other qualities of apples [15].

Main Component	Description
Proteins	**Soy proteins (SP)** are extracted from defatted protein meal. Soy protein coatings generally have poor moisture resistance and water vapor barrier properties due to the inherent hydrophilicity of the protein and the addition of hydrophilic plasticizers; whilst they are potent oxygen barriers. SP coatings are used to preserve the freshness of apple slices and/or to retard the senescence process of kiwifruit [15].
Lipids	Most **fatty acids** derived from vegetable oils are considered GRAS (generally recognized as safe). Lipid-based coatings are compatible with other coating-forming agents; in addition, they have high water vapor and gas barrier properties. However, lipid-based coatings present a greasy surface and undesirable organoleptic properties, such as waxy taste and lipid rancidity [33]. Finally, some lipid materials are unstable when subjected to temperature changes.
	Waxes (carnauba wax, beeswax, paraffin wax and others) have been used as protective coatings for blocking moisture transport, reducing surface abrasion during fruit handling, controlling the browning of the skin in fruits by improving mechanical integrity and controlling the internal gas composition of the fruits. Wax coatings are applied on fruits (citrus, apples, cucumbers) and vegetables (tomatoes, asparagus, beans, beets, carrots, celery, eggplant, peppers, potatoes, radishes, squash and turnips.
	Resin coatings are effective at reducing water loss, but are the least permeable to gases; thus, fruit can suffer undesirable changes, such as anaerobic respiration and flavor changes.

As protein films are generally brittle and susceptible to cracking (due to the strong cohesive energy density of the polymers), an improvement of their properties could be attained by adding other components; for example, the addition of compatible plasticizers could improve their extensibility [35].

Edible lipid coatings include neutral lipids, fatty acids, waxes and resins. These compounds are effective in providing a moisture barrier and improving the surface appearance. Triglycerides or neutral lipids form a continuous stable layer on the food surface, thanks to their high polarity.

The growing interest addressed toward edible coatings leads to the formulation of new compositions consisting of blends of polysaccharides, proteins and lipids. The combination could be proteins and carbohydrates, proteins and lipids and carbohydrates and lipids [13], and it is based on the fact that each polymer has characteristic functions, that, when combined with each other, enhance the final functionality of the coating [15], improving the mechanical properties and, with emulsifiers, stabilizing composite coatings and improving coating adhesion.

Edible coatings can incorporate several compounds, such as: plasticizers (glycerol, acetylated monoglyceride, polyethylene glycol and sucrose), antimicrobials (bacteriocins (nisin)), enzymes (lysozyme, peroxidase and lactoperoxidase), essential oils (cinnamon, oregano, lemongrass, clove, rosemary, tea tree, thyme and bergamot), nitrites and sulfites [36], as well as synthetic antioxidants (butylated hydroxyanisole, butylated hydroxytoluene, propyl gallate, octyl gallate, dodecyl gallate, ethoxyquin, ascorbyl palmitate and tertiary butyl hydroquinone) and natural antioxidants (tocopherols, tocotrienols, ascorbic acid, citric acid, carotenoids). In these cases, edible films and coatings act as carriers of active compounds that, applied to the surface of fruits and vegetables, lead to the extension of shelf-life, the reduction of the risk of foodborne pathogenic microorganisms' growth on cut surfaces [37] and the improvement of the quality, stability and safety of coated foods [15].

2.2. Edible Coating: A Focus on Chitosan

Physical and chemical damage accrued during minimal processing of fruits and vegetables causes disruption of the plant tissues, and the exudates become ideal substrates for the growth of several microorganisms (pathogens, molds and bacteria). Natural biodegradable compounds with antimicrobial activity are recognized as safe (generally recognized as safe (GRAS)) and environmentally friendly, and chitosan is one such compound. Chitosan is able to create a semi-permeable film on the fruit surface, which results in limiting respiration and/or transpiration and in reducing weight loss [38,39]. Furthermore, its compatibility with other substances and its capability to induce host resistance to pathogens [40] have prompted its application as a coating on fruits and vegetables [41].

Chitosan has been widely used in controlling weight-loss in fresh strawberries (*Fragaria x ananassa*) and raspberries (*Rubus idaeus*), mango (*Mangifera indica*), litchi, blueberries and other fruit and vegetables [42]. Meng *et al.* [43] and Romanazzi *et al.* [44] reported that postharvest application of chitosan coating has a good control effect on decay of grapes. Chitosan owes its antimicrobial activity to its polycationic nature, which allows the interaction and formation of polyelectrolyte complexes with acid polymers produced at the surface of microbial cells, increasing their permeability and causing cell death [45].

Several factors affect the antimicrobial activity: type of chitosan, degree of acetylation, molecular weight, concentration, target microorganism, pH of the medium and presence of other additives or food components [46].

Benhabiles *et al.* [47] demonstrated that by reducing the number of steps for the synthesis and chemical reagents, chitosan coating was effective at improving the quality of strawberries by delaying changes in weight loss and the appearance of molds. These authors used three different chitosan coatings (chitosan C1 obtained by the classical method, chitosan C2 without decoloration and chitosan C3 without the decoloration and deproteinization steps) on strawberries, and they observed the best quality for the strawberries coated with C3 (1%), which was obtained through a lesser number of steps.

In a recent work, chitosan (in acid and water solution) exhibited its antibacterial activity against *Burkholderia seminalis*, an apricot fruit rot pathogen [48]. Lou and his coworkers demonstrated that acid-solution chitosan at a concentration of 2 mg/mL inhibited *B. seminalis*, while water-solution chitosan showed limited inhibition activity.

Chitosan has been proven to be a natural compound with antifungal activity for a wide varieties of postharvest fruits [38,49]. As reported by Bautista-Banos *et al.* [38], the level of inhibition of fungi is highly correlated to chitosan concentration. It is known that the polycationic nature of this compound is the key to its antifungal properties and that the length of the polymer chain enhances its antifungal activity. An additional explanation includes the possible effect that chitosan might have on the synthesis of certain fungal enzymes. El Ghaouth *et al.* [50] studied the antifungal effect of chitosan against *Botrytis cinerea* and *Rhizopus stolonifer*. These authors hypothesized that the mechanisms by which chitosan coating reduced the decay of strawberries appear to be related to its fungistatic properties, rather than to its ability to induce defense enzymes. A further confirmation of chitosan's ability to control fungal growth was reported by Park *et al.*[51]. They obtained a reduction of 2.5 and 2 log CFU/g in the counts of *Cladosporium* sp. and *Rhizopus* sp., respectively, on strawberries coated with a chitosan-based edible film, just after the coating application. A reduction in the counts of aerobic and coliform microorganisms during storage has been also reported. Chien *et al.* [52] investigated the effects of coating with low and high molecular weight chitosan on the decay of citrus and the maintenance of its quality. A concentration of 0.2% low molecular weight chitosan (LMWC) exhibited its antifungal activity in controlling the growth of *Penicillium digitatum*and *Penicillium italicum*. LMWC coating resulted also in being able to improve firmness, titratable acidity, ascorbic acidity and the water content for citrus stored at 15 °C for 56 days.

González-Aguilar *et al.* [53] have also reported a reduction in mesophilic aerobic microorganism count when fresh-cut papayas were coated with an edible coating based on chitosan of low and medium molecular weight. These researchers also observed a complete inhibition of yeast and molds throughout the storage (14 days at 5 °C). As reported by Ali *et al.* [54], chitosan preserved papaya fruit, delaying the ripening process by reducing the respiration rate. These results could be the reason for the delayed senescence and reduced tendency to decay [40]. Chitosan had also made improvements in the taste, peel and pulp color, texture and flavor of treated papaya fruit, but the sensory features of the papaya fruits coated with a 1.5% chitosan concentration demonstrated overall superiority after five weeks of storage.

Pilon *et al.* [55] made an alternative use of chitosan. They obtained chitosan nanoparticles and demonstrated that chitosan, used as a coating based on nanoparticles, reduced the microbial growth in fresh-cut apples. The samples treated with chitosan-tripolyphosphate (CS-TPP) nanoparticles (10 nm) showed higher antimicrobial activity against mesophilic and psychrotrophic bacteria, as well as molds and yeasts than conventional chitosan coating and control [55].

Numerous papers have demonstrated that chitosan-based coatings inhibit microbial growth on fresh produce, increasing shelf-life.

In a recent paper, Benhabiles *et al.* [56] reported that chitosan coatings and a chitosan derivative (N,O-carboxymethyl chitosan (NOCC)) coating improved the quality of tomato fruits (through delaying ripening, reducing weight loss and retaining fruit firmness) and extended the shelf-life. No microbial growth was observed during storage.

Assis and Pessoa [57] and Han *et al.* [58] proposed chitosan for extending the shelf-life of sliced apples and fresh strawberries, respectively. Chien *et al.* [52] reported the effectiveness of chitosan coating (at a concentration of 0.5%, 1% and 2% (w/v)) for prolonging the quality and extending the shelf-life of sliced mango fruit through a delay in the growth of mesophilic aerobic bacteria.

Durango *et al.* [59] and Devlieghere *et al.* [60] used a chitosan-based coating to cover carrots and lettuce, respectively, observing a reduction in the respiration rate and ethylene production, as well as a decrease in firmness loss. In particular, Durango *et al.* [59] controlled the growth of mesophilic aerobes, yeasts, molds and psychrotrophs of minimally-processed carrots during the first five days of storage at 15 °C using an edible yam starch coating containing chitosan. Campaniello *et al.* [28] observed that chitosan coating in combination with low temperature and suitable packaging was able to control browning and decay in strawberry fruits. Chitosan affected the microbial growth (psychrotrophic, lactic acid bacteria and yeasts) and did not affect the visual appearance. pH and thickness values remained unchanged by chitosan coating, whereas color was positively influenced.

Pushkala *et al.* [61] investigated chitosan-based powder coating on radish shreds, demonstrating the favorable effects of two different forms of chitosan (purified chitosan (CH) and chitosan lactate (CL)) on shelf-life extension of radish shreds by a minimum of 3 d over the control. Both CH- and CL-coated samples enhanced the microbial quality and sensory acceptability of the radish shreds, exhibiting a lower degree of weight loss, respiration rate, titrable acidity, % of soluble solids, a higher content of phytochemicals, moisture and pH, compared to control samples. The treated samples also exhibited lower exudate volume, lesser browning and lower microbial load compared to the control.

Sometimes, a chitosan-only coating demonstrated certain defects (including limited inhibition to some microorganisms that led fruit to decay and a poor coating structure); thus, chitosan was combined with other substances to improve its performance [62]. Chitosan coatings combined with organic acids are easy to handle, biodegradable and cause no harm to the coated fruit and/or vegetable [62]. For example, Yu *et al.* [63] combined 1% phytic acid (known as inositol hexakisphosphate (IP6), inositol polyphosphate) with 1% chitosan to preserve fresh cut lotus root. This composite coating decreased the weight loss rate and malondialdehyde (MDA) content of fresh-cut lotus root, postponed browning, restrained the activities of peroxidase (POD), polyphenol oxidase (PPO) and phenylalanine ammonia-lyase (PAL) and maintained the content of vitamin C and polyphenol at a relatively high level.

Chitosan combined with natamycin significantly decreased fresh melon decay and weight loss caused by *Alternaria alternata* and *Fusarium semitectum*, two strains of spoilage fungi [64].

Zhang *et al.* [65] reported that chitosan was able to inhibit the growth of *Botrytis cinerea* and *Rhizopus* sp. by increasing the activities of various defense enzymes, such as β-1,3-glucanase (in orange, strawberries and raspberries) and phenylalanine ammonia-lyase (PAL) (in strawberries andtable grapes).

As expected, the antimicrobial activity of chitosan is also dependent on the food matrix; generally, the antimicrobial activity of chitosan is higher at a low pH because more of its amino groups are protonated; thus, it is able to interact with the negatively-charged surfaces inhibiting bacterial growth. Since the antimicrobial activity of chitosan is dependent on the charges on chitosan and the electrostatic forces, each food component could influence these interactions, affecting its activity.

Regarding this issue, Devlieghere *et al.* [60] published an interesting work on the interaction of chitosan and food components. The authors evaluated the effect of different food components (starch, proteins, NaCl and fat) on the antimicrobial activity of chitosan, following the growth of *Candida lambica*, a spoiling yeast strain, in a laboratory medium.

The authors reported that the higher concentrations of starch (30% (w/v)) inhibited the antimicrobial activity of chitosan, hypothesizing that it was due to a protective effect of starch or to electrostatic interactions when the starch was charged by modification. Proteins influenced the antimicrobial activity of chitosan depending on the pH of the medium, as their charges depend on the combination of the iso-electric point (IEP) of the proteins and the pH of the medium. If the pH is lower than the IEP, proteins and chitosan are positively charged; thus chitosan can exert its antimicrobial activity, as the interactions between both will be restricted. If the pH is higher than the IEP of the protein, the antimicrobial activity of chitosan is inhibited: proteins are negatively charged and neutralize most of the positive charges on the chitosan; thus, it cannot interact with the negatively-charged microbial surfaces [60].

NaCl reduces the antimicrobial activity of chitosan, because it interferes with the electrostatic forces between chitosan: Cl^- ions neutralize the positive charges on the chitosan, and on the other hand, the Na^+ ions compete with chitosan for the negative charges on the cell surface. Devlieghere *et al.* [60] observed an improved solubility of chitosan by adding NaCl at different concentrations to the medium. The authors explained that this behavior was probably due to a shielding effect of NaCl against the positive charges, which led to a coiled structure of chitosan and less interactions with components in the media. Finally, the influence of fat on the antimicrobial activity of chitosan was found negligible.

The addition of the essential oils to enhance chitosan antimicrobial action is common.

The active agents embedded into composite films may be released into or absorb substances from the packaged food or its surrounding environment. The interactions that can occur between the food product, the film or coating used as packaging and the surrounding environment are governed by different mass transfer processes: migration, adsorption, absorption and permeation [66]. For example, through these processes, the essential oils included in the coating could be transferred to the product and modify its organoleptic and nutritional characteristics by interacting with peptides, vitamins, *etc.* It is recognized that some important food components, such as vitamins, minerals and other nutrients, can also be sequestered with the consequent modification of food properties and functionality. In addition, when transferred from the coating to the food to pursue their protective action (antioxidants, antimicrobials, *etc.*), these compounds could also lose their effectiveness. It is, in fact, well recognized that generally, the efficacy of various antimicrobial compounds may be reduced by food components; for example, essential oils (EOs) and/or their components have a significant antimicrobial activity *in vitro*, but higher amounts are required (1%–3%) to achieve the same effect in foods. The presence of fat, protein, carbohydrate, water, salt, antioxidants, preservatives, other additives and pH reaction strongly influence the effectiveness of the natural compounds in foods.

High levels of fat and/or protein in foodstuffs protect bacteria from the action of EOs in some way. Some authors suggested that the fat provides a protective layer around the bacteria, or the lipid fraction could absorb the antimicrobial agent, thus decreasing its concentration and effectiveness in the aqueous phase [67]. Patrignani *et al.* [68] reported that the low fat content of vegetables may contribute to the successful use of EOs. Due to their lipophilic nature, EOs could share fat, missing the microbial target.

The pH is an important factor affecting the activity of oils. At a low pH, the hydrophobicity of some EOs increases due to their ability to dissolve more easily in the lipid phase of the bacterial membrane, thus enhancing the antimicrobial action.

Carbohydrates in foods do not protect bacteria from this action, whilst high water and/or salt level seems to facilitate the action of EOs.

The antibacterial activity of chitosan-based films combined with lemon (LO), tea tree (TTO) or bergamot essential oils (BO) was tested against two Gram-positive bacteria (*Staphylococcus aureus* and *L. monocytogenes*) and one Gram-negative bacteria (*Escherichia coli*) [69]. CH-EO composite films exhibited a significant antimicrobial activity against the three pathogens tested. The nature and concentration of the EOs, the film matrices and the interactions between CH and EOs affected the antimicrobial activity of the films. When TTO was added, CH exhibited the highest antimicrobial activity.

Chitosan-cinnamon oil coating extended the shelf-life of sweet pepper: after storage at 8 °C for 35 days, samples treated with chitosan-oil coatings showed a lower percentage of infected peppers and good sensory acceptability [70]. Sessa *et al.*[71] studied a novel approach to preserve vegetable products through modified chitosan edible coatings containing nanoemulsified natural antimicrobial compounds (lemon, mandarin, oregano or clove essential oils). A modified chitosan edible coating combined with lemon essential oil resulted in a remarkable increase in antimicrobial activity, with respect to other essential oils. This combination prolonged the shelf-life of rucola leaves from 3 to7 days, in comparison to the untreated samples.

Moreover, the modified chitosan containing the nanoemulsified antimicrobial caused a significantly longer shelf-life also in comparison to a coating made of modified chitosan or essential oil alone. Thanks to this novel treatment, it was possible to prolong the shelf life of rucola leaf vegetables to about 10–14 days, without alteration of the organoleptic properties of the product, preventing the loss of firmness and color changes and preserving palatability during storage.

More recently, Randazzo *et al.* [72] used chitosan-based and methylcellulose-based films added to several EOs derived from citrus fruits (orange, mandarin and lemon) to perform the antilisterial assay and concluded that chitosan films containing essential oil from lemon were the most effective at reducing *L. monocytogenes* counts.

A chitosan-methyl cellulose-based film was used as a coating on cantaloupe fruit. The application of a chitosan (1.5% w/v)/methyl cellulose (0.5% w/v) film on fresh-cut cantaloupe reduced populations of *E. coli* inoculated on fresh-cut cantaloupe by more than 5 log CFU/piece in 8 days at 10 °C. Furthermore, a reduction of 3 log CFU/piece of *Saccharomyces cerevisiae* in 4 days of storage at 10 °C was reported [73].

Krasaekoopt and Mabumrung [74] observed that the incorporation of 1.5% and 2% chitosan in the methylcellulose coating, applied on fresh-cut cantaloupe, produced a better microbiological quality in the final product. This coating reduced the growth of mesophilic aerobes, psychrotrophs, lactic acid bacteria, yeast and molds and prevented the multiplication of *E. coli* and *Salmonella* strains.

3. ALTERNATIVE MODIFIED ATMOSPHERES (E.G., HIGH O_2-MAP) OR THE USE OF ESSENTIAL OILS IN THE HEADSPACE

The final operation in producing minimally-processed fruits and vegetables is packaging.

Packaging performs several functions: it protects fresh-cut fruits and vegetables against deteriorative effects, contains the products, communicates to the consumers as a marketing tool and provides consumers products with ease of use and convenience. Fresh fruits and vegetables continue to respire, consuming oxygen and producing carbon dioxide and water vapor; thus, the atmosphere surrounding the package changes into another composition before sealing in vapor-barrier materials [75].

Modified atmosphere packaging (MAP) is a technique used for prolonging the shelf-life of fresh or minimally-processed foods [76]. MAP reduces the unwanted metabolic reactions and helps to protect processed products against contamination by microorganisms, thus slowing down the process of ageing [77]. MAP can be vacuum packaging (VP) and controlled atmosphere packaging (CAP). VP removes most of the air, or air is removed by vacuum or flushing and replaced with another gas mixture, before the product is packaged; the headspace atmosphere and product could change during storage, without additional manipulation of the internal environment. Differently, CAP continuously controls the environment to maintain a stable gas atmosphere within the package, as well as the temperature and humidity are monitored; it is generally used to control the ripening and spoilage of fruits and vegetables [76]. The composition, microflora, pH and the organoleptic characteristics of fruits and vegetables are extremely variable; therefore, it is impossible to define a univocal mixture of gases. It is necessary to maintain an optimum balance of oxygen and carbon dioxide; thus, an appropriate MAP must be studied for each product. Many factors affect the MAP of fresh produce and are summarized in Figure 2.

3.1. Gases Used in MAP

The main gases used in MAP are CO_2, O_2 and N_2; carbon monoxide (CO) and sulfur dioxide (SO_2) are also used. Noble gases, generally, are used for some products, such as coffee and snacks, but recently have been used also for minimally-processed apples and kiwi fruits [76]. The choice of gas depends on the food product being packed. Singly or in combination, these gases are commonly used to balance safe and shelf-life extension with optimal sensorial properties of the food. Table 2 reports the main gases used in MAP.

Low levels of oxygen and high levels of carbon dioxide reduce the produce respiration rate, delay senescence and, consequently, extend the shelf-life of fruits and vegetables. Once the package is closed, the composition of the gases inevitably changes due to produce respiration and the gas permeability of the film. If the oxygen levels are too low, fermentative processes are favored.

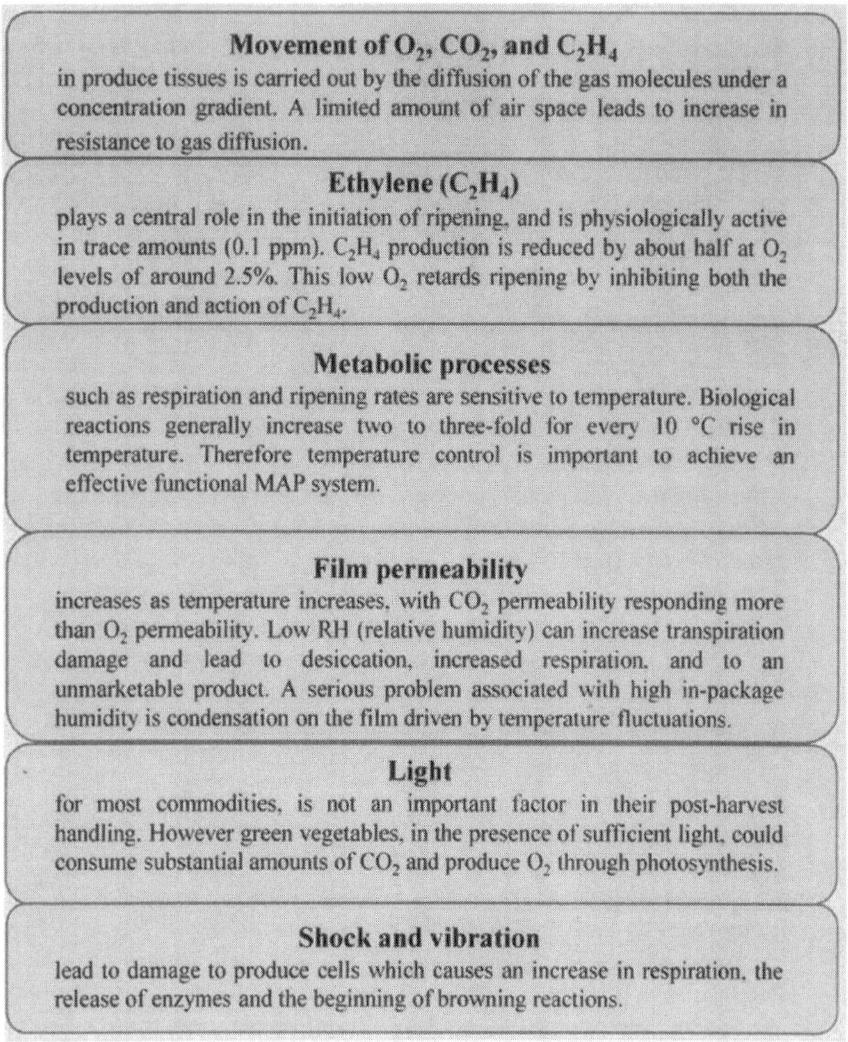

Figure 2. Factors affecting MAP of fresh products [76].

Table 2. Main gases used in MAP.

Gases	Characteristics and Functions
Carbon Dioxide (CO_2)	It is a colorless gas with a slight pungent odor at very high concentrations. It is an asphyxiant and slightly corrosive in the presence of moisture. CO_2 is lipid soluble and also dissolves readily in water to produce carbonic acid (H_2CO_3), which causes an increase of the acidity of the solution, reducing the pH; this significantly affects the microbiology of packed foods. CO_2 has bacteriostatic and fungistatic properties. This bacteriostatic effect depends on the concentration and partial pressure of CO_2, the volume of headspace gas, the type of microorganism, the age and load of the initial bacterial population, the microbial growth phase, the growth medium used, the storage temperature, the acidity, the water activity and the type of product being packaged [78–81]. As MAP contains CO_2, it is not an advisable option for these products where spoilage yeasts are stimulated by high levels of CO_2. In addition, CO_2 is not effective towards *Clostridium perfringens* and *Clostridium botulinum*. Carbon dioxide is generally effective in fresh fruits and vegetables where the normal spoilage organisms consist of aerobic, Gram-negative psychrotrophic bacteria. To guarantee the maximum antimicrobial effect, low storage temperatures for MAP are recommended, because CO_2 is most soluble at low temperatures. The fat and moisture of the product affect the absorption of CO_2. Fruits and vegetables can suffer physiological damage due to high CO_2 levels.
Oxygen (O_2)	Oxygen is a colorless and odorless gas and is highly reactive. It has a low solubility in water and promotes several deteriorative reactions in foods (fat oxidation, browning reactions and pigment oxidation). Most of the common spoilage bacteria and fungi require oxygen for growth; therefore, if the pack atmosphere contains a low concentration of oxygen, the shelf-life of foods is prolonged. As an alternative to low oxygen concentrations, superatmospheric O_2 concentrations ($\geq$70 kPa) have been proposed.
Nitrogen (N_2)	Nitrogen is an inert, tasteless, colorless and odorless gas and is relatively un-reactive. It has a lower density than air, is non-flammable and has a low solubility in water and other food constituents. N_2 is used to balance CO_2 and O_2 gases during food packaging, thus preventing packaging collapse that could occur when high concentrations of CO_2 are used. In addition, N_2 is able to delay oxidative rancidity and inhibit the growth of aerobic microorganisms without affecting the growth of anaerobic bacteria. Nitrogen can also indirectly influence the microorganisms in perishable foods by retarding the growth of aerobic spoilage microorganisms.
Carbon Monoxide (CO)	CO is a colorless, tasteless and odorless gas; it has a low solubility in water, but it is relatively soluble in some organic solvents. CO has been licensed for use in the USA to prevent browning in packed lettuce. It is highly reactive and very inflammable, thus its commercial application has been limited. Finally, CO has little inhibitory effect on microorganisms.
Noble Gases	These include helium (He), argon (Ar), xenon (Xe) and neon (Ne). These gases are not very reactive and are used in numerous food applications, e.g., potato-based snack products.

It is known that fresh-cut processing involves several forms of damage:

- Increases respiration rates;
- Causes major tissue disruption: enzymes and substrates, normally contained within the vacuole, are mixed with other cytoplasmic and nucleic substrates and enzymes;
- Increases wound-induced C_2H_4, water activity and surface area per unit volume, which lead to an accelerated water loss and favored microbial growth;

- Causes flavor loss;
- Causes cut surface discoloration or color loss;
- Causes decay;
- Causes increased rate of vitamin loss;
- Causes rapid softening;
- Causes shrinkage and a shorter storage life.

Thus, MAP facilitates the maintenance of fresh-cut products.

Actually, the safety of minimally-processed fruits and vegetables is mainly based on the correct chilling chain and hygienic practices, which seem to not be able to guarantee a sufficient degree of safety; in fact, the number of documented outbreaks of human infections associated with the consumption of these products has considerably increased. On these bases, traditional MAP is not enough to ensure the improvement of the quality and safety characteristics. In this context, many researchers have studied alternative tools, for improvement of MAP for minimally-processed fruits and vegetables; for example, the use of non-conventional atmospheres, including active packaging containing some natural antimicrobial compounds or MAP with high O_2, has been proposed.

3.2. MAP and Natural Antimicrobial Compounds

Plants and plant products are generally considered natural alternatives to improve the shelf-life and the safety of foods, since they are characterized by a wide range of GRAS volatile compounds, which are used as food flavoring agents [76]. They are able to inhibit numerous microorganisms; thus they are used as components of biological means for prolonging the shelf-life of post-harvest or minimally-processed fruits and vegetables [76].

Literature data indicate that volatile compounds can represent a useful tool to increase the shelf-life of plant products. Corbo *et al.* [82] evaluated the effects of hexanal and *trans*-2-hexenal on the shelf-life of fresh sliced apples. The authors added *trans*-2-hexenal to the gas mixture containing 70% N_2 and 30% CO_2 and found a significant extension of the shelf-life also when *Pichia subpelliculosa* (a spoilage yeast) was inoculated and abuse storage temperatures were used, although it had a weak negative effect on color retention.

Lanciotti *et al.* [83], postulated that future trends in the use of natural compounds, such as hexanal, 2-(E)-hexenal and hexyl acetate, would be focused on the use of specific active packaging able to release the active molecules in the head space slowly over time.

The authors reported that 150 ppm of hexanal, 20 ppm of 2-(E)-hexenal and 150 ppm of hexyl acetate displayed a bactericidal effect on *L. monocytogenes* and caused a significant extension of the lag phase of *E. coli* and *Salmonella* Enteritidis inoculated at levels of 10^4–10^5 CFU/g in fresh sliced apples packaged in ordinary or modified atmosphere.

Campaniello *et al.* [84] investigated the possibility of combining hexanal and MAP (65% N_2, 30% CO_2 and 5% O_2) on minimally-processed cactus pear fruits. The hexanal showed an antimicrobial effect against *Enterobacteriaceae*, normally contaminating minimally-processed fruits, both in the control and the modified atmosphere. The inclusion of the antimicrobial compound in the atmosphere determined an improvement of the original color retention; as well as a mesophilic selection favoring *Pantoea* spp., which have antagonistic activity against molds responsible for the decay of fruits during post-harvest phase.

Siroli *et al.* [85] proposed the use of several antimicrobial compounds (citron EO, hexanal, 2-(E)-hexenal, citral and carvacrol) alone or in combination in order to increase the shelf-life and quality parameters (texture and color) of sliced apples packaged in active modified atmosphere (7% O_2 and 0% CO_2), into medium permeability bags and stored at 6 °C. In all of the samples, the spoilage yeast threshold was not attained within the 35 days of storage. When treated with citral/2-(E)-hexenal and hexanal/2-(E)-hexenal, the sample showed a good color retention. This latter combination also improves on the retention of firmness, which was the best throughout 35 days of storage.

Furthermore, plant essential oils, constituted mainly by terpenoids, are used for their antimicrobial activity against many microorganisms. Most of the essential oils are GRAS; nevertheless, their use is often limited because of a high impact on the organoleptic characteristics of food products.

The activity of oils from *Labiatae* and citrus fruits, as well as the action of single constituents have been studied in order to better understand the cell targets of the molecules. Due to their antimicrobial effect, citrus essential oils could represent good candidates to improve the shelf-life and the safety of minimally-processed fruits. Essential oils, such as citrus, mandarin, cider, lemon and lime, were able to increase the shelf-life and safety of minimally-processed fruit salads without any impact on the sensory characteristics, even when the product was inoculated with spoilage or pathogenic bacterial species [76].

Citrus essential oils (EOs) exhibited their antimicrobial effect against a range of food poisoning-causing bacteria. A blend of citrus EO vapor against vancomycin-resistant (VRE) and vancomycin-susceptible (VSE) *Enterococcus faecium* and*Enterococcus faecalis* on lettuce and cucumber was assessed. Food samples were subjected to the vapor for 45 s in a 600-L vapor chamber at 25 °C. Results showed that microbial cell load was reduced and that no significant changes in taste were observed [86].

Tian *et al.* [87] tested the antifungal activity of essential oil extracted from the fruits of *Cicuta virosa L.* var. *latisecta Celak*(CVEO) against *Aspergillus niger*, *Aspergillus flavus*, *Aspergillus oryzae* and *Al. alternata* strains inoculated on cherry tomatoes. The samples were pre-treated with ethanol and wounded with a sterilized cork borer; then, each fruit was separately inoculated with 10 μL of a spore suspension containing 1×10^6 spores/mL of each fungal strain. CVEO (dissolved separately in 0.5 mL of 5% of Tween-20) was pipetted aseptically onto filter paper discs respectively placed into individual weighing bottles (without lids) to produce the concentrations of 200, 100 and 50 μL/mL. The essential oil was vaporized inside the containers spontaneously at 18 °C. The authors reported that at 200 μL/mL, CVEO showed the lowest percentages of decayed cherry tomatoes for all fungi compared to the control, as well as the highest inhibition of fungal infection.

In a recent paper, Vitoratos *et al.* [88] studied the antifungal activity of several essential oils obtained from oregano (*Origanum vulgare* L. ssp. *hirtum*), thyme (*Thymus vulgaris* L.) and lemon (*Citrus limon* L.) plants against *Bo. cinerea*inoculated in tomatoes, strawberries and cucumbers. All of the fruits were pre-treated and packaged. Different concentrations of essential oils were placed in small glass containers placed in the bottom of the package. Oregano and lemon oils were very effective at controlling disease of infected fruit by *Bo. cinerea* in tomatoes, strawberries and cucumbers. In particular, in tomatoes, *Bo. cinerea* was inhibited by oregano essential oils at 0.30 μL/mL; moreover, lemon essential oils also induced a significant reduction of grey mold. An amount of 0.05 μL/mL of lemon essential oils leads to a complete inhibition of *Bo. cinerea* in strawberries, whilst in cucumber, it leads to a reduction (39%) of the infected fruits.

3.3. Potential Applications of High O_2

It is known that low O_2 modified atmosphere packaging is universally accepted to prolong the shelf-life of minimally-processed fruits and vegetables; however, the growth of some anaerobic pathogen strains might be allowed or even stimulated. In addition, when O_2 levels are too low, this induces anaerobic respiration, leading to fermentation processes and to the production of undesirable metabolites [89].

The application of high O_2 concentrations (>70% O_2) could overcome the disadvantages of low O_2 modified atmosphere packaging for some ready-to-eat fruits and vegetables. High O_2 was found to be particularly effective at inhibiting enzymatic discoloration, preventing anaerobic fermentation reactions and inhibiting microbial growth [90]. It is hypothesized that high O_2 levels may cause substrate inhibition of the enzyme polyphenol oxidase (PPO), which is the enzyme responsible for initiating discoloration on the cut surfaces of processed products [91].

High O_2-modified atmosphere packaging has been used in some vegetables and fruits, such as minimally-processed carrots, strawberry, minimally-processed baby spinach, fresh-cut mango, apple slices, and so on [89].

Amanatidou *et al.* [92] screened numerous microorganisms (Pseudomonas fluorescens, Enterobacter agglomerans, Aureobacterium strains 27, Candida guilliermondii, Candida sake, *Salmonella* Typhimurium, *Salmonella* Enteritidis, *E. coli*,*L. monocytogenes*, *Leuconostoc mesenteroides* var. *mesenteroides*, *Lactobacillus plantarum* and *Lactococcus lactis*) associated with the spoilage and safety of minimally-processed vegetables. High oxygen levels (80%–90%) caused a reduction in the growth rate of microbial targets at 8 °C, whilst the lag phase was prolonged (it was more evident at 90% O_2 concentrations). Amanatidou *et al.* [92] and Kader and Ben Yehoshua [93] hypothesized that high O_2 concentrations lead to intracellular generation of reactive oxygen radical species (ROS, O_2^-, H_2O_2, OH^-), causing damage to the vital cellular macromolecules, thus reducing cell viability when oxidative stresses overwhelm cellular protection systems.

Jacxsens *et al.* [90] studied the effect of high oxygen-modified atmosphere packaging (*i.e.*, >70% O_2) on microbial growth and the sensorial qualities of mushroom slices, grated celeriac and shredded chicory endive and observed that high O_2atmospheres were effective at inhibiting enzymatic browning of the tested vegetables. In addition, an improvement of the microbial quality (as a reduction in yeast growth) was obtained.

In an interesting paper, Allende *et al.* [94] studied the effect of superatmospheric oxygen packaging on sensorial quality, spoilage and *L.*

monocytogenes and *Aeromonas caviae* growth in fresh processed mixed salads. An initial O_2 concentration of 95 kPa was combined with two plastic films (low and high barrier impermeability for O_2). Packaged salads were stored up to eight days at 4 °C and at temperatures simulating the chilled distribution chain.

The authors reported that superatmospheric O_2 does not affect all microorganisms in the same way: high oxygen levels affected lactic acid bacteria and *Enterobacteriaceae*, which were inhibited in both plastic films (with low and high permeability). On the contrary, the growth of yeast and *A. caviae* seemed to be stimulated, whereas the growth of psychrotrophic bacteria and *L. monocytogenes* was not affected. The general appearance was maintained for longer, and the shelf-life of the mixed salads was prolonged by using O_2 concentrations higher than 60 kPa throughout the storage period.

Chunyang *et al.* [89] studied the effect of high oxygen-modified atmosphere packaging on fresh-cut onion quality at room temperature. Onion slices were packaged in a high-barrier film package of 70 μm in thickness and stored in five different modified atmospheres (100% O_2; 95% O_2/5% CO_2; 80% O_2/20% CO_2, 75% O_2/25% CO_2; and air). Results showed that the fresh-cut onions packaged in air had a short shelf-life, because the respiration rate was quickened due to mechanical damage, and microorganisms were not inhibited. High O_2 reduced the weight loss, the respiration rate, the total reducing sugar loss and the total titrable acidity increase of the fresh-cut onions. The best modified atmosphere seemed to be 80% O_2/20% CO_2, where total bacteria counts increased slowly. High oxygen affected also the sensory characteristics: sensory quality was acceptable up to five days, while air-packaged fresh-cut onions were not acceptable after two days of storage at room temperature.

Day [91] also confirmed that the most effective high O_2 gas mixtures were found to be 80%–85% O_2/15%–20% CO_2. This had the most noticeable sensory quality and antimicrobial benefits on a range of freshly-prepared produce items.

In fact, recommended optimal headspace gas levels immediately after freshly-prepared produce package sealing are: 80%–95% O_2 and 5%–20% N_2. After package sealing, headspace O_2 levels will decline, whereas CO_2 levels will increase during storage, due to the intrinsic respiratory nature of fresh product.

The levels of O_2 and CO_2 within packages are influenced by numerous variables:

- Intrinsic produce respiration rate (which itself is affected by temperature, atmospheric composition, produce type, variety, cultivar, maturity and the severity of preparation);
- Packaging film permeability;
- Package volume;
- Surface area and fill weight;
- Produce volume/gas volume ratio;
- Degree of illumination.

It would be preferable to maintain headspace levels of $O_2 > 40\%$ and CO_2 between 10% and 25% to maximize the benefits of high O_2 MAP. This can be achieved by:

- Lowering the temperature of storage;
- Selecting produce having a lower intrinsic respiration rate;
- Minimizing cut surface tissue damage;
- Reducing the ratio of produce volume/gas;
- Using a packaging film able to maintain high levels of O_2 whilst allowing excess CO_2 to go out
- By incorporating an innovative active packaging sachet that can absorb excess CO_2 and emit an equal volume of O_2.

A further measure to maintain a gas mixture $O_2 > 40\%/CO_2$ 10%–25% could be obtained by introducing the highest level of O_2 (balanced with N_2) possible just prior to prepared produce package sealing. Generally, it is not necessary to have any levels of CO_2 in the initial gas mixture, since it will increase rapidly during chilled storage.

Numerous products (iceberg lettuce, sliced mushrooms, broccoli florets, lettuce, baby-leaf spinach, Lollo Rossa lettuce, flat-leaf parsley, cubed swede, coriander, raspberries, strawberries, grapes and oranges) processed with high O_2 MAP reported beneficial effects on sensory quality if compared to industry-standard air packing and low O_2 MAP [91].

Finally, the recommended packaging material for high O_2-modified atmosphere is 30-µm orientated polypropylene (OPP), which has sufficient O_2 barrier properties (to maintain high in-pack O_2 levels >40%), and it is sufficiently permeable to ensure that CO_2 did not rise above 25% after 7–10 days of storage at 5–8 °C.

Other packaging materials suitable for high O_2 MAP of fresh prepared produce are:

- Laminations or extrusions of OPP with low density polyethylene (LDPE);
- Ethylene-vinyl acetate (EVA);
- Polyvinyl chloride (PVC).

or other medium to very high O_2 permeability films.

4. CONDITIONING SOLUTIONS WITH ANTIMICROBIALS OR NATURAL COMPOUNDS FOR FRUIT SALAD

Coatings, dipping, spraying, *etc.*, are the main applications of antimicrobial compounds, as natural alternatives to chemical additives, whilst little is known about the use of antimicrobials as filling liquids.

An initial work by Senesi *et al.* [95] on the use of rectified apple juice as a filling liquid to increase the quality and shelf-life of fresh-cut apples was followed, more recently, by D'Amato *et al.* [96]. The authors evaluated the possibility of using a chitosan, honey and pineapple juice solution as filling liquids to prolong the microbiological shelf-life of a fruit-based salad.

"Granny Smith" apples, "Gialla" first crop cactus pear fruits and "Regina" table grapes were washed; then, apples and cactus pears were peeled and sliced, while grapes were cut without seeds. Mixed fruits were placed in conical frustum-shaped cups, filled with four different solutions (sterile distilled water as the control; 30% of acacia honey and 70% of distilled water; 50% of pineapple juice and 50% of distilled water; 50% of low molecular weight chitosan solution and 50% of sterile distilled) and stored at 4, 8 and 12 °C.

The authors observed that the use of the natural antimicrobial compounds, as a filling liquid, affected the microbiological shelf-life of salad.

At 4 °C, mesophilic and psychrotrophic bacteria did not reach the established limit (1×10^6 CFU/g) during the whole storage period (14 days), in samples with honey and chitosan solutions, whereas it ranged from about 6.34–7.81 days and from 9.38 to 11.49 for the control and samples with pineapple juice, respectively.

At 8 °C, all of the samples were stored for 10 days. For the samples treated with honey and chitosan solutions, psychrotrophic bacteria did not reach the established limit during the storage, while in control and pineapple juice, it was of 5–6 days. Concerning mesophilic bacteria, microbiological shelf-life was strongly reduced in the control samples and in pineapple juice (about 3.7–4.3 days), whilst in chitosan and honey, it was over six days. In addition, in the samples with chitosan solution, the microbiological shelf-life, calculated by using the growth of yeasts, was over seven days and, by using lactic acid bacteria, did not reach the established limit.

At 12 °C, the samples were stored for six days. The microbiological shelf-life of fruit salad without antimicrobial compounds and with pineapple juice was always very low (1.5 and 4 days, respectively).

In the samples honey added, mesophilic and psychrotrophic bacteria did not reach the limit of 1×10^6 CFU/g. For lactic acid bacteria and yeasts, microbiological shelf-life reached 3.82 and 1.81 days, respectively.

Chitosan exhibited its high antimicrobial activity towards psychrotrophic bacteria, which did not reach the established limit, while no advantage resulted for mesophilic bacteria and lactic acid bacteria and yeasts.

The authors concluded that:

- Honey inhibited mesophilic and psychrotrophic bacterial growth; the antimicrobial activity of this natural antimicrobial compound was less effective on lactic acid bacteria and yeasts at all storage temperatures;
- Chitosan exerted a high antimicrobial activity for all microbial groups considered, particularly at a low temperature of storage;
- Pineapple juice was not effective towards all microbial groups, probably due to the high amount of nutrients.

Further investigations are needed to improve the antimicrobial efficacy of honey and chitosan for potential commercial applications. The literature is poor in articles addressing this topic; thus, it would be advisable to explore the use of antimicrobial compounds as filling liquids.

5. BIOPRESERVATION AND THE USE OF PROBIOTIC COATINGS

In recent years, biological preservation has emerged as a promising strategy to extend the shelf-life and to improve the microbiological safety of foods [97], since it fits well with the diffuse desire to preserve foods by natural means. Several bacteria and yeasts have been already identified as bioprotective agents [98], and different studies have been carried out on their application to fresh-cut fruits and vegetables [99]. Table 3 summarizes the most recent studies performed on this topic.

In general, lactic acid bacteria (LAB) have shown the greatest potential as biocontrol agents of several minimally-processed foods, because they are widely used in fermented foods and have a long history of safe use [100]. However, several other bacteria and yeasts, often selected among the naturally-occurring microflora, including strains of *Pseudomonas syringae*, *Pseudomonas graminis*, *Gluconobacter asaii*, *Candida* spp., *Dicosphaerina fagi*, *Metschnikowia pulcherrima* and *C. sake*, have been proposed as biocontrol agents in fresh-cut fruits and vegetables [101,102,103,104].

Table 3. Overview of the most recent studies performed on the use of biocontrol microorganisms to biopreserve minimally-processed products.

Minimally-Processed Product	Proposed Microorganisms	Reference
Biopreservation		
Apples	*Leuconostoc mesenteroides*, *Leuconostoc citreum*	[103]
Apples	*Pseudomonas graminis*	[102,105]
Apples	*Candida sake*	[101]
Apples	*Candida* sp., *Gluconobacter asaii*, *Dicosphaerina fagi*, *Metschnikowia pulcherrima*	[106]
Apples	*Enterobacteriaceae*	[107]
Apples	*Lactobacillus plantarum*	[99]
Apples	*Lactococcus lactis*	[108]
Iceberg lettuce	*Leuconostoc mesenteroides*, *Leuconostoc citreum*	[103]
Iceberg lettuce	*Leuconostoc* spp.	[104]
Lamb's lettuce	*Lactobacillus plantarum*, *Lactobacillus casei*	[99]
Melon	*Pseudomonas graminis*	[109]
Melon	*Pseudomonas graminis*	[110]
Peaches	*Enterobacteriaceae*	[107]
Scarola salad	*Lactobacillus casei*	[111]
Probiotic Biopreservation		
Apple	*Lactobacillus rhamnosus* GG	[112]
Apple	*Lactobacillus rhamnosus* GG	[113]
Apple	*Bifidobacterium lactis* Bb-12	[114]
Papaya	*Bifidobacterium lactis* Bb-12	[114]

The success of LAB in preventing the growth and activity of undesirable microorganisms is due to a large diversity of mechanisms of action related to the production of antimicrobial compounds, organic acids, hydrogen peroxide, bacteriocins and diacetyl [103,104,115]. The combination of low pH values and antibacterial activities of organic molecules produced by LAB remains the main mechanism for biopreservation [97]. Several bacteriocin-producing LAB have been shown to be effective against spoilage and pathogenic microorganisms in minimally-processed fruits and vegetables [116,117]. However, the direct application of bacteriocins to fresh-cut products did not provide completely satisfactory results probably due to the adsorption or inactivation of the added compound into the food product [116]; on the other hand, the direct use of living bacteriocinogenic bioprotective strains could lead to a more effective protection of the food product by circumventing the mentioned problems because of the localized and constant delivery of the antibacterial compound, which will add to other advantages, like space colonization by the strain [116,117]. The inhibitory properties against contaminating foodborne pathogens and spoilage microorganisms could also consist of a mere competition for nutrients (vitamins, minerals, trace elements and peptides), and therefore, via competition or antibiosis, LAB are able to function as a hurdle to pathogen growth and survival [118].

In the last decade, different LAB species were proposed as biocontrol agents in minimally-processed fruits and vegetables (see Table 3). In 2004, Scolari and Vescovo [111] performed several challenge tests on salad leaves by simultaneously inoculating *Lactobacillus casei* and various pathogens (*S. aureus*, *Aeromonas hydrophila*, *E. coli* and *L. monocytogenes*). A significant inhibitory effect by the LAB towards all of the pathogenic strains was observed and confirmed by the same authors during a subsequent study about the influence of *Lb. plantarum* on the growth of *S. aureus* [119]. Trias *et al.* [103] found five strains of LAB (some *Leuconostoc* spp. strains) that were able to inhibit *L. monocytogenes* and *Salmonella*Typhimurium in cut iceberg lettuce leaf, but were not effective in reducing the amount of *E. coli*. More recently, Siroli *et al.*[108] showed the good performance of a nisin-producing strain of *Lc. lactis*, which was able to inhibit *L. monocytogenes*, *E. coli* and the total mesophilic species when added at a level of 7 log CFU/mL in the washing solution of minimally-processed lamb's lettuce and combined or not with thyme essential oil. In 2015, the same authors proposed two other LAB strains (*Lb. plantarum* V7B3 and *Lb. casei* V4B4) to be used as biocontrol agents alone or in combination with thyme essential oil (EO) in lamb's lettuce [99]. In this work, the use of the *Lb. plantarum* V7B3 strain (6 log CFU/mL) during the washing phase of fresh-cut lettuce increased product shelf-life and safety; *L. monocytogenes* and *E. coli* viabilities were significantly reduced over the nine days of refrigerated storage. Promising results were also obtained for biopreservation of minimally-processed golden delicious apples packaged in a modified atmosphere alone or in combination with natural antimicrobials (2-(E)-hexenal/hexanal and 2-(E)-hexenal/citral) [99]: a strain of *Lb. plantarum* (CIT3) was able to increase the safety of sliced apples, when inoculated at levels of 6–7 log CFU/g in the washing dipping solution, both alone or in combination with natural antimicrobials.

Other non-lactic acid bacteria and yeasts were also proposed as biocontrol agents (Table 3) For example, the growth of *L. monocytogenes* and *Salmonella* Enterica in fresh-cut apples has been prevented using fungal antagonists [106]. When inoculated at a low level, *L. monocytogenes* cell loads were greatly reduced (from 5.7 to 6.0 log units after seven days) by strains of *G. asaii*, *Candida* spp., *D. fagi* and *M. pulcherrima*. At high pathogen inoculum levels, only *G. asaii* and *Candida* spp. reduced the cell load of *L. monocytogenes* population to non-detectable levels. Abadias *et al.* [101] found that the application of the fungal postharvest antagonist *C. sake* CPA-1 reduced the growth of a mixture of *E. coli* strains in fresh-cut apples at 25 °C, whereas Alegre *et al.* [107] isolated a new strain of *Enterobacteriaceae* that reduced the growth of *Salmonella*, *L. monocytogenes* and *E. coli* O157:H7 on fresh-cut apples and peaches. *P. graminis* was found be able to reduce or slow down the development of foodborne pathogens on minimally-processed fresh-cut apples and peaches [105]; in this case, the inhibitory effect of the antagonists on the foodborne pathogens was not instantaneous and became apparent after 6 days at 5 °C. This strain was also proposed for fresh-cut melon, reducing *Salmonella* and *L. monocytogenes* growth during storage at 5, 10 and 20 °C [109]. The strain effectiveness depended on the pathogens' concentration and on storage temperature. At a low pathogen concentration and 20 °C, *L. monocytogenes* growth was reduced between 2.1 and 5.3 log CFU/g after two days of storage and *Salmonella* growth between 2.0 and 7.3 log CFU/g.

Although research on the use of biocontrol agents in minimally-processed fruits and vegetables has increased in recent decades, the standardization of a biopreservative approach is still difficult to realize. Independent of the species and strains proposed, in fact, the various studies available in the literature clearly highlight that the efficacy of biocontrol agents is affected by different factors, such as the inoculation level, the presence of other bacteria, the physico-chemical and compositional features of the products and the storage conditions. Further investigations are required, especially considering that process conditions have to be taken into account, during the scaling up at the industrial level.

An interesting modern challenge is to incorporate probiotic bacteria into coated processed fruits and vegetables in order to improve their shelf-life (biopreservation), while providing new non-dairy functional foods. As is well accepted, edible films/coatings may serve as carriers of food additives, such as anti-browning agents, antimicrobials, colorants, flavors, nutrients and spices [24]. Some authors [114,120] proved that the coatings were also good carriers for antioxidant agents, such as cysteine, glutathione and ascorbic and citric acids; thus, the addition of LAB to obtain functional edible films and coatings could have successful implications. Several studies were already conducted on different fruits, as is shown in Table 3. This new trend has arisen from more considerations. As just mentioned, the inclusion into coatings of cultures with inhibitory properties could improve the shelf-life and safety of minimally-processed products, while reducing the need to use increasing levels of chemical additives [118]. In addition, the steps of minimally-processing vegetables, such as peeling and cutting, promote the release of cellular content rich in minerals, sugars, vitamins and other nutrients, creating ideal conditions for microbial growth: this characteristic allows the use of fruit and vegetable food products as probiotic carriers. According to Soccol *et al.* [121], in fact, minimally-processed fruits and vegetables are very good matrices providing ideal substrates for probiotics, since they contain minerals, vitamins, antioxidants and fibers. Moreover, due to their cellulose content, fruits, such as apples and pears, may also exert a protective effect on the probiotic microorganisms during passage through the intestinal tract [122], allowing these microorganisms to reach the colon and benefit the host. It has been reported that the optimum probiotic growth temperature is between 35 and 40 °C, and the best pH is between 6.4 and 4.5, ceasing when a pH of 4.0–3.6 is reached [123]. This situation could be solved by using some supports, such as agar, polyacrylamide, calcium pectate gel, chemically-modified chitosan beads and alginates, to provide a physical barrier against unfavorable conditions [124,125,126].

One of the first reports about this issue is the study of Tapia *et al.* [114]; fresh-cut apple and papaya cylinders were coated with 2% (w/v) alginate or gellan film-forming solutions containing viable Bifidobacteria, namely *Bifidobacterium lactis* Bb-12. The bifidus-containing coatings were more permeable to water vapor than the corresponding films without probiotics, and gellan coatings were more resistant to water transfer than the alginate ones. The most important result was that the edible coatings were efficient in supporting *Bb. lactis* Bb-12 on fresh-cut apple and papaya. In 2010, Rößle *et al.* [112] applied a probiotic microorganism (*Lactobacillus rhamnosus* GG; LGG) to fresh-cut apple wedges (cultivar Braeburn). All samples were able to maintain a probiotic load of about *ca.* 10^8 CFU/g over 10 storage days, which is sufficient for a probiotic effect, and this is comparable to the counts of probiotic bacteria in commercially-available dairy products. It is also important to underline that the physico-chemical properties of the apple wedges containing LGG compared to the control (without probiotics) remained stable over the observation period. In a subsequent work [113], the effectiveness as a biocontrol agent of the same probiotic strain against *Salmonella* and *L. monocytogenes* on minimally-processed apples throughout storage, as well as its effect on apple quality and natural microflora was evaluated. The obtained results showed that *Salmonella* was not affected by co-inoculation with LGG, whereas the *L. monocytogenes* population was 1-log unit lower in the presence of probiotic population maintained over recommended levels for probiotic action (10^6 CFU/g) along 14 days.

Although still unexplored, this new challenge appears to be highly advantageous, since minimally-processed fruits and vegetables are a food category rich in nutrients, intended for consumption by all individuals and widely accepted among consumers [127]. Thus, edible coatings, including probiotic bacteria, when applied to minimally-processed products, open new possibilities to improve their shelf-life and safety while providing innovative functional foods.

6. CONCLUDING REMARKS

The use of non-conventional tools to preserve and prolong the quality of minimally-processed fruits and vegetables seems to be the future trend to obtain improved final products. In particular, the application of the antimicrobial compounds into coatings and/or filling liquids should be further investigated due to their environmental friendliness and versatility. A combined approach of these technologies, through additive and synergistic interactions, could be useful to obtain innovative and safe minimally-processed products.

AUTHOR CONTRIBUTIONS

The authors contributed equally to this paper.

CONFLICTS OF INTEREST

The authors declare no conflict of interest.

REFERENCES

1. Allende, A.; Tomas-Barberan, F.A.; Gil, M.I. Minimal processing for healthy traditional foods. Trends Food Sci. Tech.2006, 7, 179–186.

2. Lamikanra, O. Preface. In Fresh-Cut Fruits and Vegetables; Lamikanra, O., Ed.; CRC Press: Boca Raton, FL, USA, 2002.

3. Gonzales-Aguilar, G.A.; Ayala-Zavala, J.F.; de La Rosa, L.A.; Ivarez-Parrilla, E.A. Preserving quality of fresh-cut product using safe technology. J. Verbrauch. Lebensm. 2010, 5, 65–72.

4. Boesso Oriani, V.; Molina, G.; Chiumarelli, M.; Pastore, G.M.; Dupas Hubinger, M. Properties of cassava starch-based edible coating containing Essential Oils. J. Food Sci. 2014, 79, E189–E194.

5. McHugh, T.H.; Senesi, E. Apple wraps: A novel method to improve the quality and extend the shelf life of fresh-cut apples. J. Food Sci. 2000, 65, 480–485.

6. Nisperos-Carriedo, M.O.; Baldwin, E.A.; Shaw, P.E. Development of an edible coating for extending postharvest life of selected fruits and vegetables. Proc. Fla. State Hort. Soc. 1991, 104, 122–125.

7. Lerdthanangkul, S.; Krochta, J.M. Edible coating effects on post harvest quality of green bell peppers. J. Food Sci.1996, 61, 176–179.

8. Avena-Bustillos, R.J.; Krochta, J.M.; Saltveit, M.E. Water vapor resistance of red delicious apples and celery sticks coated with edible caseinate-acetylated monoglyceride films. J. Food Sci. 1997, 62, 351–354.

9. Guilbert, S.; Gontard, N.; Gorris, L.G.M. Prolongation of the shelf-life of perishable food products using biodegradable films and coatings. LWT—Food Sci. Technol. 1996, 29, 10–17.

10. Smith, S.; Geeson, J.; Stow, J. Production of modified atmospheres in deciduous fruits by the use of films ad coatings.Hort. Sci. 1987, 22, 772–776.

11. Krochta, J.M.; Mulder-Johnston, C.D. Edible and biodegradable polymer films: Challenges and opportunities. Food Technol. 1997, 51, 61–74.

12. Souza, B.W.S.; Cerqueira, M.A.; Teixeira, J.A.; Vicente, A.A. The use of electric fields for edible coatings and films development and production: A review. Food Eng. Rev. 2010, 2, 244–255.

13. Campos, C.A.; Gerschenson, L.N.; Flores, S.K. Development of edible films and coatings with antimicrobial activity.Rev. Food Bioprocess. Technol. 2010, 4, 849–875.

14. Allen, L.; Nelson, A.I.; Steinberg, M.P.; McGill, J.N. Edible corn-carbohydrate food coatings. Development and physical testing of starch-algin coating. Food Technol. 1963, 17, 1437–1442.

15. Lin, D.; Zhao, Y. Innovations in the development and application of edible coatings for fresh and minimally processed fruits and vegetables. Compr. Rev. Food Sci. Food Saf. 2007, 6, 60–75.

16. Diab, T.; Biliaderis, C.G.; Gerasopoulos, D.; Sfakiotakis, E. Physicochemical properties and application of pullulan edible films and coatings in fruit preservation. J. Sci. Food Agric. 2001, 81, 988–1000.

17. Synowiec, A.; Gniewosz, M.; Kraśniewska, K.; Przybył, J.L.; Bączek, K.; Węglarz, Z. Antimicrobial and antioxidant properties of pullulan film containing sweet basil extract and an evaluation of coating effectiveness in the prolongation of the shelf life of apples stored in refrigeration conditions. Inn. Food Sci. Emer. Technol. 2014, 23, 171–181.

18. Vargas, M.; Pastor, C.; Chiralt, A.; McClements, D.J.; González-Martínez, C. Recent advances in edible coatings for fresh and minimally processed fruits. Crit. Rev. Food Sci. 2008, 48, 496–511.

19. Coma, V.; Deschamps, A.; Martial-Gros, A. Bioactive packaging materials from edible chitosan polymer-antimicrobial assessment on dairy related contaminants. J. Food Sci. 2003, 68, 2788–2792.

20. Li, P.; Barth, M.M. Impact of edible coatings on nutritional and physiological changes in lightly processed carrots.Postharvest Biol. Tec. 1998, 14, 51–60.

21. Amanatidou, A.; Slump, R.A.; Gorris, L.G.M.; Smid, E.J. High oxygen and high carbon dioxide modified atmospheres for shelf life extension of minimally processed carrots. J. Food Sci. 2000, 65, 61–66.

22. Hershko, V.; Nussinovitch, A. Relationships between hydrocolloid coating and mushroom structure. J. Agric. Food Chem. 1998, 46, 2988–2997.

23. Bryan, D.S. Prepared citrus fruit halves and method of making the same. U.S. Patent 3,707,383, 26 December 1972.

24. Lee, J.Y.; Park, H.J.; Lee, C.Y.; Choi, W.Y. Extending shelf-life of minimally processed apples with edible coatings and antibrowning agents. Lebens. Wissen. Technol. 2003, 36, 323–329.

25. Arnold, F.W. Infrared Roasting of Coated Nutmeats. U.S. Patent 3,383,220, 14 May 1968.

26. Maftoonazad, N.; Ramaswamy, H.S.; Moalemiyan, M.; Kushalappa, A.C. Effect of pectin-based edible emulsion coating on changes in quality of avocado exposed to Lasiodiplodia theobromae infection. Carbohyd. Polym. 2007, 68, 341–349.

27. Vargas, M.; Albors, A.; Chiralt, A.; Gonzalez-Martinez, C. Quality of cold-stored strawberries as affected by chitosan-oleic acid edible coatings. Postharvest Biol. Tec. 2006, 41, 164–171.

28. Campaniello, D.; Bevilacqua, A.; Sinigaglia, M.; Corbo, M.R. Chitosan: Antimicrobial activity and potential applications for preserving minimally processed strawberries. Food Microbiol. 2008, 25, 992–1000.

29. Krochta, J.M.; Pavlath, A.E.; Goodman, N. Edible films from casein-lipid emulsions for lightly-processed fruits and vegetables. In Engineering and food, Preservation Processes and Related Techniques; Spiess, W.E., Schubert, H., Eds.; Elsevier Science Publishers: New York, NY, USA, 1990; pp. 329–340.

30. Trezza, T.A.; Krochta, J.M. Application of edible protein coatings to nut and nut-containing food products. In Protein-Based Films and Coatings; Gennadios, A., Ed.; CRC Press: Boca Raton, FL., USA, 2002; pp. 527–549.

31. Chan, M.A.; Krochta, J.M. Grease and oxygen barrier properties of whey-protein isolate coated paperboard. Solutions2001, 84, 57.

32. Lin, S.Y.; Krochta, J.M. Plasticizer effect on grease barrier properties of whey protein concentrate coatings on paperboard. J. Food Sci. 2003, 68, 229–233.

33. Guilbert, S. Technology and application of edible protective films. In Food Packaging and Preservation; Mathlouthi, M., Ed.; Elsevier Applied Science Publishers: London, UK, 1986; pp. 371–394.

34. Gennadios, A.; McHugh, T.H.; Weller, G.L.; Krochta, J.M. Edible coatings and films based on proteins. In Edible Coatings and Films to Improve Food Quality; Krochta, J.M., Baldwin, E.A., Nisperos-Carriedo, M.O., Eds.; Technomic Publishing Co., Inc.: Lancaster, PA, USA, 1994; pp. 201–277.

35. Lim, L.T.; Mine, Y.; Britt, I.J.; Tung, M.A. Formation and properties of egg white protein films and coatings. In Protein-Based Films and Coatings; Gennadios, A., Ed.; CRC Press: Boca Raton, FL, USA, 2002; pp. 233–252.

36. Franssen, L.R.; Krochta, J.M. Edible coatings containing natural antimicrobials for processed foods. In Natural Antimicrobials for Minimal Processing of Foods; Roller, S., Ed.; CRC Press: Boca Raton, FL, USA, 2003; pp. 250–262.

37. Martín-Belloso, O.; Rojas-Graü, M.A.; Soliva-Fortuny, R. Delivery of flavor and active ingredients using edible films and coatings. In Edible Films and Coatings for Food Applications; Embuscado, M.E., Huber, K.C., Eds.; Springer: New York, NY, USA, 2009; pp. 295–313.

38. Bautista-Banos, S.; Hernandez-Lauzardo, A.N.; Velazquez-del Valle, M.G.; Hernandez-Lo, M.; Ait Barka, E.; Bosquez-Molina, E.; Wilson, C.L. Chitosan as a potential natural compound to control pre and postharvest diseases of horticultural commodities. Crop. Prot. 2006, 25, 108–118.

39. Jiang, Y.M.; Li, J.R.; Jiang, W.B. Effects of chitosan coating on shelf life of cold stored litchi fruit at ambient temperature. LWT-Food Sci. Technol. 2005, 38, 757–761.

40. Shiekh, R.A.; Malik, M.A.; Al-Thabaiti, S.A.; Shiekh, M.A. Chitosan as a novel edible coating for fresh fruits. Food Sci. Technol. Res. 2013, 19, 139–155.

41. Shahidi, F.; Arachchi, J.K.V.; Jeon, Y. Food application of chitin and chitosan. Trends Food Sci. Tech. 1999, 10, 37–51.

42. Moncayo, D.; Buitrago, G.; Algecira, N. The surface properties of biopolymer-coated fruit: A review. Ing. Invest. 2013,33, 11–16.

43. Meng, X.H.; Li, B.Q.; Liu, J.; Tian, S.P. Physiological responses and quality attributes of table grape fruit to chitosan preharvest spray and postharvest coating during storage. Food Chem. 2008, 106, 501–508.

44. Romanazzi, G.; Nigro, F.; Ippolito, A.; di Venere, D.; Salerno, M. Effects of pre- and postharvest chitosan treatments to control storage grey mold of table grapes. J. Food Sci. 2002, 67, 1862–1867.

45. Ziani, K.; Fernandez Pan, I.; Royo, M.; Maté, J. Antifungal activity of films and solutions based on chitosan. Food Hydrocoll. 2009, 23, 2309–2314.

46. Aider, M. Chitosan application for active bio-based films production and potential in the food industry: Review. LWT Food Sci. Technol. 2010, 43, 837–842.

47. Benhabiles, M.S.; Drouiche, N.; Lounici, H.; Pauss, A.; Mameri, N. Effect of shrimp chitosan coatings as affected by chitosan extraction processes on postharvest quality of strawberry. Food Meas. 2013, 7, 215–221.

48. Lou, M.M.; Zhu, B.; Muhammad, I.; Li, B.; Xie, G.L.; Wang, Y.L.; Li, H.Y.; Sun, G.C. Antibacterial activity and mechanism of action of chitosan solutions against apricot fruit rot pathogen Burkholderia seminalis. Carbohydr. Res.2011, 346, 1294–1301.

49. Liu, L.; Liu, C.K.; Hicks, K.B. Composite films from pectin and fish skin gelatin or soybean flour protein. J. Agric. Food Chem. 2007, 55, 2349–2355.

50. El Ghaouth, A.; Arul, J.; Grenier, J.; Asselin, A. Antifungal activity of chitosan on two post-harvest pathogens of strawberry fruits. Phytopathology 1992, 82, 398–402.

51. Park, S.I.; Stan, S.D.; Daeschel, M.A.; Zhao, Y. Antifungal coatings on fresh strawberries (Fragaria × ananassa) to control mold growth during cold storage. J. Food Sci. 2005, 70, 202–207.

52. Chien, P.J.; Sheu, F.; Lin, H.R. Coating citrus (Murcott tangor) fruit with low molecular weight chitosan increases postharvest quality and shelf life. Food Chem. 2007, 100, 1160–1164.

53. Gonzalez-Aguilar, G.A.; Valenzuela-Soto, E.; Lizardi-Mendoza, J.; Goycoolea, F.; Martinez-Tellez, M.A.; Villegas-Ochoa, M.A.; Monroy-Garcia, I.N.; Ayala-Zavala, J.F. Effect of chitosan coating in preventing deterioration and preserving the quality of fresh-cut papaya “Maradol”. J. Sci. Food Agric. 2009, 89, 15–23.

54. Ali, A.; Muhammad, M.T.M.; Sijam, K.; Siddiqui, Y. Effect of chitosan coatings on the physicochemical characteristics of Eksotika. II papaya (Carica papaya L.) fruit during cold storage. Food Chem. 2011, 124, 620–626.

55. Pilon, L.; Poliana, C.; Spricigo, M.M.; de Moura, M.R.; Assis, O.B.G.; Mattoso, L.H.C.; Ferreira, M.D. Chitosan nanoparticle coatings reduce microbial growth on fresh-cut apples while not affecting quality attributes. Int. J. Food Sci. Technol. 2015, 50, 440–444.

56. Benhabiles, M.S.; Tazdait, D.; Abdi, N.; Lounici, H.; Drouiche, N.; Goosen, M.F.A.; Mameri, N. Assessment of coating tomato fruit with shrimp shell chitosan and N,O-carboxymethyl chitosan on postharvest preservation. Food Meas.2013, 7, 66–74.

57. Assis, O.B.; Pessoa, J.D. Preparation of thin films of chitosan for use as edible coatings to inhibit fungal growth on sliced fruits. Braz. J. Food Technol. 2004, 7, 7–22.

58. Han, C.; Lederer, C.; McDaniel, M.; Zhao, Y. Sensory evaluation of fresh strawberries (Fragaria ananassa) coated with chitosan-based edible coatings. J. Food Sci. 2005, 70, S172–S178.

59. Durango, A.M.; Soares, N.F.F.; Andrade, N.J. Microbiological evaluation of an edible antimicrobial coating on minimally processed carrots. Food Control 2006, 17, 336–341.

60. Devlieghere, F.; Vermeulen, A.; Debevere, J. Chitosan: Antimicrobial activity, interactions with food components and applicability as a coating on fruit and vegetables. Food Microbiol. 2004, 21, 703–714.

61. Pushkala, R.; Raghuram, P.K.; Srividya, N. Chitosan based powder coating technique to enhance phytochemicals and shelf life quality of radish shreds. Postharvest Biol. Tec. 2013, 86, 402–408.

62. Jianglian, D.; Shaoying, Z. Application of chitosan based coating in fruit and vegetable preservation: A review. J. Food Process. Technol. 2013, 4, 227:1–227:4.

63. Yu, Y.W.; Li, H.; Jinhua, D.; Ren, Y.Z. Study of natural film with chitosan combining phytic acids on preservation of fresh-cutting lotus root. J. Chin. Inst. Food Sci. Technol. 2012, 12, 131–136.

64. Cong, F.; Zhang, Y.; Dong, W. Use of surface coatings with natamycin to improve the storability of Hami melon at ambient temperature. Postharvest Biol. Technol. 2007, 46, 71–75.

65. Zhang, H.; Li, R.; Liu, W. Effects of chitin and its derivative chitosan on postharvest decay of fruits: A Review. Int. J. Mol. Sci. 2011, 12, 917–934.

66. Salgado, P.R.; Ortiz, C.M.; Musso, Y.S.; di Giorgio, L.; Mauri, A.N. Edible films and coatings containing bioactives.Curr. Opin. Food Sci. 2015, 5, 86–92.

67. Speranza, B.; Corbo, M.R. Essential Oils for Preserving Perishable Foods: Possibilities and Limitations. In Application of Alternative Food Preservation Technologies to Enhance Food Safety and Stability; Bevilacqua, A., Corbo, M.R., Sinigaglia, M., Eds.; Bentham Publisher: Saif Zone, Sharjah, 2010; pp. 35–57.

68. Patrignani, F.; Siroli, L.; Serrazanetti, D.I.; Gardini, F.; Lanciotti, R. Innovative strategies based on the use of essential oils and their components to improve safety, shelf-life and quality of minimally processed fruits and vegetables.Trends Food Sci. Technol. 2015.

69. Sánchez-González, L.; Cháfer, M.; Hernández, M.; Chiralt, A.; González-Martínez, C. Antimicrobial activity of polysaccharide films containing essential oils. Food Control. 2011, 22, 1302–1310.

70. Xing, Y.; Li, X.; Xu, Q.; Yun, J.; Lu, Y.; Tang, Y. Effects of chitosan coating enriched with cinnamon oil on qualitative properties of sweet pepper (Capsicum annuum L.). Food Chem. 2011, 124, 1443–1450.

71. Sessa, M.; Ferraria, G.; Donsìb, F. Novel edible coating containing Essential Oil nanoemulsions to prolong the shelf life of vegetable products. Chem. Eng. Trans. 2015.

72. Randazzo, W.; Jimenez-Belenguer, A.; Settanni, L.; Perdones, A.; Moschetti, M.; Palazzolo, E.; Guarrasi, V.; Vargas, M.; Germanà, M.A.; Moschetti, G. Antilisterial effect of citrus essential oils and their performance in edible film formulations. Food Control. 2016, 59, 750–758.

73. Sangsuwan, J.; Rattanapanone, N.; Rachtanapun, P. Effect of chitosan/methyl cellulose films on microbial and quality characteristics of fresh-cut cantaloupe and pineapple. Postharvest Biol. Technol. 2008, 49, 403–410.

74. Krasaekoopt, W.; Mabumrung, J. Microbiological evaluation of edible coated fresh-cut cantaloupe. Kasetsart. J. Nat. Sci. 2008, 42, 552–557.

75. Corbo, M.R.; Speranza, B.; Campaniello, D.; D'Amato, D.; Sinigaglia, M. Fresh-cut fruits preservation: Current status and emerging technologies. In Current Research, Technology and Education topics in Applied Microbiology; Mendez-Villas, A., Ed.; Formatex Research Center: Badajoz, Spain, 2011; pp. 1143–1154.

76. Corbo, M.R.; Bevilacqua, A. Alternative Modified Atmosphere for Fresh Food Packaging. In Application of Alternative Food Preservation Technologies to Enhance Food Safety and Stability; Bevilacqua, A., Corbo, M.R., Sinigaglia, M., Eds.; Bentham Publisher: Saif Zone, Sharjah, 2010; pp. 196–204.

77. Siddiqui, M.W.; Chakraborty, I.; Ayala-Zavala, J.F.; Dhua, R.S. Advances in minimal processing of fruits and vegetables: A review. J. Sci. Ind. Res. 2011, 70, 823–834.

78. Church, N. Developments in modified-atmosphere packaging and related technologies. Trends Food Sci. Technol. 1994,5, 345–352.

79. Farber, J.M. Microbiological Aspects of Modified-Atmosphere Packaging Technology—A Review. J. Food Protect. 1991,54, 58–70.

80. Phillips, C.A. Review: Modified Atmosphere Packaging and Its Effects on the Microbiological Quality and Safety of Produce. Int. J. Food Sci. Technol. 1996, 31, 463–479.

81. Church, I.J.; Parsons, A.L. Modified atmosphere packaging technology: A Review. J. Sci. Food Agric. 1995, 67, 143–152.

82. Corbo, M.R.; Lanciotti, R.; Gardini, F.; Sinigaglia, M.; Guerzoni, M.E. Effects of hexanal, (E)-2-hexenal, and storage temperature on shelf life of fresh sliced apples. J. Agric. Food Chem. 2000, 48, 2401–2408.

83. Lanciotti, R.; Gianotti, A.; Patrignani, F.; Belletti, N.; Guerzoni, M.E.; Gardini, F. Use of natural aroma compounds to improve shelf-life and safety of minimally processed fruits. Trends Food Sci. Technol. 2004, 15, 201–208.

84. Campaniello, D.; D'Amato, D.; Corbo, M.R.; Sinigaglia, M. Combined action of hexanal and MAP on minimally processed cactus pear fruit. Int. J. Food Sci. 2005, 1–10.

85. Siroli, L.; Patrignani, F.; Serrazanetti, D.I.; Tabanelli, G.; Montanari, C.; Tappi, S.; Rocculi, P.; Gardini, F.; Lanciotti, R. Efficacy of natural antimicrobials to prolong the shelf-life of minimally processed apples packaged in modified atmosphere. Food Control. 2014, 46, 403–411.

86. Fisher, K.; Phillips, C.; McWatt, L. The use of an antimicrobial citrus vapour to reduce Enterococcus sp. on salad products. Int. J. Food Sci. Technol. 2009, 44, 1748–1754.

87. Tian, J.; Ban, X.; Zeng, H.; He, J.; Huang, B.; Wang, Y. Chemical composition and antifungal activity of essential oil from Cicuta virosa L. var. latisecta Celak. Int. J. Food Microbiol. 2011, 145, 464–470.

88. Vitoratos, A.; Bilalis, D.; Karkanis, A.; Efthimiadou, A. Antifungal Activity of Plant Essential Oils against Botrytis cinerea, Penicillium italicum and Penicillium digitatum. Not. Bot. Horti Agrobot. 2013, 41, 86–92.

89. Chunyang, H.; Xiqing, Y.; Fei, L.; Binxin, S. Effect of High Oxygen Modified Atmosphere Packaging on Fresh-Cut Onion Quality. In Proceedings of the 17th IAPRI World Conference on Packaging, Tianjin, China, 12–15 October 2010.

90. Jacxsens, L.; Devlieghere, F.; Van der Steen, C.; Debevere, J. Effect of high oxygen modified atmosphere packaging on microbial growth and sensorial qualities of fresh-cut produce. Int. J. Food Microbiol. 2001, 71, 197–210.

91. Day, B.P.F. New modified atmosphere techniques fresh prepared fruit and vegetables. In Fruit and Vegetables Processing; Jongen, W., Ed.; Woodhead Publishing Limited: Cambridge, UK, 2002; pp. 310–330.

92. Amanatidou, A.; Smid, E.J.; Gorris, L.G.M. Effect of elevated oxygen and carbon dioxide on the surface growth of vegetable-associated micro-organisms. J. Appl. Microbiol. 1999, 86, 429–438.

93. Kader, A.; Ben-Yehoshua, S. Effects of superatmospheric oxygen levels on postharvest physiology and quality of fresh fruits and vegetables. Postharvest Biol. Technol. 2000, 20, 1–13.

94. Allende, A.; Jacxsens, L.; Devlieghere, F.; Debevere, J.; Artés, F. Effect of Superatmospheric Oxygen Packaging on Sensorial Quality, Spoilage, and Listeria monocytogenes and Aeromonas caviae Growth in Fresh Processed Mixed Salads.J. Food Protect. 2002, 65, 1565–1573.

95. Senesi, E.; De Regibus, P. Use of rectified apple juice as filling liquid to increase quality and shelf-life of fresh-cut apples (Succo di mela rettificato come liquido di governo per prolungare la curabilita' di spicchi di mela di IV gamma). Ind. Aliment. 2002, 41, 139–146.

96. D'Amato, D.; Sinigaglia, M.; Corbo, M.R. Use of chitosan, honey and pineapple juice as filling liquids for increasing the microbiological shelf life of a fruit-based salad. Int. J. Food Sci. Technol. 2010, 45, 1033–1041.

97. Galvez, A.; Abriouel, H.; Benomar, N.; Lucas, R. Microbial antagonists to food-borne pathogens and biocontrol. Curr. Opin. Biotechnol. 2010, 21, 142–148.

98. Vermeiren, L.; Devlieghere, F.; Debevere, J. Evaluation of meat born lactic acid bacteria as protective cultures for the biopreservation of cooked meat products. Int. J. Food Microbiol. 2004, 96, 149–164.

99. Siroli, L.; Patrignani, F.; Serrazanetti, D.I.; Tabanelli, G.; Montanari, C.; Gardini, F.; Lanciotti, R. Lactic acid bacteria and natural antimicrobials to improve the safety and shelf-life of minimally processed sliced apples and lamb's lettuce. Food Microbiol. 2015, 47, 74–84.

100. Carr, F.J.; Chill, D.; Maida, N. The lactic acid bacteria: A literature survey. Crit. Rev. Microbiol. 2002, 28, 281–370.

101. Abadias, M.; Usall, J.; Alegre, I.; Torres, R.; Vinas, I. Fate of Escherichia coli in apple and reduction of its growth using the postharvest biocontrol agent Candida Sake CPA-1. J. Sci. Food Agric. 2009, 89, 1526–1533.

102. Alegre, I.; Vinas, I.; Usall, J.; Anguera, M.; Altisent, R.; Abadias, M. Antagonistic effect of Pseudomonas graminis CPA-7 against foodborne pathogens in fresh-cut apples under simulated commercial conditions. Food Microbiol. 2013, 33, 139–148.

103. Trias, R.; Badosa, E.; Montesinos, E.; Baneras, L. Bioprotective Leuconostoc strains against Listeria monocytogenes in fresh fruits and vegetables. Int. Microbiol. 2008, 127, 91–98.

104. Trias, R.; Baneras, L.; Badosa, E.; Montesinos, E. Bioprotection of Golden Delicious apples and Iceberg lettuce against foodborne bacterial pathogens by lactic acid bacteria. Int. J. Food Microbiol. 2008, 123, 50–60.

105. Alegre, I.; Vinas, I.; Usall, J.; Anguera, M.; Figge, M.J.; Abadias, M. Control of foodborne pathogens on fresh-cut fruit by a novel strain of Pseudomonas graminis. Food Microbiol. 2013, 34, 390–399.

106. Leverentz, B.; Conway, W.S.; Janisiewicz, W.; Abadias, M.; Kurtzman, C.P.; Camp, M.J. Biocontrol of the food-borne pathogens Listeria monocytogenes and Salmonella Enterica serovar poona on fresh-cut apples with naturally occurring bacterial and yeast antagonists. Appl. Environ. Microb. 2006, 72, 1135–1140.

107. Alegre, I.; Vinas, I.; Usall, J.; Anguera, M.; Figge, M.J.; Abadias, M. An Enterobacteriaceae species isolated from apples controls foodborne pathogens on fresh-cut apples and peaches. Postharvest Biol. Technol. 2012, 74, 118–124.

108. Siroli, L.; Patrignani, F.; Salvetti, E.; Torriani, S.; Gardini, F.; Lanciotti, R. Use of a nisin-producing Lactococcus lactisstrain, combined with thyme essential oil, to improve the safety and shelf-life of minimally processed lamb's lettuce. In Proceeding of the 11th International Symposium on lactic acid bacteria, Egmond aan Zee, The Netherlands, 31 August–4 September 2014.

109. Abadias, M.; Altisent, R.; Usall, J.; Torres, R.; Oliveira, M.; Vinas, I. Biopreservation of fresh-cut melon using the strain Pseudomonas graminis CPA-7. Postharvest Biol. Technol. 2014, 96, 69–77.

110. Plaza, L.; Altisent, R.; Alegre, I.; Vinas, I.; Abadias, M. Changes in the quality and antioxidant properties of fresh-cut melon treated with the biopreservative culture Pseudomonas graminis CPA-7 during refrigerated storage. Postharvest Biol. Technol. 2016, 111, 25–30.

111. Scolari, G.; Vescovo, M. Microbial antagonism of Lactobacillus casei added to fresh vegetables. Int. J. Food Sci. 2004, 16, 465–475.

112.Rößle, C.; Auty, M.A.E.; Brunton, N.; Gormley, R.T.; Butler, F. Evaluation of fresh-cut apple slices enriched with probiotic bacteria. Innov. Food Sci. Emerg. 2010, 11, 203–209.

113.Alegre, I.; Viñas, I.; Usall, J.; Anguera, M.; Abadias, M. Microbiological and physicochemical quality of fresh-cut apple enriched with the probiotic strain Lactobacillus rhamnosus GG. Food Microbiol. 2011, 28, 59–66.

114.Tapia, M.S.; Rojas-Graü, M.A.; Rodríguez, F.J.; Ramirez, J.; Carmona, A.; Martín-Belloso, O. Alginate and Gellan-based edible films for probiotic coatings on fresh-cut fruits. J. Food Sci. 2007, 72, 190–196.

115.Cleveland, J.; Montville, T.J.; Nes, I.F.; Chikindas, M.L. Bacteriocins: Safe, natural antimicrobials for food preservation. Int. J. Food Microbiol. 2001, 71, 1–20.

116.Allende, A.; Martınez, B.; Selma, V.; Gil, M.I.; Suarez, J.E.; Rodrıguez, A. Growth and bacteriocin production by lactic acid bacteria in vegetable broth and their effectiveness at reducing Listeria monocytogenes in vitro and in fresh-cut lettuce. Food Microbiol. 2007, 24, 759–766.

117.Randazzo, C.L.; Pitino, I.; Scifo, G.O.; Caggia, C. Biopreservation of minimally processed iceberg lettuces using a bacteriocin produced by Lactococcus lactis wild strain. Food Control. 2009, 20, 756–763.

118.Schuenzel, K.M.; Harrison, M.A. Microbial antagonists of foodborne pathogens on fresh, minimally processed vegetables. J. Food Protect. 2002, 65, 1909–1915.

119.Scolari, G.; Vescovo, M.; Zacconi, C.; Bonade, A. Influence of Lactobacillus plantarum on Staphylococcus aureus growth in a fresh vegetable model system. Eur. Food Res. Technol. 2004, 218, 274–277.

120.Rojas-Graü, M.A.; Sobrino-López, A.; Tapia, M.S.; Martín-Belloso, O. Browning inhibition in fresh-cut Fuji apple slices by natural antibrowning agents. J. Food Sci. 2006, 71, 59–65.

121.Soccol, C.R.; Vandenberghe, L.P.S.; Spier, M.R.; Medeiros, A.B.P.; Yamaguishi, C.T.; Lindner, J.D.; Pandey, A.; Soccol, V.T. The potential of probiotics. Food Tech. Biotechnol. 2010, 48, 413–434.

122.Kourkoutas, Y.; Kanellaki, M.; Koutinas, A.A. Apple pieces as immobilization support of various microorganisms.LWT-Food Sci. Technol. 2006, 39, 980–986.

123.Shah, N.P. Functional cultures and health benefits. Int. Dairy J. 2007, 17, 1262–1277.

124. Kailasapathy, K. Microencapsulation of probiotic bacteria: Technology and potential application. Curr. Issues Intest. Microbiol. 2002, 3, 39–49. [PubMed]

125. Guerin, D.; Vuillemard, J.C.; Subirade, M. Protection of Bifidobacteria encapsulated in polysaccharide-protein gel beads against gastric juice and bile. J. Food Protect. 2003, 66, 2076–2084.

126. Kourkoutas, Y.; Xolias, V.; Kallis, M.; Bezirtzoglou, E.; Kanellaki, M. Lactobacillus casei cell immobilization on fruit pieces for probiotic additive, fermented milk and lactic acid production. Process. Biochem. 2005, 40, 411–416.

127. Saad, S.M.I.; Bedani, R.; Mamizuka, E.M. Benefícios à Saude dos Probióticos e Prebióticos. In Probióticos e Prebióticos em Alimentos: Fundamentos e Aplicações Tecnológicas; Saad, S.M.I., Cruz, A.G., Faria, J.A.F., Eds.; Editora Varela: São Paulo, Brazil, 2011; pp. 51–84, In Portuguese.

Chapter 4

Effect of Storage Conditions on the Sensory Quality, Colour and Texture of Fresh-Cut Minimally Processed Cabbage with the Addition of Ascorbic Acid, Citric Acid and Calcium Chloride

Eleni Manolopoulou[1], Theodoros Varzakas[2*]

[1]Department of Plant Production, School of Agricultural Technology, Higher Institute of Kalamata, Kalamata, Greece; [2]Department of Food Technology, School of Agricultural Technology, Higher Institute of Kalamata, Kalamata, Greece.

ABSTRACT

The aim of this study was to investigate the effect of storage conditions on the sensory quality, colour and texture of fresh-cut cabbage during the addition of ascorbic acid, citric acid and calcium chloride. Ascorbic acid maintained the overall quality for 14 days at 0°C and 7 days at 5°C; no difference, however, was observed regarding browning of cut surface compared to the control sample at both storage temperatures. Calcium chloride maintained the overall quality and cut surface browning for 14 days at both storage temperatures. It was also found that citric acid 1% can be used for minimally processed cabbage. Soaking with citric acid helped retain the color and increased the overall acceptance and organoleptic quality of fresh cut cabbage; it reduced browning of the cut surface and protected against formation of black specks. Citric acid treatment combined with low temperature storage (0°C) prolonged the shelf life of minimally processed cabbage for 22 days, time sufficient for acceptable marketing of the product. The lightness of minimally processed cabbage decreased linearly from 70.94 ± 6 to 63.8 ± 8.5 - 61.3 ± 8 units for the chemical treatments during 22 days of storage at 0˚C. Hue angle values during storage time were also significantly influenced by chemical treatments mainly at 0˚C.

Keywords: storage, sensory quality, colour, texture, fresh-cut minimally processed cabbage, ascorbic acid, citric acid, calcium chloride

1. INTRODUCTION

Minimally processed vegetables have become widely accepted in restaurants, catering services, salad bars and consumer packs for home use.

The primary quality attributes of a fresh-cut vegetable include colour, texture, flavour and odour [1].

Minimally processing operations such as peeling, coring and cutting damage the tissue integrity resulting in cellular decompartmentation. Contact of enzymes and substrates lead to biochemical injuries such as browning, off-flavours and texture breakdown. Enzymatic browning and textural defects reduce product quality and shelf-life [2,3].

Degree of browning depends on phenol content, polyphenol oxidase (PPO) activity [4] and oxygen concentrations.

Brown discolorations in foods are complex because of the large number of secondary reactions that may occur [5].

Various approaches are now applied to limit enzymatic browning such as modified atmosphere [6], chemicals inhibitors and gaseous chlorine dioxide [7,8] or novel decontamination methods such as neutral electrolysed oxidising water [9]. Chemicals exhibit different modes of action on preventing enzymatic browning; the following groups may be used: reducing agents, chelating agents, inorganic salts.

Ascorbic acid is a moderate reducing component, acidic in nature. It forms neutral salts with bases and is water soluble. Ascorbic acid is generally recognized as safe (GRAS) for use in fruits and vegetables [10] and is probably the most widely used.

Citric acid is the best known chelating agent for use in fruits and vegetables [11,12] and is a phenolase Cu-chelating agent. The inhibition of polyphenol oxides (PPO) was attributed to the chelating action [13]. Citric acid is usually used in combination with other anti-browning agents [14].

Salts of calcium have been tested as anti-browning agents. Van Rensburg and Engelbrecht [15] studied the effect of calcium salts on susceptibility to browning of avocados. Varella et al. [16] proposed the use of minimally processed apples "Fuji" in a treatment with 1% $CaCl_2$ for 16 days storage at 8°C - 10°C. Kim and Klieber [17] reported that calcium chloride reduce browning in minimally processed Chinese cabbage.

Cabbage is an important member of cruciferous vegetables being also an important dietary vegetable consumed as fresh cut salad or cooked. Recently it was widely marketed as a minimally processed product. Brassica vegetables (cabbage, cauliflower, broccoli, Brussels sprouts, Kale) possess antioxidant and anticarcinogenic properties [18] and a large group of glycosinolates. Their hydrolysis products can protect against cancer [19].

Finally Rinaldi et al. [20] studied the shelf life of minimally processed cabbage stored under controlled atmospehere (CA) and found that CA did not prolong the shelf life of 10 days at 5°C. Rinaldi et al. [21] also reported that LDPE packaging was more adequate for fresh-cut cabbage storage.

The objective of this study was to investigate the potential of ascorbic acid, citric acid and calcium chloride, individually, to extend the shelf life and maintain the quality of fresh-cut cabbage.

2. MATERIALS AND METHODS

2.1. Raw Material and Treatments

White Cabbage (Brassica oleracea L. var. capitata) var. Bunner harvested in March and April from a farm located in Messinia (Peloponnese, Greece) was minimally processed using the best preparation techniques. The leaves were cut into quarters and then sliced into 3 cm wide strips with a sharp stainless steel knife sterilized in ethanol. The shredded cabbage was washed for 4 min in chlorinated (100 ppm NaOCl) water at 5°C (at a solid/ liquid ratio of 1:3) to remove released nutrients. Then the cabbage was rinsed with tap water for 4 min and centrifuged for 1 min at a speed of 500 rpm.

The treatments were conducted in parallel and prepared for the same batch of product. To minimize product heterogeneity, processed cabbage was pooled, mixed and subsequently divided in four equal parts. The first part was imbibed in distilled water and was used as a control. The second, third and fourth parts were imbibed in ascorbic acid, citric acid and calcium chloride solution, respectively.

Imbibition treatments with solutions of ascorbic acid (Serva, Germany), citric acid (Merck, Germany) and calcium chloride (Carlo Erba, Italy) were applied separately for 3 min, after draining. All dipping solutions were made with distilled water at 5°C and the concentrations were 1% (w/v) at all cases.

100 ± 2 g cabbage slices treated with ascorbic acid, citric acid, calcium chloride or distilled water (control) were placed in clear polystyrene trays (15.5 × 22.5 × 3 cm) covered with PVC film (13 μm) and stored at 0°C and 5°C and 90% E.R.H. in darkness.

Mass loss, instrumental texture, instrumental colour, total soluble solids (TSS), visual quality were evaluated at days 0, 7, 14 and 22. Six packs (replicates) were prepared for each sampling date and treatment. For mass loss ten packs per treatment were prepared.

Mass loss (%) was calculated as percentage of the initial mass of each sample using an electronic scale with an accuracy of ±0.01 g. Weighing was carried out in air conditioned area to avoid package condensation.

2.2. Texture Analysis

Firmness was evaluated using a Texture Analyser TAXT2i (SMS, UK) with a 100 N load cell attached. A Kramer shear cell with a five probe attached to the instrument was used. The speed of the probe was set to 1 $mm \cdot s^{-1}$. Firmness was evaluated in terms of maximum force (rupture force in N) recorded in a typical forcedistance diagram.

2.3. Colour

Colour was measured with a Minolta CR-300 chromameter (Minolta Corp. Japan) on CIE L*, a*, b* chromatic space. The degree of browning was analysed in terms of changes in L*, and hue angle values (h°).

Total soluble solids (TSS) was measured using a digital refractometer (SR-400) in a 10 g homogeneous and filtered cabbage pulp sample.

Quality changes (cut surface browning, black speck, microbial spoilage and overall visual acceptance) were evaluated in 6 packs per treatment, by sensory evaluation performed by 6 semi trained panellists using a score system from 1 to 5 (**Table 1**) with the score of 3 as the cut off for marketing.

2.4. Statistical Analysis

The experiment was performed according to a full factorial design (treatments × storage time) and subjected to an analysis of variance (ANOVA) using the Statgraphics

Table 1. Estimation of visual quality characteristics of minimally processed cabbage (Adapted from Kim and Klieber, 1997).

Characteristics	Score				
	5	4	3	2	1
Cut surface browning	None	Traces in <5 pieces	Very light in <10 pieces	Light in >10 pieces	Dark in >10 pieces
Black speck	None	Few small specks on <5 pieces	Some small specks on <10 pieces	Some large specks on <20 pieces	Many large specks on >20 pieces
Microbial spoilage	None	Traces in 1 piece	Traces in 2 - 5 pieces	Spoilage on 6 - 10 pieces	Spoilage on >10 pieces
Overall visual quality	Very fresh	Fresh	Marketable	Edible non marketable	Non edible.

Plus 5.1 (Statpoint Technologies, Inc, VA, USA) statistical software. Mean values were subjected to Fisher's Least Significant Difference test (LSD) at p = 0.05. This statistical test is liberal with respect to the comparison wise error rate, however, it is powerful for detecting true difference [22]. Data shown in the figures and tables are the mean values of two experimental series since no significant difference between the series was found.

3. RESULTS AND DISCUSSION

3.1. Weight Loss

The multifactor ANOVA indicated that storage temperature, chemical treatments and storage time affected (p = 0.05) weight loss. After 22 days at 0˚C in air, weight loss (**Table 2**) was similar for the control and the treatments with ascorbic acid and calcium chloride (2.28% - 2.81%). The treatment with citric acid showed significantly higher weight loss (3.17%). Similar results were presented at 5˚C. A linear correlation was found between weight loss and storage time (R^2 = 0.985 - 0.995).

Leafy vegetables are particularly vulnerable to rapid water loss [23]. The loss in moisture results in a reduction of fresh weight accompanied by the loss of freshness, appearance and texture. The maximum acceptable weight loss of cabbage is 7% [23,24]. In the present study, weight loss was below 7% at both storage temperatures (0˚C, 5˚C).

3.2. Texture

Firmness changes of minimal processed cabbage during storage at 0°C and 5°C depend upon the soaking solution and the time of storage (**Figure 1**).

The rupture values for minimally processed cabbage treated with different chemicals as well as these of control increased over the 22 days storage period. This increase varied between 2.2% - 5% at 0°C for the control and the treatments with ascorbic and citric acid. At 5°C the increase varied between 6% - 8% for the control and treatment with citric acid and 13% for the treatment with ascorbic acid. Calcium chloride had a greater firming effect ($p = 0.05$) at both storage temperatures compared to other treatments. This was due to cross-linking of both cell wall and middle-lamella pectin by calcium ions [25]. The texture of a plant food is affected by the genetic background, morphology, cell turgor, water content, biochemical components and growing conditions [26,27]. Treatments with ascorbic and citric acid did not result in tissue softening, results not agreeable to those reported by Gil et al. [28]. Cocci et al. [29] showed a structural breakdown with consequent softening in apple tissue.

Table 2. Weight loss of minimal processing cabbage treated with different chemicals after 22 days of storage at 0°C and 5°C.

Temperature	Ascorbic acid	Citric acid	Calcium chloride	control
0°C	2.45^{Aa}	3.17^{Ab}	2.28^{Aa}	2.81^{Aa}
5°C	3.63^{Ba}	4.2^{Bb}	3.35^{Ba}	3.22^{Ba}

Values followed by different superscript **lowercase** letters in the same row show statistically significant differences. $p = 0.05$; Different superscript **uppercase** letters show statistically significant differences in the same column (among temperatures). $p = 0.05$; Mean values. N = 2 repetitions × 10 samples per repetition = 20.

3.3. Colour

The L* colour value and hue angle (h°) were selected as the most suitable parameters to measure the colour change of cut surface of minimally processed cabbage. When the colour became darker, a decrease in L* and hue angle (h°) was observed. The hue value represents true colour which is effective for visualizing the colour appearance of food products [30]. The storage time and the interaction between storage time and temperature had significant effect on L*. Among the parameters which were studied (temperature, chemical treatments, storage time and their interactions) only temperature did not have a significant effect ($p = 0.05$) on hue values.

The lightness of minimally processed cabbage decreased linearly (**Figure 2**) from 70.94 ± 6 to 63.8 ± 8.5 −61.3 ± 8 for the chemical treatments during 22 days of storage at 0˚C. The control sample presented a decrease of L* values to 59.4 ± 5.4.

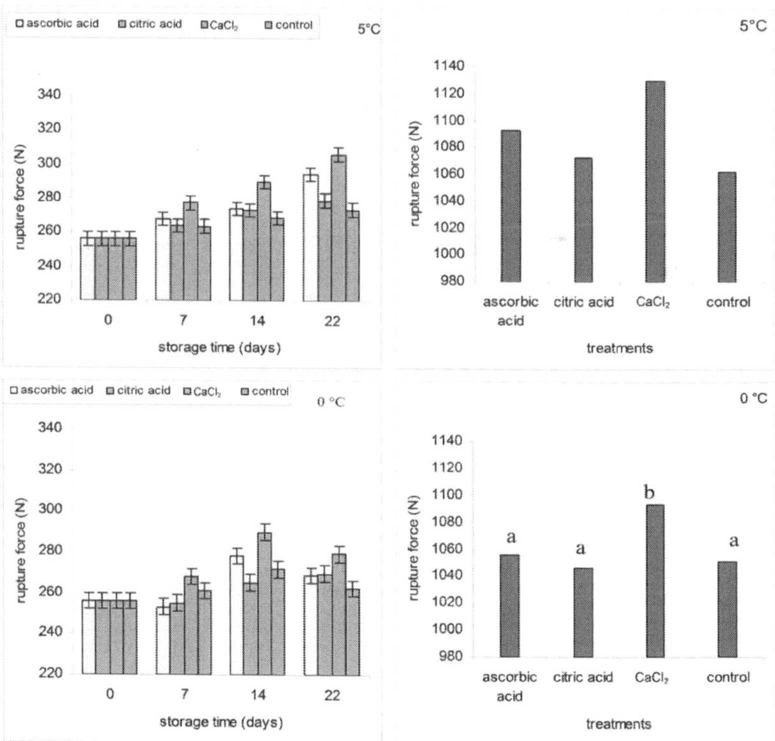

Figure 1. Effects of treatment with chemicals on the firmness of minimal processed cabbage stored at 0˚C and 5˚C. Data observed on the left set of histograms show the changes during the 22 days of storage (Mean values. N = 2 repetitions × 6 samples = 12). Vertical bars represent LSD at the 5% level. The data observed in the right set of histograms are the mean of all times.

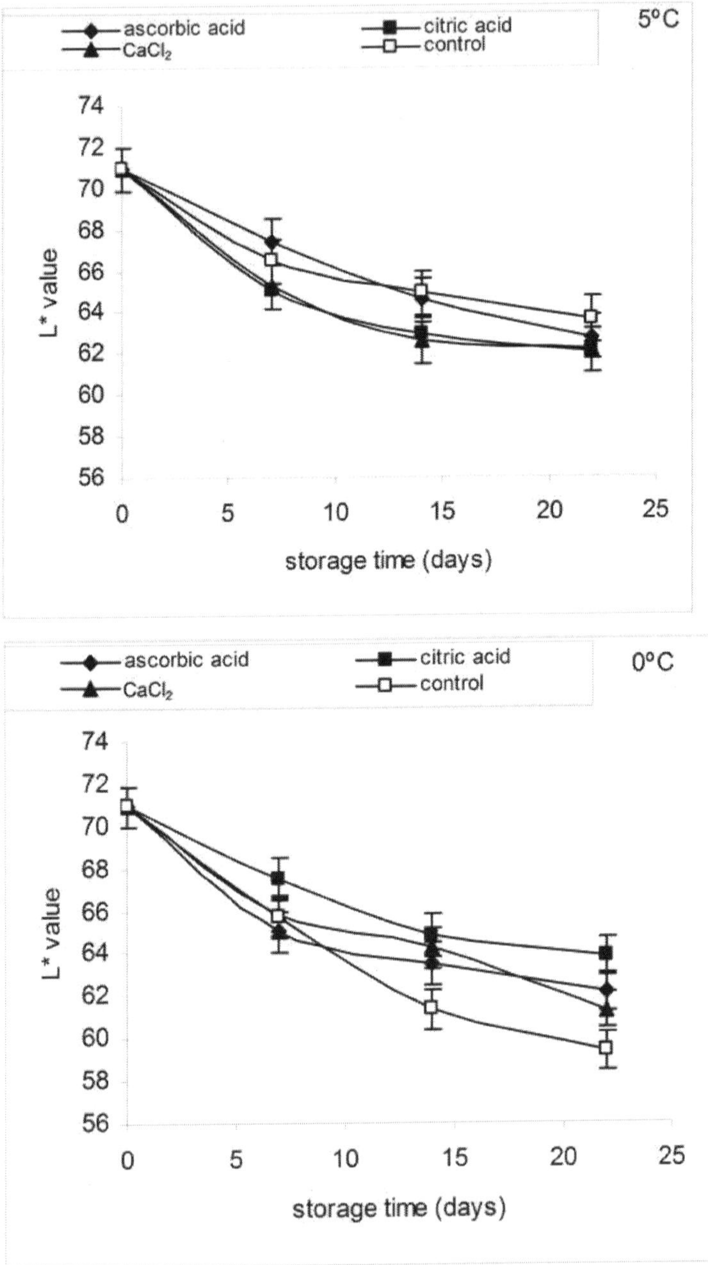

Figure 2. Effects of chemical treatments on changes in L* value of minimal processed cabbage stored at 0˚C and 5˚C for 22 days. Vertical bars represent LSD at the 5% level (Mean values. N = 2 repetitions × 10 samples = 20).

At 5˚C the lightness decreased from 70.94 ± 6 to 63.6 ± 7.9 - 62.07 ± 8.2 at all cases during 22 days of storage. By day 22, at 0˚C the L* value was lower in non treated cabbage (control) than in treated cabbage with chemicals. Among the chemical treatments citric acid was the best inhibitor showing the least change in lightness. At 5˚C, no significant difference in L* values between the control and treated samples was observed after 22 days of storage.

Hue angle values (**Figure 3**) during storage time were significantly affected by chemical treatments mainly at 0˚C.

Hue angle values of control decreased rapidly at 0˚C. Among the chemical treatments, citric acid retained better the color of fresh-cut cabbage. Citric acid has been reported extensively for its inhibitory activity on polyphenol oxidase, PPO [31] and its antibrowning activity in minimally processed fruits and vegetables [32]. Our results come in agreement with Kim and Klieber [17] who worked on minimally processed Chinese cabbage and Kaur and Kapoor [33] working on minimally processed Indian cabbage. Ascorbic acid and calcium chloride are intermediate inhibitors as seen in their hue angle (h˚) values (**Figure 3**). This result was similar to results of other researchers [16,34] working on banana slices and minimally processed apples. Ascorbic acid is a reducing agent widely used as an antibrowning agent for processing of fruits and vegetables. Reductants are irreversibly oxidized during the reaction, which means that the protection they confer is only temporary since they are consumed in the reaction [34]. The action of calcium chloride might be due to the PPO inhibition by chloride ions [35].

3.4. Soluble Solids Contents (SSC)

The soluble solids content in minimally processed cabbage during storage at 0˚C and 5˚C are presented in **Table 3**.

At 0˚C the control exhibited higher SSC and the cabbage treated with calcium chloride showed the smallest. At 5˚C the treatments with ascorbic acid and citric acid exhibited the highest SSC. Similar results were reported by Jiang et al. [36] in Chinese water chestnut. Gonzalez-Aguilar et al. [37] reported an appreciable increase in fructose and glucose in fresh-cut mangoes treated with antibrowning agents and stored at 10˚C.

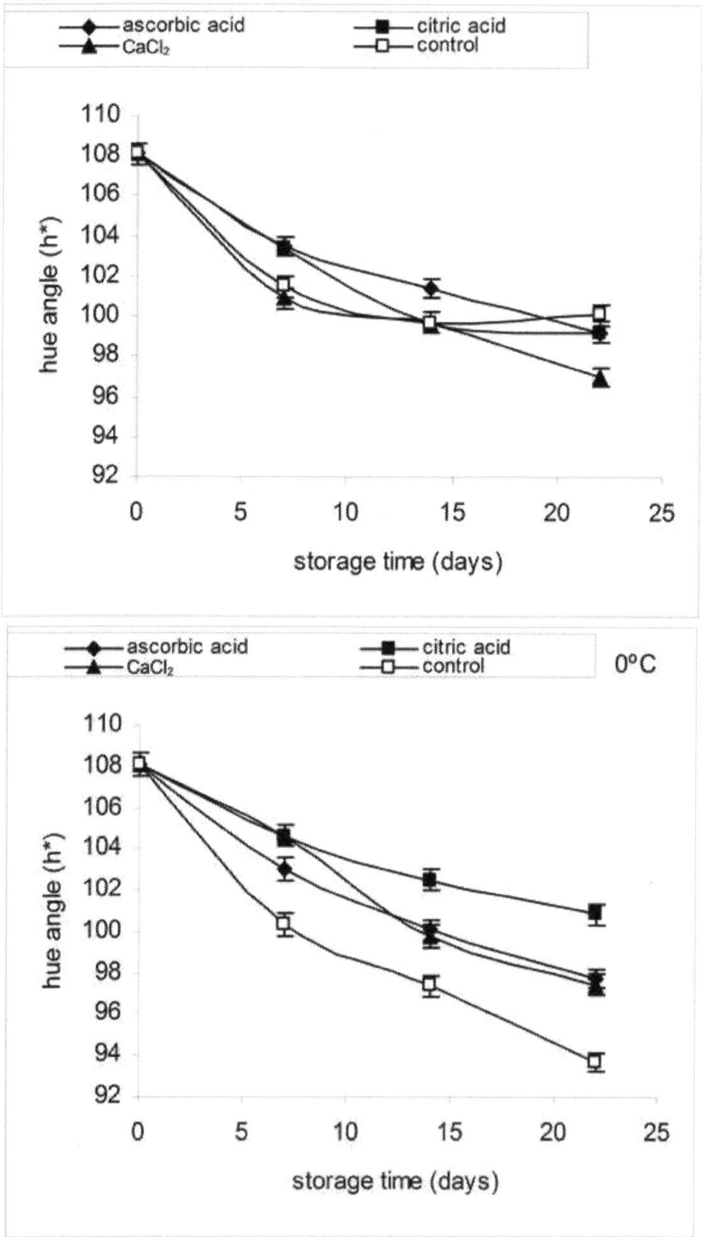

Figure 3. Hue angle changes of minimal processing cabbage stored at 0˚C and 5˚C for 22 days. Vertical bars represent LSD at the 5% level (Mean values. N = 2 repetitions × 10 samples = 20).

Table 3. Soluble solids content of minimal processed cabbage treated with different chemicals after 22 days of storage at 0˚C and 5˚C.

Temperature (˚C)	Time of storage (days)				
	0 day	22 days			
		Ascorbic acid	Citric acid	Calcium chloride	control
0	5.88	4.5^{aA}	4.4^{aA}	4.0^{bA}	4.9^{cA}
5	5.88	4.8^{aB}	4.4^{aA}	3.8^{bA}	3.9^{bB}

Values followed by different superscript **lowercase** letters in the same row show statistically significant differences. $p = 0.05$; Different superscript **uppercase** letters show statistically significant differences in the same column (among temperatures). $p = 0.05$; Mean values. N = 2 repetitions × 6 samples per repetition = 12.

Table 4. Changes in organoleptic quality properties of minimally processed cabbage with different chemical treatments during storage at 0˚C and 5˚C.

Treatments	Temperature 0˚C				Temperature 5˚C			
	Storage time (days)				Storage time (days)			
	0	7	14	22	0	7	14	22
Overall acceptance								
Ascorbic acid	5^{aA}	4.75^{aA}	3.8^{bA}	2.1^{cA}	5^{aA}	4^{bA}	2^{cA}	1.1^{dA}
Citric acid	5^{aA}	4.9^{aA}	4^{bA}	3.3^{cB}	5^{aA}	4.8^{aB}	3.3^{bB}	2^{cB}
Calcium chloride	5^{aA}	4.9^{aA}	3.4^{bA}	2.1^{cA}	5^{aA}	3.4^{bC}	3^{bB}	1.13^{cA}
control	5^{aA}	3.9^{bB}	2.8^{cB}	2.1^{dA}	5^{aA}	3.3^{bC}	2.4^{cC}	1.8^{dB}
Cut surface browning								
Ascorbic acid	5^{aA}	4.6^{aA}	2.5^{bA}	1.6^{cA}	5^{aA}	3.9^{bA}	2.5^{cA}	1.4^{dA}
Citric acid	5^{aA}	4.9^{aB}	3.5^{bB}	3^{cB}	5^{aA}	4.6^{aB}	3.1^{bB}	1.8^{cB}
Calcium chloride	5^{aA}	4.5^{aA}	3.5^{bB}	2.4^{cC}	5^{aA}	2.9^{bC}	3^{bB}	1^{cC}
control	5^{aA}	4.4^{bA}	2.6^{bA}	2.1^{cC}	5^{aA}	3.8^{bA}	2.5^{cA}	1.8^{dB}
Black speck								
Ascorbic acid	5^{aA}	5^{aA}	2.1^{bA}	1.9^{cA}	5^{aA}	4.8^{aA}	3.3^{bA}	1.6^{cA}
Citric acid	5^{aA}	5^{aA}	3.8^{bB}	3.1^{cB}	5^{aA}	5^{aA}	3.9^{bB}	2.8^{cB}
Calcium chloride	5^{aA}	5^{aA}	3.3^{bC}	2.6^{cC}	5^{aA}	5^{aA}	3.1^{bA}	1^{cC}
control	5^{aA}	5^{aA}	3^{bC}	2^{cA}	5^{aA}	5^{aA}	2.8^{bC}	2.3^{cD}
Microbial spoilage								
Ascorbic acid	5^{aA}	5^{aA}	3.6^{bA}	1.9^{cA}	5^{aA}	5^{aA}	3.3^{bA}	2^{cA}
Citric acid	5^{aA}	5^{aA}	4.6^{bB}	3.4^{cB}	5^{aA}	5^{aA}	3.8^{bB}	3.3^{cB}
Calcium chloride	5^{aA}	5^{aA}	4.1^{bA}	1^{cC}	5^{aA}	5^{aA}	3.3^{bA}	2.9^{cC}
control	5^{aA}	5^{aA}	3.9^{bA}	2.8^{cD}	5^{aA}	5^{aA}	2.9^{bA}	2^{cA}

Values followed by different superscript **lowercase** letters in the same row show statistically significant differences. $p = 0.05$; Different superscript **uppercase** letters show statistically significant differences in the same column (among treatments). $p = 0.05$; Mean values. N = 2 repetitions × 6 samples per repetition × 6 panellists = 72.

3.5. Overall Organoleptic Quality

Organoleptic quality evaluation was carried out by human volunteers in conjunction with quantifiable physical and chemical measurements to determine if a shelf-life extension treatment has merit.

Overall acceptance together with cut surface browning determined the shelf life of control samples which was 7 days at both storage temperatures (**Table 4**) using a score of 3 as the limit. Treatment with citric acid extended the shelf life of shredded cabbage stored at 0˚C up to 22 days. During marketing the predominant temperature is 5˚C and at this temperature treatment with citric acid extended the shelf life from 7 days (control) to 14 days. Ascorbic acid maintained the overall quality for 14 days at 0˚C and 7 days at 5˚C, however, no difference was observed regarding cut surface browning compared to the control sample at both storage temperatures. Similar results were reported by Rosen and Kader [38] working with pear slices. Calcium chloride maintained the overall quality and cut surface browning for 14 days at both storage temperatures. Similar results in apples were reported by Drake and Spayed [39]. However, Bolin and Huxoll [40] reported that calcium chloride is not effecttive in preventing browning in peaches. Therefore, 1% citric acid can be used for minimally processed cabbage.

Citric acid is the most commonly used acidifier in combination with other types of antibrowning agents to control browning in fruits and vegetables.

During storage, the quality of minimally processed cabbage was affected by black speck. Black speck is a non-pathogenic disorder appearing during storage. It consists of very small to moderate size discolored lesions on the midrib and veins of the leaves [41]. At 0˚C citric acid reduced the effects of black speck at the end of shelf-life (22 days) resulting in a better quality. Citric acid may reduce black speck through its inhibitory effect on polyphenoloxidase by lowering the pH and resulting in the chelation of copper at the active site of the enzyme [42]. At 5˚C the chemical treatments affected black speck and reduced the incidence at the end of shelf-life (14 days).

Citric acid at 0˚C and all chemical treatments at 5˚C prevented microbial spoilage at the end of shelf-life (22 and 14 days, respectively).

4. CONCLUSIONS

The results presented indicate that soaking with citric acid better retains the color and the overall acceptance and organoleptic quality of fresh cut cabbage. It can reduce the cut surface browning and protect from black speck. Citric acid treatment combined with low temperature storage (0˚C) prolonged the shelf life of minimally processed cabbage for 22 days, time sufficient for acceptable marketing.

REFERENCES

1. E. Watada and L. Qi, "Quality of Fresh-Cut Produce," Postharvest Biology and Technology, Vol. 15, No. 3, 1999, pp. 201-205.

2. D. King and H. R. Bolin, "Physiological and Microbiological Storage Stability of Minimally Processed Fruits and Vegetables," Food Technology, Vol. 43, No. 2, 1989, pp. 132-139.

3. J. K. Brecht, "Physiology of Lightly Processed Fruits and Vegetables," Horticultural Science, Vol. 30, No. 1, 1995, pp. 18-22.

4. J. Zawistowshy, C. G. Biliaderis and N. A. M. Eskin, "Polyphenol Oxidase," In: D. S. Robinson and N. A. M. Eskin, Eds., Oxidative Enzymes in Foods, Elsevier, London, 1991, pp. 217-273.

5. G. M. Sapers, "Browning of Foods: Control by Sulfites. Antioxidants and Other Means," Food Technology, Vol. 47, No. 10, 1993, pp. 75-84.

6. M. E. Pirovani, D. R. Güemes, A. M. Piagentini and J. H Di Pentina, "Storage Quality of Minimally Processed Cabbage Packaged in Plastic Films," Journal of Food Quality, Vol. 20, No. 5, 1997, pp. 381-389.

7. M. Friedman, "Food Browning and Its Prevention—A Review," Journal of Agricultural and Food Chemistry, Vol. 44, No. 3, 1996, pp. 631-653.

8. V. M. Gomez-Lopez, P. Ragaert, V. Jeyachchandran, J. Debevere and F. Devlieghere, "Shelf-Life of Minimally Processed Lettuce and Cabbage Treated with Gaseous Chlorine Dioxide and Cysteine," International Journal of Food Microbiology, Vol. 121, No. 1, 2008, pp. 74-83.

9. V. M. Gomez-Lopez, P. Ragaert, J. Ryckeboer, V. Jeyachchandran, J. Debevere and F. Devlieghere, "Shelf-Life of Minimally Processed Cabbage Treated with Neutral Electrolysed Oxidizing Water and Stored under Equilibrium Modified Atmosphere," International Journal of Food Microbiology Vol. 117, No. 1, 2007, pp. 91-98.

10. L. D. Alvarez and A. Chiralt, "Color of Minimally Processed Fruits and Vegetables as Affected by Some Chemical and Biochemical Changes," In: S. M. Alzamora, M. S. Tapia and A. Lopez-Malo Eds., Minimally Processed Fruits and Vegetables, Aspen Publishers, Frederick, 2000, pp. 111-126.

11. F. Pizzocaro, D. Torreggiani and G. Gilardi, "Inhibition of Apple Polyphenol Oxidase by Ascorbic Acid, Citric Acid and Sodium Chloride," Journal of Food Processing and Preservation Vol. 17, No. 1, 1993, pp. 21-30.

12. M. A. Rojas-Graü, Solina-Fortuny and O. Martin-Belloso, "Effect of Natural Antibrowning Agents on Color and Related Enzymes in Fresh-Cut 'Fuji' Apples as an Alternative to the Use Ascorbic Acid," Journal of Food Science, Vol. 73, 2008, pp. 267-272.

13. Y. M. Fu, J. R. Jiang, G. Zanberman and Y. Fuchs, "Purification of Polyphenol Oxidase and the Browning Control of Litchi Fruit by Glutathiose and Citric Acid," Journal of the Science of Food and Agriculture, Vol. 79, No. 7, 1999, pp. 950-954.

14. H. E. Moline, J. G. Buta and I. M. Newman, "Prevention of Browning of Banana Slices Using Natural Produce and Their Derivatives," Journal of Food Quality, Vol. 22, No. 5, 1999, pp. 499-511.

15. E. Van Rensburg and A. H. P. Engelbrecht, "Effect of Calcium Salts on Susceptibility to Browning of Avocados Fruit," Journal of Food Science, Vol. 51, No. 4, 1986, pp. 1067-1068.

16. P. Varela, A. Salvador and S. M. Fiszman, "The Use of Calcium Chloride in Minimally Processed Apples," European Food Research and Technology, Vol. 224, No. 4, 2007, pp. 461-467.

17. S. Kim and A. Klieber, "Quality Maintenance of Minimally Processed Chinese Cabbage with Low Temperature and Citric Acid Dip," Journal of the Science of Food and Agriculture, Vol. 75, No. 1, 1997, pp. 31-36.

18. Y. F. Chu, J. Sun, X. Wu and R. H. Liu, "Antioxidant and Anti Proliferative Activities of Common Vegetables," Journal of Agricultural and Food Chemistry, Vol. 50, No. 23, 2002, pp. 6910-6916.

19. Y. S. Keum, W. S. Jeong and A. N. T. Kong, "Chemoprevention by Isothiocyanates and Their Underlying Molecular Signalling Mechanisms," Mutation Research, Vol. 555, No. 1-2, 2004, pp. 191-202.

20. M. M. Rinaldi, B. C. Benedetti and C. L. Moretti, "Respiration Rate, Ethylene Production and Shelf Life of Minimally Processed Cabbage under Controlled Atmosphere," Acta Horticulturae, Vol. 876, 2010a, pp. 279- 286.

21. M. M. Rinaldi, C. I. G. L. Sarantopoulos, B. C. Benedetti and C. L. Moretti, "Storage of Minimally Processed Cabbage in Different Packaging Systems," Acta Horticulturae, Vol. 877, 2010b, pp. 597-602.

22. M. C. Gacula and J. Singh, "Statistical Methods in Food and Consumer Research," Academic Press Inc., New York, 1984.

23. S. J. Kays, "Postharvest Physiology of Perishable Plant Products," AVI Publisher, New York, 1991, pp. 356-357.

24. H. M. Kang, K. W. Park and M. E. Saltveit, "Elevated Growing Temperatures during the Day Improve the Postharvest Chilling Tolerance of Greenhouse—Grown Cucumber," Postharvest Biology and Technology Vol. 24, 2002, pp. 49-57.

25. W. Poovaich, "Role of Calcium in Prolonging Storage Life of Fruits and Vegetables," Food Technology, Vol. 40, No. 5, 1986, pp. 86-89.

26. F. R. Harker, R. J. Redgwell, I. C Hallet and S. H. Murray, "Texture of Fresh Fruit," Horticultural Reviews, Vol. 20, 1997, pp. 121-224.

27. E. Sams, "Preharvest Factors Affecting Postharvest Texture," Postharvest Biology and Technology, Vol. 15, No. 3, 1999, pp. 249-254.

28. M. I. Gil, J. R. Gorny and A. A. Kader, "Responses of 'Fuji' Apple Slices to Ascorbic Acid Treatments and Low Oxygen Atmospheres," Horticultural Science, Vol. 33, No. 2, 1998, pp. 305-309.

29. Cocci, P. Rocculi, S. Romani and M. Dalla Rosa, "Changes in Nutritional Properties of Minimally Processed Apples during Storage," Postharvest Biology and Technology, Vol. 39, No. 3, 2006, pp. 265-271.

30. R. G. Mcguire, "Reporting of Objective Colour Measurements," Horticultural Science, Vol. 27, No. 12, 1992, pp. 1254-1255.

31. T. T. Langdon, "Preventing of Browning in Fresh Prepared Potatoes without the Use of Sulfiting Agents," Food Technology, Vol. 41, No. 64, 1987, pp. 66-67.

32. R. Ahvenainen, "New Approaches in Improving the Shelf Life of Minimally Processed Fruit and Vegetables,". Trends in Food Science and Technology, Vol. 7, No. 6, 1996, pp. 179-187.

33. C. Kaur and H. C. Kapoor, "Antioxidant Activity and Quality of Minimally Processed Indian Cabbage (Brassica Oleracea var. Capitata)," Journal of Food Processing and Preservation, Vol. 25, No. 5, 2001, pp. 367-380.

34. M. Apintanapong, K. Cheachumluang, P. Suansawan and N. Thongpraset, "Effect of Antibrowning Agents on Banana Slices and Vacuum-Fried Slices," Journal of Food Agriculture and Environment, Vol. 5, No. 3-4, 2007, pp. 151-157.

35. Garcia and D. M. Barette, "Preservation Treatments for Fresh Cut Fruits and Vegetables," In: O. Lamikanra, Ed., Fresh Cut Fruits and Vegetables: Science, Technology and Market, CRC Press, Boca Raton, 2002, pp. 267-303.

36. Y. M. Jiang, L. T. Peng and J. R. Li, "Use of Citric Acid for Shelf Life and Quality Maintenance of Fresh-Cut Chinese Water Chestnut," Journal of Food Engineering, Vol. 63, No. 3, 2004, pp. 325-328.

37. A. Gonzalez-Aguilar, C. Y. Wang and J. C. Buta, "Maintaining Quality of Fresh-Cut Mangoes Using Antibrowning Agents and Modified Atmosphere Packaging," Journal of Agricultural and Food Chemistry, Vol. 48, No. 9, 2000, pp. 4204-4208.

38. J. C. Rosen and A. A. Kader, "Postharvest Physiology and Quality Maintenance of Sliced Pear and Strawberry Fruits," Journal of Food Science, Vol. 54, No. 3, 1989, pp. 656-659.

39. S. R. Drake and S. E. Spayed, "Influence of Calcium Treatment on 'Golden Delicious' Apple Quality," Journal of Food Science, Vol. 48, No. 2, 1983, pp. 403-405.

40. R. Bolin and C. C Huxoll, "Storage Stability of Minimally Processed Fruits," Journal of Food Processing and Preservation, Vol. 31, No. 4, 1989, pp. 281-292.

41. R. Walsh, E. C. Lougheed and P. M. A. Toivenon, "The Effect of Benomyl. Sodium Hypochlorite and Controlled Atmospheres upon the Incidence of Black Speck of Stored Cabbage," Journal of American Society of Horticultural Sciences, Vol. 108, No. 4, 1983, pp. 533-536.

42. R. Lyengar and A. J. Mcevily, "Anti-Browning Agents: Alternatives to the Use of Sulfites in Foods," Trends in Food Science and Technology, Vol. 3, 1992, pp. 60-64.

Chapter 5

Edible Coating as Packaging Strategy to Extend the Shelf-life of Fresh-Cut Fruits and Vegetables

Raffaele Porta[1*], Giovanna Rossi-Marquez[1], Loredana Mariniello[1], Angela Sorrentino[2], C. Valeria L. Giosafatto[1], Marilena Esposito[1] and Prospero Di Pierro[1]

[1]*Department of Chemical Sciences, University of Naples "Federico II", Naples, Italy*

[2]*Department of Agriculture, University of Naples "Federico II", Naples, Italy*

Food market continuously generates novel products by following the trends of consumers whose preferences are increasing towards ready-to-use foods. Among these, fresh-cut fruits and vegetables are at the top of the list and are stimulating a great effort of research in this area. In fact, their market has grown rapidly in recent years as a result of the changes in consumer habits. Therefore, there is a real need to find new methods for preservation of processed food products able to achieve widespread acceptance by the industry.

Food appearance remains the most required attribute, strongly affecting consumer decision to buy it or not. In addition, food texture is also a fundamental feature determining product acceptability, appearance and texture changes being very tightly linked markers of food deterioration [1]. Both appearance and texture of a fruit or vegetable tissue strictly depend on genetic, environmental, postharvest handling and storage factors. Therefore, the precise knowledge of the processes leading to the appearance and textural modifications is of crucial importance in developing effective approaches to counteract them and, hence, to improve quality and shelf-life of these products.

Most of the changes occurring in fresh-cut fruits and vegetables represent an extension of the normal ripening events leading to their softening, although they are also specifically influenced by the tissue cutting and wounding. Product maturity at the time of processing mostly affects the shelf-life of fruits since they become increasingly soft and susceptible to transport and handling damage during ripening [2-4]. Precise understanding of ripening-related softening, hence, must be carefully taken into consideration when fresh-cut fruits are studied. Conversely, fresh-cut vegetables normally present minor problems because they generally have a much greater proportion of cells with thickened secondary walls and, consequently, are much firmer and less susceptible to deterioration.

However, the consequences of both fruit and vegetable cutting and post-cutting processes have not been extensively investigated. In this respect, it should be enphasized that fresh-cut products are "wounded" tissues and, thus, they deteriorate more rapidly because their behavior significantly differs from that of intact fruits and vegetables. In fact, fresh product cutting may result in several biochemical events such as browning, off-flavor spillage and loss of nutritional value due to release of the intracellular content at the sites of wounding [5]. Also the subcellular compartmentalization is shattered at the cut surfaces, and the mixing of some substrates and enzymes, which are normally separated, gives rise to reactions that in the intact cells do not occur. For instance, fruit injury results in an increased rate of respiration and production of ethylene, with visible effects within minutes to few hours, whereas the development of hardening and lignification is an undesireable effect of vegetable wounding [6]. A further problem in many fresh-cut fruits and vegetables -particularly evident in fruits with a white flesh, such as apples and pears- is the cut-edge browning due to the polyphenol oxidase and phenol peroxidase interaction with polyphenols and oxygen [6]. Furthermore, other two phenomena, namely drying and microbial colonization, may influence fresh-cut fruit and vegetable appearance and texture. Both these factors, having different causes and effects, also lead to unattractive products. Water loss and osmotic changes are the major reasons of postharvest deterioration determining a decay of food crispness and turgor. They occur very rapidly following the exposure of the internal tissues of fresh-cut products [7,8]. It has been demonstrated that, at a given temperature, the rate of water loss from the products strictly depends on the environment relative humidity [9], and it is well known that whereas temperature management is a common practice to delay fruit and vegetable deterioration, post harvest maintenance of the recommended RH is not frequently carried out [10]. Finally, food surface microbial stability is another major determinant of fresh-cut product quality and safety during storage and distribution.

To prolong the shelf-life of fresh-cut fruits and vegetables different technologies have been so far used [11]. Among these numerous chemical and physical preservation strategies were explored to reduce enzymatic browning and tissue softening after cutting. However, almost all of the chemical treatments (sulfite, citric acid, ascorbic acid derivatives, cinnamate, benzoate, and cyclodextrins) confere off flavors, and many of the most effective substances added are recognized as unsafe. Conversely, the strategy based on the manipulation of both temperature and headspace atmosphere is more feasible and reliable. Modified Atmosphere Packaging (MAP), a dynamic system with two gas fluxes, allows the respiration of the fresh products and the gas exchange through a packaging film [12]. MAP gas composition, generally low in O2 and high in CO2, primarily depends on package total surface area, product weight and respiration rate, film gas transmission capacity and storage temperature. However, food packaging by MAP at low temperature is usually not sufficient to significantly extend the shelf-life of the majority of the fresh-cut fruits and vegetables.

The application of edible coatings is increasingly demonstrating to be a relatively new and simple technology effective in preventing the appearance and textural deterioration of several products. The use of different types of films based on a variety of single biopolymers, or on their combinations, results to be extremely advantageous even though all data obtained so far indicate that the coatings need to be tailored and optimized for each kind of foods [13-18]. More than twenty years ago Nisperos-Carriedo et al. [19] demonstrated that edible coatings based on cellulose gums effectively delay ripening in some climacteric fruits like mangoes, papayas, and bananas and significantly reduce enzymatic browning on sliced mushrooms. Ten years later Le Tien et al. [20] showed that milk protein coatings are able to counteract oxidative degradation of cut fruits and vegetables through polyphenol oxidase inhibition. From then several polysaccharides (chitosan, alginate, methylcellulose or pectin) and proteins (casein, collagen, gelatin, phaseolin, zein, soy or whey proteins), or mixtures of them, were shown to give rise to edible films effective as water vapor and gas barriers for a wide range of food products and as carriers for antimicrobials [17,21-25]. In particular, antimicrobial containing films are recently gaining potential interest in reducing the deleterious effects caused by also minimal cut processing of fresh fruits and vegetables [26,27]. Some of the most commonly used antimicrobials include benzoic acid, sorbic acid, lysozyme, bacteriocins and plant-derived secondary metabolites, such as essential oils and phytoalexins. In particular, Mastromatteo et al. [28] recently found that the combined use of ethanol pretreatment and edible coating prolonged shelf-life of minimally processed kiwi fruits.

Moreover, edible coatings may serve as carriers also for a wide range of food additives, including anti-browning agents, colorants, flavors, nutrients, spices that not only extend the shelf-life of the products, but also improve their safety and acceptability [29-32]. An example derives from the fresh-cut carrots, one of the most consumed vegetables, whose marketing is limited by their fast physiological changes during storage, developing characteristic odors of anaerobic catabolism due to higher respiration rate and microbiological deterioration. Thus, minimally processed carrots quickly lose their firmness and bright orange color and develop a white blush on the surface, thereby reducing consumer acceptability [6,33,34]. It was recently demonstrated that dipping into a hydro-alcoholic solution after fresh-cut carrot coating with sodium alginate, followed by a final packaging into a microperforated polypropylene film under passive or active MAP, prevents dehydration and microbial proliferation, delays the respiratory activity and enhances the quality of the product stored at 4°C [35]. In addition, from the sensory point of view, coated carrots were appreciated for about two weeks, whereas all the uncoated ones were refused after only two days.

In addition, the effectiveness of soy protein coatings in reducing oxidative browning and moisture loss of cut apples and potatoes during their storage at 4°C was recently reported [36]. This kind of coating was shown to have antioxidative activity and the addition of carboxymethyl cellulose to the formulation was found to significantly improve its effect. In fact, browning in control cut potatoes was increased by 106.6% in contrast to 35.2% for soy protein-coated samples, and moisture barrier effect was significantly improved after treatment. Soy protein coatings reduced moisture loss in apples and potatoes by 21.3 and 29.6%, respectively, over the control.

Similar results were recently obtained in our laboratories (unpublished data) by coating fresh-cut apples, carrots and potatoes with a pectin/whey protein film containing the enzyme transglutaminase. The ability of such hydrocolloid film to act as an effective barrier to water vapor allowed coated fruit and vegetables to markedly decrease their water loss and to counteract the microbial growth during their storage at 4°C for 10 days. In addition, food texture was assessed both at the harvest of the products and after storage by acquiring their mechanical profiles. An almost stable physiological situation was depicted between the two stages (at harvest and after storage), with a practically unmodified pattern of both product hardness and chewiness, when apples, potatoes and carrots were coated with the edible films. The picturebelow clearly shows the different browning and softening degrees betweenuncoated and coated apple and potato samples after 10 days of storage at 4°C, thus suggesting that milk whey, often considered only as a polluting dairy by-product, could effectively be used as a natural component of antioxidative and moisture barrier coatings to enhance the shelf-life of fresh-cut fruits and vegetables (Figure 1).

In conclusion, we want to highlight that a relatively simple technology, such as the application of edible coatings obtainable from inexpensive raw materials, is effective against both browning development and textural deterioration happening during the management and storage of fresh-cut products. The use of various combinations of treatments may be certainly advantageous, even though they should be optimized for each plant species and sometimes probably even for each cultivar.

REFERENCES

1. Cantwell MA, Suslow TV (2002) Post harvest handling systems: fresh-cutfruits and vegetables. In: Postharvest Technology of Horticultural Crops, Kader AA (3rd edn), Agric Natural Res Publ, CA,USA.

2. Beaulieu JC, Ingram DA, Lea JM, Bett-Garber KL (2004) Effect of harvest maturity on the sensory characteristics fresh-cut cantaloupe. J Food Sci 69: S250–S258.

3. Soliva-Fortuny RC, Alos-Saiz N, Espachs-Barroso A, Martin-Belloso O (2004) Influence of maturity at processing on quality attributes of fresh-cut conference pears. J Food Sci 69: S290–S294.

4. Lana MM, Tijskens LMM, van Kooten O (2005) Effects of storage temperature and fruit ripening on firmness of fresh cut tomatoes. Post harvest BiolTechnol 35:87–95.

5. Saltveit ME (2003) Fresh-cut vegetables. In: Post harvest Physiology and Pathology of Vegetables, Bartz JA, Brecht JK, Marcel Dekker Inc, USA

6. Toivonen PMA, Brummell DA (2008) Biochemical bases of appearance and texture changes in fresh-cut fruit and vegetables. Postharvest BiolTechnol 48:1–14.

7. Shackel KA, Greve C, Labavitch JM, Ahmadi H (1991) Cell turgor changes associated with ripening in tomato pericarp tissue. Plant Physiol 97: 814-816.

8. Saladié M, Matas AJ, Isaacson T, Jenks MA, Goodwin SM, et al. (2007) A reevaluation of the key factors that influence tomato fruit softening and integrity. Plant Physiol 144: 1012-1028.

9. Kays SJ (1991) Post harvest physiology of perishable plant products. Van Nostrand Reinhold, New York.

10. Paull RE (1999) Effect of temperature and relative humidity on fresh commodity quality. Postharvest BiolTechnol 15:263–277.

11. Rojas-Graü MA, Oms-Oliu G, Soliva-Fortuny R, Martín-Belloso O (2009) The use of packaging techniques to maintain freshness in fresh-cut fruits and vegetables: a review. Int J Food SciTechnol, 44: 875-889.

12. Van de Velde K, Kiekens P (2002) Biopolymers: overview of several properties and consequences on their applications. Polymer Testing 21:433–442.

13. Olivas GI, Barbosa-Cánovas GV (2005) Edible coatings for fresh-cut fruits. Crit Rev Food Sci Nutr 45: 657-670.

14. Shon J, Eo JH, Eun JB (2010) Effect of soy protein isolate coating on quality attributes of cut raw Han-Woo (Korean cow) beef, aerobically packaged and held refrigerated. J Food Quality 33:42-60.

15. Plackett D (2011) Biopolymers: New materials for sustainable films and coatings. John Wiley & Sons eds.

16. Porta R, Di Pierro P, Sorrentino A, Mariniello L (2011) Promising perspectives of transglutaminase in "bioplastics" production. J Biotechnol Biomater.

17. Porta R, Mariniello L, Di Pierro P, Sorrentino A, Giosafatto CV (2011) Transglutaminase crosslinked pectin- and chitosan-based edible films: a review. Crit Rev Food Sci Nutr 51: 223-238.

18. Rojas-Graü MA, Salvia-Trujillo L, Soliva-Fortuny R, Martín-Belloso (2012) Edible films and coatings. In: Decontamination of fresh and minimally processed produce,Vicente M. Gómez-López, John Wiley & Sons Inc.

19. Nisperos-Carriedo MO, Baldwin EA, Shaw PE (1991) Development of an edible coating for extending postharvest life of selected fruits and vegetables. Proc Florida State Hort. Soc 104:122-125.

20. Le Tien C, Vachon C, Mateescu MA, Lacroix M (2001) Milk protein coatings prevent oxidative browning of apples and potatoes. J Food Sci 66:512-516.

21. Shon J, Haque ZU (2007) Efficacy of sour whey as a shelf-life enhancer: use in antioxidative edible coatings of cut vegetables and fruit. J Food Quality 30:581-593.

22. Shon J, Chin KB (2008) Effect of whey protein coating on quality attributes of low-fat, aerobically packaged sausage during refrigerated storage. J Food Sci 73: C469-475.

23. Di Pierro P, Sorrentino A, Mariniello L, Giosafatto CVL, Porta R (2011) Chitosan/wheyprotein film asactivecoating to extend Ricotta cheese shelf-life. LWT-Food Sci Technol 44:2324-2327.

24. Porta R, Mariniello L, Di Pierro P, Sorrentino A, Giosafatto VC, Rossi Marquez G, Esposito M (2012) Water barrier edible coatings of fried foods. J BiotechnolBiomater.

25. Rossi Marquez G, Di Pierro P, Esposito M, Mariniello L, Porta R (2013) Application of transglutaminase-crosslinked whey protein/pectin films as water barrier coatings in fried and baked foods. Food Bioprocess Technol.

26. Conte A, Scrocco C, Brescia I, Del Nobile MA (2009) Packaging strategies to prolong the shelf life of minimally processed lampascioni (Muscari comosum). J Food Eng 90:199–206.

27. Del Nobile MA, Conte A, Scrocco C, Laverse J, Brescia I, et al. (2009). New packaging strategies to preserve fresh-cut artichoke quality during refrigerated storage. Innov Food Sci EmergTechnol10:128–133.

28. Mastromatteo M, Mastromatteo M, Conte A, Del Nobile MA (2010) Combined effect of active coating and MAP to prolong the shelf life of minimally processed kiwi fruit (Actinidia deliciosa cv. Hayward). Food Res Int.

29. Cagri A, Ustunol Z, Ryser ET (2004) Antimicrobial edible films and coatings. J Food Prot 67: 833-848.

30. Pranoto Y, Salokhe V, Rakshit KS (2005) Physical and antibacterial properties of alginate-based edible film incorporated with garlic oil. Food Res Int 38:267–272.

31. Eswaranandam S, Hettiarachchy NS, Meullenet JF (2006) Effect of malic and lactic acid incorporated soy protein coatings on the sensory attributes of whole apple and fresh cut cantaloupe. J Food Sci 71:S307-313.

32. Rojas-Graü MA, Soliva-Fortuny R, Martín-Belloso (2009) Edible coatings to incorporate active ingredients to fresh-cut fruits: a review. Trends Food SciTechnol 20: 438-447.

33. Peiyin L, Barth MM (1998) Impact of edible coatings on nutritional and physiological changes in lightly-processed carrots. Postharvest BiolTechnol14:51–60.

34. Barry-Ryan C, Pacussi JM, O'Beirne D (2000) Quality of shredded carrots as affected by packaging film and storage temperature. J Food Sci 65:726–730.

35. Mastromatteo M, Conte A, Del Nobile MA (2012) Packaging strategies to prolong the shelf life of fresh carrots (Daucus carota L.). Innov Food Sci Emerg Technol 13:215–220.

36. Shon J, Choi Y-H (2011) Effect of edible coatings containing soy protein isolate (SPI) on the browning and moisture content of cut fruit and vegetables. J Appl Biol Chem 54:190-196.

Chapter 6

Value-Added Fruit Processing for Human Health

H.P. Vasantha Rupasinghe[1] and Li Juan Yu[1]

[1] *Faculty of Agriculture, Dalhousie University, Truro, Nova Scotia, Canada*

1. INTRODUCTION

Fruits are staple food in human diet. There has been a growing interest in the connection of fruit and vegetable consumption and improved health. Research have shown that biologically active components in plant-based foods, particularly phytochemicals such as polyphenolics and carotenoids, have important role in reducing the risks of chronic diseases, including cancer, cardiovascular disease, diabetes and Alzheimer's disease, among others. The first part of the chapter provides a brief update of the links between fruit-based antioxidants and other biologically active compounds and potential health benefits.

Fruit production is increasing globally. Despite the increasing fruit production at the global level, a significant amount of fruit produced is lost or wasted due to poor post-harvest management. The second part of the chapter provides information on current status of post-harvest losses in selected fruits and methods to prevent these losses. Therefore, processing fruits into value-added products is one of the strategies to reduce post-harvest losses and promote consumption of fruits.

Fresh-cut fruits, also called minimally processed fruits, are products that are partially prepared, maintain a fresh-like state and ready for use and eating. Recently, fresh-cut fruits have become popular because they meet the consumer demand for convenient ready-to-eat foods with fresh-like quality. However, fresh-cut fruits are more perishable than whole fruits. The third part of the chapter covers some recently developed approaches for the value addition of fresh-cut fruits with respect to the use of natural antimicrobials, anti-browning agents, edible coating, modified atmosphere packaging (MAP), 1-methylcyclopropene (1-MCP) application and vacuum impregnation (VI).

2. FRUITS AND HUMAN HEALTH

Consumption of fruits and vegetables is increasing because of strong evidence that many beneficial effects for human health are associated with the dietary intake of fruits and vegetables (Kaur & Kapoor 2001, Rupasinghe et al. 2012). As suggested by epidemiological studies, the consumption of fruit and vegetables may lead to prevention of many chronic diseases, including cardiovascular disease (Weichselbaum 2010, Al-Dosari et al. 2011; Thilakarathna and Rupasinghe 2012), type II diabetes (Johnston et al. 2002, Yu et al. 2012b) and some cancers (De Mejía & Prisecaru 2005, Lala et al. 2006, Sun & Liu 2008, Lippi & Targher 2011). These disease prevention effects of fruits could be due to the presence of health promoting phytochemicals such as carotenoids (Chichili et al. 2006), flavonoids (Yu et al. 2012a), other phenolic compounds (Masibo & He 2008) and vitamins (Lippi & Targher 2011, Gutierrez 2008). Furthermore, the health-protective effects may be rather produced by complex mixtures of interacting natural chemicals than a single component in these plant-derived foods (Lila 2007). Table 1 gives a summary of selected fruit-based antioxidants and other health promoting compounds for disease prevention.

3. FRUIT PRODUCTION AND POST-HARVEST LOSS

3.1. Fruit Production

Fruit production is increasing dramatically worldwide. According to the FAO, the total world fruit production in 2008 was 572.4 million tons, and the number climbed to 609.2 million tons in 2010 (FAO 2010). Among these fruits, thirty percent of which were tropical fruits, with water melon occupied of 59.2%, mongo and guavas of 20.5% and pineapple of 11.4% (Rawson et al. 2011).

Despite the increasing food production at the global level, about one-third of the food produced in the world is lost or wasted (Prusky 2011), among which, post-harvest stage losses and marketing stage losses are major losses.

3.2. Post-Harvest Loss Of Fruits

Despite of food production is increasing globally, a significant amount of the food for human consumption is lost or wasted, especially perishable foods such as fruits and vegetables (Prusky 2011). The amount of food lost each year is equivalent to more than half of the world's annual cereals production (2.3 billion tonnes in 2009/2010) (Gustavsson et al. 2011).

Table 1. Fruit-based health promoting compounds and postulated disease prevention

Source	Active component	Prevention mechanism	Disease	References
Grape	Anthocyanins	Anti-proliferative	Cancer	Lala et al. 2006
	Flavonoids	Inhibition of HNR-adduct formation	Macular degeneration and cataract	Yu et al. 2012a
	Resveratrol	Antioxidant, anti-inflammatory, activation of SIRT1	Alzheimer's	Sun et al. 2010
	Resveratrol	Normalize iron and Ca^{2+}, increase SOD activity	Cardiotoxicity	Mokni et al. 2012
	Resveratrol	Enhance insulin secretion	Diabetes	Yu et al. 2012b
Apple	Polyphenols	Antioxidant, cell cycle modulation	Cancer	Sun & Liu 2008
	Polyphenols	Antioxidant, multiple mechanisms	Cardiovascular	Weichselbaum 2010
	Phloridzin	Anti-inflammatory, bone resorption	Bone protection	Puel et al. 2005
	Polyphenols	Reduce amyloid-β formation	Alzheimer's	Chan & Shea 2009
	Phloretin-2'-O-Glucoside	Delay glucose absorption	Diabetes	Johnston et al. 2002
Banana	Lectins (Bioactive protein)	Cell cycle arrest, apoptosis	Cancer	De Mejía & Prisecaru 2005
	Polyphenols	Antioxidant, reduce LDL modification	cardiovascular	Yin et al. 2008
	Polyphenols	Antioxidant	Alzheimer's	Heo et al. 2008
Pineapple	Bromelain	Proteolytic enzyme regulation	Anti-inflammatory	Hale et al. 2010
Mango	Phenolic compounds	Antioxidant, multiple mechanisms	Degenerative diseases	Masibo & He 2008

It is hard to give precise information on the amount of fruit losses generated globally, because fruit losses vary greatly among varieties, countries, and climatic regions, and there is no universally applied method for measuring losses. As a consequence, the food loss data during post-harvest are mostly estimated and the variations are from 10% to 40% (Prusky 2011). Table 2 lists some examples of post-harvest losses of selected fruits in India, Egypt and United States.

Table 2. Post-harvest losses in selected fruits

Source	Post-harvest loss %				References
	Farm	Wholesale	Retail	Total	
Grape	15.1	6.9	6.0	28.0	Kader 2010
Grape	7.3	4.2	2.9	14.4	Murthy et al. 2009
Grape	N/A	N/A	7.6	N/A	Buzby et al. 2009
Mango	15.6	8.9	5.3	29.7	Murthy et al. 2009
Mango	N/A	N/A	14.5	N/A	Buzby et al. 2009
Banana	5.5	6.7	16.7	28.8	Murthy et al. 2009
Banana	N/A	N/A	8.0	N/A	Buzby et al. 2009
Papaya	N/A	N/A	54.9	N/A	Buzby et al. 2009
Pineapple	N/A	N/A	14.6	N/A	Buzby et al. 2009
Kiwi	N/A	N/A	12.7	N/A	Buzby et al. 2009
Apple	N/A	N/A	8.6	N/A	Buzby et al. 2009
Avocado	N/A	N/A	9.3	N/A	Buzby et al. 2009
Tomato	9.0	17.9	16.3	43.2	Kader 2010

3.3. Prevention And Reduction Of Post-Harvest Loss

Methods of preventing losses of fruits and vegetables could be found from papers and fact sheet written by Singh and Goswami (2006), Sonkar et al. (2008), Prusky (2011) and DeEll and Murr (2009). These methods include selection of new cultivars with firm fruits and longer postharvest life, minimizing physical damage during harvesting and postharvest handling, control and monitoring of temperature and relative humidity, use of controlled or modified atmosphere storage, use of pre- and post-harvest fungicides (hydrogen peroxide) before and after harvest and use of physical treatment such as ozonation technology. Table 3 gives examples of use of controlled atmosphere storage of selected fruits.

Table 3. Controlled atmosphere storage conditions of selected fruits

Source	Temp (°C)	RH (%)	O_2 (%)	CO_2 (%)	Storage life	References
Grape	0-5	90-95	5-10	15-20	> two weeks	Singh and Goswami 2006
Mango	10-15	90	3-7	5-8	> two weeks	Singh & Goswami 2006
Banana	12-16	90	2-5	2-5	> two weeks	Singh & Goswami 2006
Papaya	10-15	90	2-5	5-8	> two weeks	Singh & Goswami 2006
Pineapple	8-13	90	2-5	5-10	> two weeks	Singh & Goswami 2006
Kiwi	0-5	90	1-2	3-5	> two weeks	Singh & Goswami 2006
Avocado	5-13	90	2-5	3-10	> two weeks	Singh & Goswami 2006
Apple (Empire)	1-2	N/A	1.5-2.5	1.5-2.0	5-8 months	DeEll & Murr 2009
Apple (Gala)	0	N/A	1.5-2.5	1.5-2.5	5-8 months	DeEll & Murr 2009
Apple (Golden Delicious)	0	N/A	1.5-2.5	1.5-2.5	5-8 months	DeEll & Murr 2009
Apple (McIntosh)	3	N/A	1.0-2.5	0.5-2.5	5-8 months	DeEll & Murr 2009

* Temp: temperature; RH: Relative humidity

4. FRUIT PROCESSING AND PRESERVATION

Processed fruit products generally include minimally processed fruit products such as fresh-cut fruit, fermented fruit products such as cider, wine and vinegar, traditional thermally processed fruit products such as jam, jelly, juice and beverage, novel non-thermal processed fruit products such as juice and beverage, etc. A comprehensive review has been given by the same authors on novel non-thermal processed fruit product preservation including juices and beverages (Rupasinghe & Yu 2012). At the same time, fresh-cut fruit stands out to be a promising food that meets the demand of consumers for convenient and ready-to-eat fruits with a fresh-like quality. In this case, this part of the chapter would give emphasis on fresh-cut fruit processing and preservation.

4.1. Fresh-Cut Fruit Processing

The sales of fresh-cut produce have grown from approximately $5 billion in 1994 to $10–12 billion in 2005, which is about 10% of total produce sales (Rupasinghe et al. 2005). Fresh-cut fruits and vegetables are products that are partially prepared, maintain a fresh-like state and no additional preparation is necessary for use and eating (Watada & Qi 1999). Figure 1 shows the flowchart of fresh-cut fruit processing. It generally includes washing, and/or peeling, cutting, and/or slicing or wedging and packaging. Dipping solutions or edible coating materials could be applied during dipping or coating process.

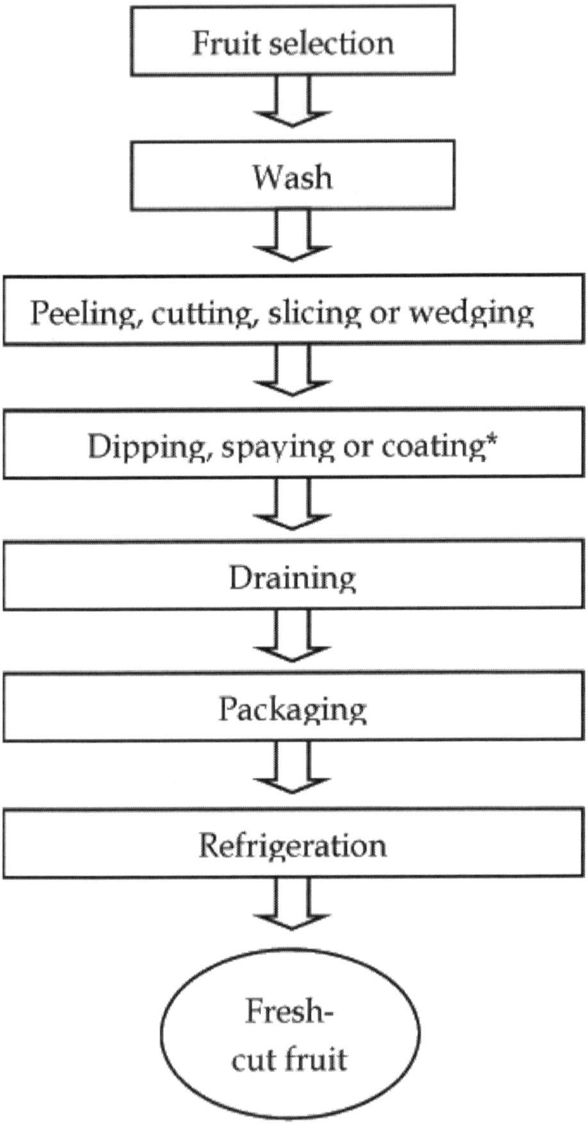

Figure 1. Major steps for fresh-cut fruit processing (revised from Corbo et al. 2010) * During this process, natural preservatives or edible coating materials could be applied

4.2. Fresh-Cut Fruit Preservation

Fresh-cut fruits are more perishable than whole fruits, because the tissue integrity of fruits is more easily altered during processing. Post-cut quality of fresh-cut fruits suffers from wound induced biochemical and physiological changes such as water loss, accelerated respiration and cut-surface browning as well as microbiological spoilage (Kader 2002, Chiabrando & Giacalone 2012). Therefore, preservation of fresh-cut fruits needs combinative efforts of antimicrobial agents, anti-browning substances as well as packaging strategies. A detailed review was given by Oms-Oliu et al. (2010) about recent approaches for preserving quality of fresh-cut fruits.

4.2.1. Antimicrobial Agent

During the preparatory steps of fresh-cut fruit processing, the natural protection of fruit is removed and chances of contamination may increase. Damage of tissues allows the growth and fermentation of some species of yeasts such as *Saccharomyces cerevisiae* and the attack by pathogenic microorganisms such as *Listeria monocytogenes*, *Salmonella* spp., *Staphylococcus aureus* and *Escherichia coli* O157:H7 (Martin-Belloso et al. 2006). Therefore, the searching for methods to retard microbial growth is of great interest to researchers and fresh-cut industry.

Traditionally, the most commonly used antimicrobials are potassium sorbate and sodium benzoate. However, consumer demand for natural origin, safe and environmental friendly food preservatives is increasing. Natural antimicrobials such as organic acids, herb leaves extracts and oils, chitozan and*bacteriocins* have shown feasibility for use in some food products including fresh-cut fruits (Gould 2001, Corbo et al. 2009). Some of them have been considered as Generally Recognized As Safe (GRAS) additives in foods. Selected natural antimicrobials and their status for GRAS additives are listed in Table 4.

Cinnamon as an antimicrobial agent has been investigated in fresh-cut apple slices (Muthuswamy et al. 2008). Ethanol extract of cinnamon bark (1% to 2% w/v) and cinnamic aldehyde (2 mM) could reduce*E. coli* O157:H7 and *L. innocua in vitro*. Ethanol extract of cinnamon bark (1% w/v) reduced significantly the aerobic growth of bacteria inoculated in fresh-cut apples during storage at 6°C up to 12 days. It was also found that cinnamic aldehyde has greater antimicrobial activity than potassium sorbate (Muthuswamy et al. 2008).

Table 4. Selected natural antimicrobial agents and their status for GRAS additives*

Name	Origin	GRAS status
Rosemary	Plant	Yes
Cinnamon	Plant	Yes
Cinnamic acid	Plant	Yes
Clove	Plant	Yes
Lactoperoxidase	Animal	No
Lemon (peel, balm, grass)	Plant	Yes
Lime	Plant	Yes
Nisin	Microorganism	Yes
Chitozan	Animal	No
Carvacrol	Plant	Yes
Citric acid	Plant	Yes
Ascorbic acid	Plant	Yes
Vanillin	Plant	Yes

* Revised from USFDA (2006): Food Additive Status List

Carvacrol and cinnamic acid could delay microbial spoilage of fresh-cut melon and kiwifruit. Dipping of fresh-cut kiwifruit in carvacrol solutions at up to 15 mM reduced total viable counts from 6.6 to less than 2 *log* CFU/g for 21 days of storage at 4°C. Also, treatment with 1 mM of carvacrol or cinnamic acid reduced viable counts on kiwifruit by 4 and 1.5 *log* CFU/g for 5 days of storage at 4°C and 8°C, respectively (Roller & Seedhar 2002).

Vanillin was also proved to be a practical preservative for processing fresh-cut mango and apples under refrigerated conditions. Fresh-cut mango slices were dipped for 1 min in solutions containing 80 mM vanillin before being packaged. Results indicated that treatment with 80 mM vanillin significantly delayed ($P < 0.05$) the development of total aerobic bacteria and yeast and mold populations of fresh-cut mangoes stored at 5 and 10 ◦C for up to 14 and 7 d, respectively (Ngarmsak et al. 2006). Also, a dip of 12 mM vanillin incorporated with a commercial anti-browning dipping solution (calcium ascorbate, NatureSeal™) inhibited the total aerobic microbial growth by 37% and 66% in fresh-cut 'Empire' and 'Crispin' apples, respectively, during storage at 4 °C for 19 days. Furthermore, vanillin (12 mM) did not influence the control of enzymatic browning and softening by NatureSeal (Rupasinghe et al. 2006).

4.2.2. Anti-browning agents

Enzymatic browning is also a major concern on the extension of shelf-life of fresh-cut fruit (Oms-Oliu et al. 2010). It is caused by the enzymatic oxidation of phenols to quinones by enzymes, typically polyphenoloxidases, in the presence of oxygen. Quinones are then subjected to further reactions, leading to the formation of browning pigments (Ozoglu & Bayindirli 2002, Jeon & Zhao 2005). Traditionally, sulfites have been used for browning prevention. However, their use on fresh-cut fruit and vegetables was banned in 1986 by the FDA owing to their potential hazards to health (Buta et al. 1999). Therefore, various alternative substances, such as honey, citric acid, ascorbic acid, calcium chloride, calcium lactate and calcium ascorbate, among others, have being used to retard browning in fresh-cut fruit (Jeon & Zhao 2005, Oms-Oliu et al. 2010). These anti-browning products are not often used alone because it is difficult to achieve efficient browning inhibition, and combination of them would give preferable results. Table 5 gives examples of anti-browning treatment on fresh-cut Apples. Examples of anti-browning treatment on other fresh-cut fruits including banana, kiwifruits, mango, among others could be found in Table 3 in Oms-Oliu et al. (2010)'s paper.

Table 5. Examples of anti-browning treatments assessed on fresh-cut fruits

Cultivar of apple	Anti-browning agent	Storage conditions	References
Gala	10% honey solution with vacuum impregnating	3°C for 14 days	Jeon & Zhao (2005)
Granny Smith	1% w/v of citric acid/CaCl2 and 1% of ascorbic acid/CaCl2	4°C for 5 days	Chiabrando & Giacalone (2012)
Granny Smith	0.05% w/v of sodium chlorite and 1% of calcium propionate	10°C for 14 days	Guan & Fan (2010)
Golden Delicious	1% w/v of citric acid/CaCl2 and 1% of ascorbic acid/CaCl2	4°C for 5 days	Chiabrando & Giacalone (2012)
Golden Delicious	80 mg/L acidic electrolyzed water (AEW) followed by 5%calcium ascorbate	4°C for 11 days	Wang et al. (2007)
Granny Smith	1% w/v of citric acid/CaCl2 and 1% of ascorbic acid/CaCl2	4°C for 5 days	Chiabrando & Giacalone (2012)
Red Delicious	300 mg /L sodium chlorite (SC) and 300 mg /L citric acid	5°C for 14 days	Luo et al. (2011)
Scarlet Spur	1% w/v of citric acid/CaCl2 and 1% of ascorbic acid/CaCl2	4°C for 5 days	Chiabrando & Giacalone (2012)

4.2.3. Edible coating

The incorporation of antimicrobial and anti-browning agents to fresh-cut fruits could be done by dipping, spaying or edible coating treatment. Dipping or spraying aqueous solutions to fruit pieces containing antimicrobial agents, antioxidants, calcium salts or functional ingredients such as minerals and vitamins are widely used to improve quality of fresh-cut fruit. However, the effectiveness of these compounds could be better improved with their incorporation into edible coatings. The application of edible coatings to deliver active ingredients is one of the recent progresses made for shelf-life extension of fresh-cut fruits. Detailed information on edible coating for fresh-cut fruits could be found in review papers from Vargas et al. (2008), Rojas-Graü et al. (2009) and Valencia-Chamorro et al. (2011).

Edible coatings may be defined as a thin layer of material that covers the surface of the food and can be eaten as a part of the whole product. Therefore, the composition of edible coatings has to be food grade or GRAS. Furthermore, the coating materials need to be transparent, odourless, permeable for water vapour and selectively permeable to gases and volatile compounds (Kester & Fennema 1986).

Ingredients that can be used to form edible coatings include polysaccharides such as cellulose, starch, alginate, chitosan, pectin, carrageenan, gum Arabic, guar gum and xanthan gum, proteins such as zein, gluten, soy, whey protein, lipids such as beeswax, lecithin, cocoa butter and fatty acids (Vargas et al. 2008). Examples of edible coating treatment on fresh-cut apples are listed in Table 6.

4.2.4. Modified atmosphere packaging (map) and 1-methylcyclopropene (1-mcp)

The respiration rate of fresh-cut fruits is greater than that of intact fruits (Kader 1986). The increased respiration rate can induce the ethylene synthesis, increase enzymatic activity, promote oxidation of phenolic compounds and microbial growth, and therefore contributes to quality losses such as color and firmness. In this case, the control of respiration is essential for maintaining quality and prolonging the shelf life of fresh-cut fruits (Rocha & Morais 2003).

Modified atmosphere packaging (MAP) is a technology which offers the optimum gas conditions around the product by adjusting the barrier properties of the packaging film (Simpson and Carevi 2004). Various approaches to prolong the shelf life of fresh-cut products, such as edible coatings and refrigeration could be applied in combination with MAP (Rupasinghe 2005).

Table 6. Examples of edible coating treatment on fresh-cut apples

Cultivar of apple	Functional ingredients	Concentration (%)	Coating materials	References
Gala	N/A	N/A	Cassava starch, glycerol, carnauba wax, stearic acid	Chiumarelli & Hubinger 2012
Gala	N/A	N/A	Chitosan	Wu et al. 2005
Fuji	Oregano oil	0.1 – 0.5 (v/v)	Apple puree, alginate	Rojas-Graü et al. 2007
Fuji	Lemongrass	1.0 – 1.5 (v/v)	Apple puree, alginate	Rojas-Graü et al. 2007
Fuji	Vanillin	0.3 – 0.6 (v/v)	Apple puree, alginate	Rojas-Graü et al. 2007
Fuji	Cinnamon	0.7 (v/v)	Alginate	Raybaudi-Massilia et al. 2008
Fuji	Clove	0.7 (v/v)	Alginate	Raybaudi-Massilia et al. 2008
Fuji	Lemongrass	0.7 (v/v)	Alginate	Raybaudi-Massilia et al. 2008
Fuji	Cinnamaldehyde	0.5 (v/v)	Alginate	Raybaudi-Massilia et al. 2008
Fuji	Citral	0.5 (v/v)	Alginate	Raybaudi-Massilia et al. 2008
Fuji	Ascorbic acid, $CaCl_2$	1.0 (w/v)	Carrageenan	Lee et al. 2003
Fuji	Ascorbic acid, $CaCl_2$	1.0 (w/v)	Whey protein concentrate	Lee et al. 2003
Fuji	Ascorbic acid, $CaCl_2$	1.0 (w/v)	Whey protein concentrate	Lee et al. 2003
Golden Delicious	Ascorbic acid	0.5-1.0 (w/v)	Whey protein concentrate, beeswax	Perez-Gago et al. 2006
Golden Delicious	Cysteine	0.1-0.5 (w/v)	Whey protein concentrate, beeswax	Perez-Gago et al. 2006
Granny Smith	Ascorbic acid, citric acid	0.5 (w/v)	Pectin, apple purée	McHugh & Senesi 2000

1-Methylcyclopropene (1-MCP) may retard or inhibit the generation of ethylene, the natural ripening hormone which is undesirable in terms of storage of certain fruits. Therefore, 1-MCP is becoming a commercial tool (SmartFresh, AgroFresh Inc., Philadelphia) for extending the shelf-life and quality of certain fruits and plant products (Rupasinghe et al. 2005). 1-MCP can be applied immediately after harvest (Aguayo et al. 2006; Mao & Fei 2007), just before fresh-cut processing or at both steps (Calderón-López et al. 2005; Vilas-Boas & Kader 2007). However, treatment of intact fruit with 1-MCP before fresh-cut processing is easier and more convenient than after processing.

Moreover, the increase in ethylene production promoted by peeling, slicing or wedging could be prevented by the pre-use of 1-MCP (Rupasinghe et al., 2005).

4.2.5. Vacuum impregnation

Osmotic treatments have been traditionally used as a pre-treatment step in freezing, canning and frying to improve the quality of the final produce (Alzamora et al., 2000). Among developments in osmotic treatments of fruit products, vacuum impregnation (VI) may be the latest (Zhao & Xie 2004). The VI technique is performed by applying a vacuum pressure in a tank or oven containing the immersed product for a short time and then restoring the atmospheric pressure with the product remains immersed (Martínez-Monzó et al., 1998). The process of VI is a hydrodynamic mass transfer process based on an exchange between internal gas or liquid and an external liquid phase (Zhao & Xie, 2004). VI technique could be used to develop novel minimally processed fruit products with value-addition since nutritional and bioactive ingredients could be incorporated into the fruit based products during VI process (Xie & Zhao, 2003; Guillemin et al., 2008, Rößle 2011) and which gives a bright future for VI application in fresh-cut fruits. Table 7 gives examples of VI treatment on fresh-cut fruits.

Table 7. Examples of VI treatment conditions on fresh-cut fruits or value-added products

Type of fruit	VI treatment conditions				References
	VI solution	VI pressure (mmHg)	VI time (min)	Restoration Time (min)	
Apple (Gala)	20% (w/w) of HFCS, Ca, Zn	50	15	30	Xie & Zhao 2003
Apple (Gala)	10% (w/w) of honey	75	15	30	Jeon & Zhao 2005
Strawberry	8°Brix of glucose solution	37.5	5	5	Castelló et al. 2006
Apple (Empire)	15°Brix of grape juice, 1.6% of $CaCl_2$ (w/v), 0.05% of NaCl (w/v), 0.1% of vitamin E (v/v)	152.4	10	22	Joshi et al. 2010
Apple (Empire)	20-40 % (v/v) of maple syrup, 1.6% of $CaCl_2$ (w/v), 0.05% of NaCl (w/v)	152	10	22	Joshi et al. 2011
Apple (Granny Smith)	50% (v/v) of honey	525	10	10	Rößle et al. 2011

HFCS: High fructose corn syrup

5. CONCLUSION

Fruits are not only consumed as stable food but also provide desirable health benefits beyond their basic nutrition. However, the quantitative and qualitative losses of fruits are significant during post-harvest, marketing, processing and storage. Prevention of these losses during post-harvest management could be done by multiple steps and methods such as controlled or modified atmosphere packaging and application of ozonation technology.

On the other hand, promotion of minimally processed fruit products such as fresh-cut fruit into the commercial market is a practical, economical, and consumer and environmental friendly approach compared with traditional processing methods. However, fresh-cut fruits are more perishable than whole fruits in terms of biochemical and physiological changes such as water loss, accelerated respiration and cut-surface browning as well as microbiological spoilage. Therefore, preservation of fresh-cut fruits needs combinative efforts of antimicrobial agents, anti-browning substances as well as packaging strategies.

Natural or GRAS additives have been the popular ingredients used as antimicrobial agents and anti-browning agents, or bioactive ingredients. The incorporation of antimicrobial and anti-browning agents to fresh-cut fruits could be done by dipping, spaying or edible coating treatment. The application of edible coatings to deliver active ingredients is one of the recent progresses made for shelf-life extension of fresh-cut fruits. It could be used in combination with modified atmosphere packaging (MAP), 1-methylcyclopropene (1-MCP) and refrigeration for better results.

In addition for edible coating, vacuum impregnation (VI) may be another practical approach for incorporation of health promoting natural ingredients into fresh-cut fruits. VI technique could be used to develop novel minimally processed fruit products with value-addition through incorporation of nutritional and bioactive ingredients.

REFERENCES

1. Aguayo, E., Jansasithorn, R. & Kader, A. A. (2006). Combined effects of 1-Methylcyclopropene, calcium chloride dip, and/or atmospheric modification on quality changes in fresh-cut strawberries, *Postharvest Biology and Technology* 40(3): 269-278.

2. Alzamora SM, Tapia MS, Leunda A, Guerrero SN, Parada-Arias E. 2000. Minimal Presser vation of fruits: a cited project, *in*: Lozano, J.E., Anon, C., Parada-Arias, E. & Borabosa-Canovas GV (ed.), *Trend in Food*

Engineering, Technomic Publishing Company,Pennsylvania, USA, pp. 205-225.

3. Buta, J. G., Moline, H. E., Spaulding, D. W. & Wang, C. Y. (1999). Extending storage life of fresh-cut apples using natural products and their derivatives, *Journal of Agricultural and Food Chemistry* 47(1): 1-6.

4. Buzby, J. C.,Wells, H. F., Axtman, B. & Mickey, J. (2009). Supermarket loss estimates for fresh fruit, vegetables, meat, poultry, and seafood and their use in the ERS loss- adjusted food availability data, *United States Department of Agriculture,* Economic Research Service, Economic Information Bulletin, 44: 20.

5. Calderon-Lopez, B., Bartsch, J. A., Lee, C. Y. & Watkins, C. B. (2005). Cultivar effects on quality of fresh cut apple slices from 1-methylcyclopropene (1-MCP)-treated apple fruit, *Journal of Food Science* 70(3): S221-227.

6. Castelló, M. L., Fito, P. J. & Chiralt, A. (2006). Effect of osmotic dehydration and vacuum impregnation on respiration rate of cut strawberries,*LWT - Food Science and Technology* 39(10): 1171-1179.

7. Chan, A. & Shea, T. B. (2009). Dietary supplementation with apple juice decreases endo- genous amyloid-β levels in murine brain, *Journal of Alzheimer's Disease* 16(1): 167- 171.

8. Chiabrando, V. & Giacalone, G. (2012). Effect of anti-browning agents on color and related enzymes in fresh-cut apples during cold storage, *Journal of Food Processing and Preservation* 36(2): 133-140.

9. Chiumarelli, M. & Hubinger, M. D. (2012). Stability, solubility, mechanical and barrier properties of cassava starch - carnauba wax edible coatings to preserve fresh-cut apples, *Food Hydrocolloids* 28(1): 59-67.

10. Corbo, M. R., Bevilacqua, A., Campaniello, D., D'Amato, D., Speranza, B. & Sinigaglia, M. (2009). Prolonging microbial shelf life of foods through the use of natural compounds and non-thermal approaches - A review, *International Journal of Food Science and Technology* 44(2): 223-241.

11. Corbo, M.R., Speranza, B., Campaniello, D., D'Amato, D. & Sinigaglia, M. (2010). Fresh-cut fruits preservation: current status and emerging technologies, *in* Mendez-Vilas, A (Ed.), *Current research, Technology and Education Topics in Applied Microbiology and Microbial Biotechnology,* Formatex Research Center, Badajoz, Spain, pp. 1143-1154.

12. DeEll J. R. & Murr, D. P. (2009). CA storage guidelines and recommendations for apples, *http://www.omafra.gov.on.ca/english/crops/facts/03-073.htm*

13. De Mejía EG & Prisecaru VI. (2005). Lectins as bioactive plant proteins: a potential in cancer Treatment, *Critical Reviews in Food Science and Nutrition* 45 (6): 425-445.

14. FAO (2010). World fruit production in metric tonnes. *http://faostat.fao.org/default.aspx*.

15. Gould, G. W. (2001). Symposium on 'nutritional effects of new processing technologies' – new processing technologies: An overview, *Proceedings of the Nutrition Society* 60(4): 463-474.

16. Guan, W. & Fan, X. (2010). Combination of sodium chlorite and calcium propionate reduces enzymatic browning and microbial population of fresh-cut "granny smith" apples, *Journal of Food Science* 75920, M72-M77

17. Guillemin, A., Guillon, F., Degraeve, P. et al. (2008). Firming of fruit tissues by vacuum- infusion of pectin methylesterase: visualisation of enzyme reaction, *Food Chemistry* 109, 368–378.

18. Gustavsson, J., Cederberg, C., Sonesson, U., van Otterdijk, R. & Meybeck, A. (2011). Global food losses and food waste: extent causes and prevention, Food and Agriculture Organization of the United Nations. International congress Save Food!, *International packaging industry fair Interpack 2011*, Dusseldorf, Germany.

19. Hale, L. P., Chichlowski, M., Trinh, C. T. & Greer, P. K. (2010). Dietary supplementation with fresh pineapple juice decreases inflammation and colonic neoplasia in IL-10- deficient mice with colitis, *Inflammatory Bowel Diseases* 16(12): 2012-2021.

20. Heo, H.J., Choi, S.J., Choi, S-G, Lee. J.M. & Lee, C.Y. (2008). Effects of banana, orange, and apple on oxidative stress-induced neurotoxicity in PC12 cells, *Journal of Food Science* 73(2), H28-H32.

21. Jeon, M. & Zhao, Y. (2005). Honey in combination with vacuum impregnation to prevent enzymatic browning of fresh-cut apples, *International Journal of Food Sciences and Nutrition* 56(3): 165-176.

22. Johnston, K. L., Clifford, M. N. & Morgan, L. M. (2002). Possible role for apple juice phenolic compounds in the acute modification of glucose

tolerance and gastrointestinal hormone secretion in humans, *Journal of the Science of Food and Agriculture* 82(15): 1800-1805.

23. Joshi, A. P. K., Rupasinghe, H. P. V. & Pitts, N. L. (2010). Sensory and nutritional quality of the apple snacks prepared by vacuum impregnation process, *Journal of Food Quality* 33(6): 758-767.

24. Joshi, A. P. K., Rupasinghe, H. P. V. & Pitts, N. L. (2011). Comparison of non-fried apple snacks with commercially available fried snacks, Food*Science and Technology International* 17(3): 249-255.

25. Kader, A.A., (1986). Biochemical and physiological basis for effects of controlled and Modified atmospheres on fruits and vegetables, *Food Technol.* 40: 99-104.

26. Kader, A. (2002). Quality parameters of fresh-cut fruit and vegetable products, in Lamikanra, O. (ed.), *Fresh-Cut Fruits and Vegetables, Science, Technology and Market*, CRC Press, Boca Raton, Florida, USA, pp. 11–19.

27. Kader, A. A. (2010). Handling of horticultural perishables in developing vs. developed countries, *Acta Horticulturae* 877: 121–126.

28. Kaur, C. & Kapoor, H. C. (2001). Antioxidants in fruits and vegetables - the millennium's health, *International Journal of Food Science and Technology* 36(7): 703-725.

29. Kester, J.J. & Fennema O. R. (1986). Edible films and coatings: A review, *Food Technol.* 40: 47– 49.

30. Lala, G., Malik, M., Zhao, C., He, J., Kwon, Y., Giusti, M. M. & Magnuson, B. A. (2006). Anthocyanin-rich extracts inhibit multiple biomarkers of colon cancer in rats, *Nutrition and Cancer* 54(1): 84-93.

31. Lee, J. Y., Park, H. J., Lee, C. Y. & Choi, W. Y. (2003). Extending shelf-life of minimally processed apples with edible coatings and anti-browning agents, *Lebensmittel Wissenschaft und Technologie* 36: 323-329.

32. Lila, M. A. (2007). From beans to berries and beyond: Teamwork between plant chemicals for protection of optimal human health, *Annals of the New York Academy of Sciences* 1114: 372-380.

33. Luo, Y., Lu, S., Zhou, B. & Feng, H. (2011). Dual effectiveness of sodium chlorite for enzymatic browning inhibition and microbial inactivation on fresh-cut apples, *LWT-Food Science and Technology* 44(7): 1621-1625.

34. Mao, L. & Fei, G. (2007). Application of 1-methylciclopropene prior to cutting reduces wound responses and maintains quality in cut kiwifruit, *J. Food Eng*. 78: 361–365.

35. Martin-Belloso, O., Soliva-Fortuny, R. & Oms-Oliu, G. (2006). Fresh-cut fruits, in: Hui, Y.H. (Ed.), *Handbook of Fruits and Fruit Processing,*Blackwell Publishing, Oxford, pp. 129– 144.

36. Martínez-Monzó, J., Martínez-Navarrete, N., Chiralt, A. & Fito, P. (1998). Mechanical and structural changes in apple (var. granny smith) due to vacuum impregnation with cryoprotectants, *Journal of Food Science* 63(3): 499-503.

37. Masibo, M. & He, Q (2008). Major mango polyphenols and their potential significance to human health, *Comprehensive Reviews in Food Science and Food Safety* 7(4): 309-319.

38. McHugh, T. H. & Senesi, E. (2000). Apple wraps: a novel method to improve the quality and extend the shelf life of fresh-cut apples, *Journal of Food Science* 65: 480-485.

39. Mokni, M., Hamlaoui-Guesmi, S., Amri, M., Marzouki, L., Limam, F. & Aouani, E. (2012). Grape seed and skin extract protects against acute chemotherapy toxicity induced by doxorubicin in rat heart, *Cardiovascular Toxicology* 1-8.

40. Murthy, D. S., Gajanana, T. M., Sudha, M. & Dakshinamoorthy, V. (2009). Marketing and post-harvest losses in fruits: Its implications on availability and economy, *Indian Journal of Agricultural Economics* 64(2): 259-275.

41. Muthuswamy, S., Rupasinghe, H. P. V. & Stratton, G. W. (2008). Antimicrobial effect of cinnamon bark extract on *Escherichia coli* O157:H7, *Listeria innocua* and fresh-cut apple slices, *Journal of Food Safety* 28(4): 534-549.

42. Ngarmsak, M., Delaquis, P., Toivonen, P., Ngarmsak, T., Ooraikul, B. & Mazza, G., (2006). Antimicrobial activity of vanillin against spoilage microorganisms in stored fresh- cut mangoes, *J. Food Prot*. 69: 1724–1727.

43. Oms-Oliu, G., Rojas-Graü, M. A., González, L. A., Varela, P., Soliva-Fortuny, R., Hernando, M. I. H., . . . Martín-Belloso, O. (2010). Recent approaches using chemical treatments to preserve quality of fresh-cut fruit: A review, *Postharvest Biology and Technology* 57(3): 139-148.

44. Ozoglu, H. & Bayindirli, A. (2002). Inhibition of enzymatic browning in cloudy apple juice with selected anti-browning agents, *Food Control* 13: 213-221.

45. Perez-Gago, M. B., Serra, M. & del Rio, M. A. (2006). Color change of fresh-cut apples coated with whey protein concentrate-based edible coatings,*Postharvest Biology and Technology* 39: 84-92.

46. Prusky, D. (2011). Reduction of the incidence of postharvest quality losses, and future prospects, *Food Security* 3(4): 463-474.

47. Puel, C., Quintin, A., Mathey, J., Obled, C., Davicco, M. J., Lebecque, PCoxam, V. (2005). Prevention of bone loss by phloridzin, an apple polyphenol, in ovariectomized rats under inflammation conditions, *Calcified Tissue International* 7(5): 311-318.

48. Rawson, A., Patras, A., Tiwari, B. K., Noci, F., Koutchma, T. & Brunton, N. (2011). Effect of thermal and non-thermal processing technologies on the bioactive content of exotic fruits and their products: Review of recent advances, *Food Research International*, 4(7): 1875-1887.

49. Raybaudi-Massilia, R. M., Rojas-Grau, M. A., Mosqueda-Melgar, J. & Martín -Belloso, O. (2008). Comparative study on essential oils incorporated into an alginate-based edible coating to assure the safety and quality of fresh-cut Fuji apples, *J. Food Prot*. 71: 1150–1161.

50. Rößle, C., Brunton, N., Gormley, T. R. & Butler, F. (2011). Quality and antioxidant capacity of fresh-cut apple wedges enriched with honey by vacuum impregnation, *International Journal of Food Science and Technology* 46(3): 626-634.

51. Rocha, A. M. C. N. & Morais, A. M. M. B. (2003). Shelf life of minimally processed apple (cv. Jon gold) determined by colour changes, *Food Control*14(1): 13-20.

52. Rojas-Graü, M. A., Raybaudi-Massilia, R. M., Soliva-Fortuny, R. C., Avena-Bustillos, R. J., McHugh, T. H. & Martín-Belloso, O. (2007). Apple puree-alginate edible coating as carrier of antimicrobial agents to prolong shelf-life of fresh-cut apples, *Postharvest Biol. Technol.* 45: 254–264.

53. Rojas-Graü, M. A., Soliva-Fortuny, R. & Martín-Belloso, O. (2009). Edible coatings to incorporate active ingredients to fresh-cut fruits: A review,*Trends in Food Science and Technology* 20(10): 438-447.

54. Roller, S& Seedhar, P. (2002). Carvacrol and cinnamic acid inhibit microbial growth in fresh- cut melon and kiwifruit at 4° and 8°C, *Letters in Applied Microbiology* 35: 390–394.

55. Rupasinghe, H.P.V., Murr, D.P., DeEll, J.R. & Odumeru, J. A. (2005). Influence of 1-methylcyclopropene (1-MCP) and NatureSealTM on the quality of fresh-cut 'Empire' and 'Crispin' apples, *J. Food Qual.* 28: 289-307.

56. Rupasinghe, H.P.V., Boulter-Bitzer, J, Ahn, T. & Odumeru, J.A. (2006). Vanillin inhibits pathogenic and spoilage microorganisms in vitro and aerobic microbial growth on fresh-cut apples, *Food Res. Intern.* 39: 575-580.

57. Rupasinghe, H. and Yu, L. J. 2012. Emerging preservation methods for fruit juices and beverages, in El-Samragy Y. (Eds.), *Food Additives*, InTech - Open Access Publisher, Rijeka, Croatia. ISBN: 978-953-51-0067-6. pp: 65-82.

58. Rupasinghe, H.P.V., Thilakarathna, S. & Nair, S. (2012). Polyphenols of Apples and Their Potential Health Benefits, in: J. Sun, K. N. Prasad, A. Ismail, B. Yang, X. You and L. Li (Eds.), *Polyphenols: Chemistry, Dietary Sources and Health Benefits,* Nova Science Publishers, Inc. Hauppauge, NY, USA,

59. Simpson, R. & Carevic, E. (2004). Designing a modified atmosphere packaging system for foodservice portions on non-respiring foods: optimal gas mixture and food/headspace ratio, *Foodservice Research International* 14: 257-272.

60. Singh, A. K. & Goswami, T. K. (2006). Controlled atmosphere storage of fruits and vegetables: A review, *Journal of Food Science and Technology*43(1): 1-7.

61. Sonkar, R. K., Sarnaik, D. A., Dikshit, S. N., Saroj, P. L. & Huchche, A. D. (2008). Post-harvest management of citrus fruits: A review, *Journal of Food Science and Technology* 45(3): 199-208.

62. Sun, J. & Liu, R. H. (2008). Apple phytochemical extracts inhibit proliferation of estrogen-dependent and estrogen-independent human breast cancer cells through cell cycle modulation, *Journal of Agricultural and Food Chemistry* 56(24): 11661-11667.

63. Sun, A. Y., Wang, Q., Simonyi, A. & Sun, G. Y. (2010). Resveratrol as a therapeutic agent for neurodegenerative diseases, *Molecular Neurobiology*41(2-3): 375-383.

64. Thilakarathna, S.H. & H.P.V. Rupasinghe. 2012. Anti-atherosclerotic effects of fruit bioactive compounds: A review of current scientific evidence,*Can J. Plant Sc.* 92: 407-419.

65. U.S. Food and Drug Administration. (2006). Food Additive Status List. http://www.cfsan.fda.gov/~dms/opa-appa.html

66. Valencia-Chamorro, S. A., Palou, L., Delŕio, M. A. & Pérez-Gago, M. B. (2011). Antimicrobial edible films and coatings for fresh and minimally processed fruits and vegetables: A review, *Critical Reviews in Food Science and Nutrition* 51(9): 872-900.

67. Vargas, M., Pastor, C., Chiralt, A., McClements, D. J. & González-Martínez, C. (2008). Recent advances in edible coatings for fresh and minimally processed fruits, *Critical Reviews in Food Science and Nutrition* 48(6): 496-511.

68. Vilas-Boas, E. & Kader, A. (2007). Effect of 1-MCP on softening of fresh-cut kiwifruit, mango and persimmon slices, *Post. Biol. Technol.* 43: 238–244.

69. Wang, H., Feng, H. & Luo, Y. (2007). Control of browning and microbial growth on fresh-cut apples by sequential treatment of sanitizers and calcium ascorbate, *Journal of Food Science* 72(1), M1-M7.

70. Watada, A.E. & Qi, L. (1999). Quality of fresh-cut produce, *Postharvest Biol. Technol.* 15: 201–205.

71. Weichselbaum, E., Wyness, L. & Stanner, S. (2010). Apple polyphenols and cardiovascular disease - a review of the evidence, *Nutrition Bulletin*35(2): 92-101.

72. Wu, T., Zivanovic, S., Draughon, F.A., Conway, W. S. & Sams, C. E. (2005). Physicochemical properties and bioactivity of fungal chitin and chitosan,*J. Agric. Food Chem.* 53: 3888–3894.

73. Xie, J. & Zhao, Y. (2003). Nutritional enrichments of fresh apple (Royal Gala) by vacuum impregnation, *International Journal of Food Sciences and Nutrition* 54: 387–398.

74. Yin, X., Quan. J. & Kanazawa T. (2008). Banana prevents plasma oxidative stress in healthy individuals, *Plant Foods in Human Nutrition* 63 (2): 71-76.

75. Yu, C., Nandrot, E. F., Dun, Y. & Finnemann, S. C. (2012a). Dietary antioxidants prevent age-related retinal pigment epithelium actin

damage and blindness in mice lacking αvβ5 integrin, *Free Radical Biology and Medicine* 52(3): 660-670.

76. Yu, W., Fu, Y. -. & Wang, W. (2012b). Cellular and molecular effects of resveratrol in health and disease, *Journal of Cellular Biochemistry* 113(3): 752-759.

77. Zhao, Y. & Xie, J. (2004). Practical applications of vacuum impregnation in fruit and vegetable processing, *Trends in Food Science & Technology* 15; 434–451.

Chapter 7

Irrigation Water Quality for Leafy Crops: A Perspective of Risks and Potential Solutions

Ana Allende [1] and James Monaghan [2,*]

[1]*Research Group on Quality, Safety and Bioactivity of Plant Foods, Department of Food Science and Technology, CEBAS-CSIC, Campus Universitario de Espinardo, 30100 Murcia, Spain*

[2]*Fresh Produce Research Centre, Department of Crop and Environment Sciences, Harper Adams University, Newport, Shropshire, TF10 8NB, UK*

ABSTRACT

There is increasing evidence of the contribution of irrigation water in the contamination of produce leading to subsequent outbreaks of foodborne illness. This is a particular risk in the production of leafy vegetables that will be eaten raw without cooking. Retailers selling leafy vegetables are increasingly targeting zero-risk production systems and the associated requirements for irrigation water quality have become more stringent in regulations and quality assurance schemes (QAS) followed by growers. Growers can identify water sources that are contaminated with potential pathogens through a monitoring regime and only use water free of pathogens, but the low prevalence of pathogens makes the use of faecal indicators, particularly *E. coli*, a more practical approach. Where growers have to utilise water sources of moderate quality, they can reduce the risk of contamination of the edible portion of the crop (*i.e.*, the leaves) by treating irrigation water before use through physical or chemical disinfection systems, or avoid contact between the leaves and irrigation water through the use of drip or furrow irrigation, or the use of hydroponic growing systems. This study gives an overview of the main problems in the production of leafy vegetables associated with irrigation water, including microbial risk and difficulties in water monitoring, compliance with evolving regulations and quality standards, and summarises the current alternatives available for growers to reduce microbial risks.

Keywords: leafy vegetables; irrigation water; food safety; QAS; GAP; water disinfection treatment; hydroponics

1. INTRODUCTION

Water of inadequate quality has the potential to be a direct source of contamination and a vehicle for spreading localized contamination in the field, facility or transportation environments in the production of fresh produce crops [1]. A number of comprehensive review articles have been published which highlight irrigation water as a source of pathogenic microorganisms in produce [2,3,4,5,6,7,8]. Pathogenic microorganisms associated with irrigation water include bacteria, viruses and parasites (protozoa and helminths). Although viruses and parasites are of paramount importance and can be transmitted to fresh produce via irrigation water, this paper focuses on pathogenic bacteria.

The increasing evidence of contamination of produce from irrigation water has been reviewed recently by Uyttendaele *et al.* [9] and contamination events were identified where water is a risk factor in the production and harvesting of fresh produce. However, the authors made clear that identification of the implicated food vehicle and/or the location of the point of food contamination in fresh produce-associated outbreaks are a recurrent challenge.

Despite this evidence, research focused on microbial quality of agricultural water is relatively limited. Most research regarding microbial quality of water (e.g., pathogen prevalence or indicator organisms) has been conducted for objectives related to reclaimed water, drinking and recreational water supplies and the effects of agriculture on the environment [10]. In 2011, Pachepsky *et al.* [5] highlighted that no databases on microbial quality of irrigation water had been compiled. However, there is an increasing number of recent research papers that focus on the evaluation of the microbial quality of irrigation water used for the production of fresh produce and its significance as a source of contamination e.g., [11,12,13,14,15,16,17,18,19,20]. Several of these published papers show the results obtained within the frame of a research project consortium of the FP7 project Veg-i-Trade (www.vegitrade.org) focused on the impact of climate change and globalisation on safety of fresh produce.

Growers use a variety of water sources for field operations and irrigation and much knowledge is needed to relate risk factors associated with the transfer coefficients for pathogens by source, concentration and use [10]. Quantitative microbial risk assessment (QMRA) is being applied to establish the links between concentrations of pathogenic microorganisms in agricultural water and the probability of illness [21,22,23]. Lately, as a result of several longitudinal microbiological surveys, different scenarios have been evaluated to estimate the potential impact of agricultural practices, including water treatments and the application of potable water, on the *E. coli* levels and prevalence of pathogenic microorganisms in fresh produce [24,25].

Based on this evidence, standards and guidelines for Quality Assurance Schemes (QAS) have been designed to cultivate prevention and control of food safety hazards [26] but QAS have become not only a prevention strategy but also a marketing strategy used by the industry, which is pushing towards zero-risk for many factors including irrigation water sources, in response to consumers concerns over food safety [27]. In many cases, current QAS require growers to develop and implement management systems and to risk assess the irrigation water sources to reduce microbial risks. Increasingly QAS are requiring stringent testing regimes of water coming into contact with the edible portion of leafy vegetables [28]. However, some studies highlight that the required monitoring programs can be self-defeating and create additional uncompensated monitoring costs [26].

Growers should also consider other alternatives to guarantee the safety of the produce they produced under any circumstance. Surface water represents one of the most risky water sources, thus where growers are utilising surface water for agricultural applications, particularly for leafy greens or produce that is intended to be consumed raw, and mitigation strategies may be required [29]. Several strategies have been proposed to reduce the risk of produce contamination with pathogens during irrigation [5]. The main aims of these strategies is to reduce the use of uncontrolled sources of water and to establish distance limits for water resources used for irrigation from livestock housing, stored effluents, and/or land spread with manure [30].

Water treatment is also a feasible strategy to guarantee the microbial quality of irrigation water. Treating water during storage and while in the delivery systems may represent a good alternative to high frequency microbial testing [8] and probably the only possibility for growers using irrigation water with microbial loads above regulatory thresholds. Another option for growers using water sources of poor microbiological quality is the use of different production systems, such as hydroponic systems and drip irrigation, which avoid contact between the edible part of the crop and the irrigation water.

This study gives an overview of the main problems associated with irrigation water, including microbial risk and difficulties in water monitoring, compliance with evolving regulations and quality standards, and summarises the current alternatives available for growers to reduce microbial risks.

2. IRRIGATION WATER AS A RISK FACTOR

Agricultural water has been defined as a major risk factor in the contamination of leafy crops eaten raw as salads [30]. When available, fresh water is consumed for agricultural production but water scarcity is becoming a major threat to the sustainability of agriculture, which needs to rely much more on marginal water sources, including treated wastewater [31]. However, little is known regarding the microbial quality of irrigation water used for leafy crop production. Systematic sampling focused on determining the risks of microbial contamination associated with different water sources are rare and most of the existing information comes from U.S. The available literature indicates that generic *E. coli* levels and prevalence of pathogenic foodborne bacteria in irrigation water significantly varies depending on several factors including seasonality, geographical location and weather conditions among others (reviewed by [8,9]). The Center for Produce Safety [10] have published a report to provide a summary of the scientific and technical information related to factors that affect the microbiological safety of agricultural water. This report summarizes the most recent research on irrigation water and gives a good overview of the needs for future research including sampling strategies that provide an estimate of the true underlying distribution of bacteria in a water system, correlation of field and water system management practices with pathogen prevalence in agricultural water samples, as well as a better understanding of risk factors leading to survival and/or growth of pathogens on fresh produce following application of contaminated water.

Current Situation in Europe

Recently, a comprehensive review has summarized the main irrigation water sources used in Europe highlighting municipal water, rainwater, groundwater and surface water as the most relevant [9]. In a temperate production area, such as the UK, the primary sources of water for irrigation

Table 1. Enumeration of microbial indicators and prevalence of foodborne pathogens in water used to irrigate fresh produce in Europe.

Country	Produce	Water Source	Microorganisms	Average cfu/100 mL	Prevalence	Reference
Belgium	Strawberry	Groundwater	STEC	–	0/22	Delbeke *et al.*, 2015
			E. coli spp.	1	4/22	
		Rainfall water collected in ponds	STEC	–	11/56	
			E. coli spp.	40–45	40/56	
Belgium	Lettuce	Rainfall water collected in open wells and bore hole water	STEC	–	6/68	Holvoet *et al.*, 2014
			Campylobacter spp.	–	37/120	
			Salmonella spp.	–	1/68	
			E. coli spp.	30–35	90/120	
Spain	Baby spinach	Surface water collected in water reservoirs	STEC	–	0/50	Castro-Ibañez *et al.*, 2015
			Salmonella spp.	–	1/50	
			E. coli spp.	5–10	72/250	
Spain	Tomatoes	Surface water	STEC	–	0/16	López-Gálvez *et al.*, 2014
			Salmonella spp.	–	1/16	
			E. coli spp.	20–25	6/32	
			Listeria spp.	30–35	26/30	
		Reclaimed water	STEC	–	0/16	
			Salmonella spp.	–	2/16	
			E. coli spp.	240–280	31/32	
			Listeria spp.	350–400	26/30	
Italy	Tomatoes	Tap water	*E. coli* spp.	–	0/30	Forslund *et al.*, 2012
		Reclaimed water	*E. coli* spp.	10,300	11/30	
Crete	Tomatoes	Tap water	*E. coli* spp.	400	2/31	
		Reclaimed water	*E. coli* spp.	596	4/31	

Table 2. Characteristics, advantages and disadvantages of water treatment technologies for irrigation water.

Water Treatment	Active Agent	Recommended Dose	Reported Microbial Reductions Range (Log cfu/mL)	Advantages	Disadvantages
Sodium hypochlorite	Hypochlorous acid	2–5 mg/L	0.2–4.0	High bactericidal action Liquid ready to use Low operating costs	Organic matter reduces its efficacy Influenced by pH Storage of large volumes By-product formation (trihalomethanes and chlorates among others)
Calcium hypochlorite	Hypochlorous acid	2–5 mg/L	0.2–4.0	High bactericidal action Tablets ready to use Low operating costs	Organic matter reduces its efficacy Influenced by pH By-product formation (trihalomethanes and chlorates among others)
Chlorine dioxide	Chlorine dioxide molecule	0.1–5.0 mg/L	0.5–5.0	High bactericidal action Effective at a wide pH range (4–10) Does not react with organic matter as chlorine	"*in situ*" generation or use of stabilized solutions By-product formation (mostly chlorates)
Ultrasound	Cavitation	20–40 kHz	3.0	Not affected by pH Easy to use No formation of disinfection by-products	Lack of residual bactericidal action
UV-C	DNA damage	1–200 mJ/cm^2	0.5–5.0	High bactericidal action Not affected by pH Easy to use No formation of disinfection by-products Low operational costs	Water turbidity affects efficacy Lack of residual bactericidal action
Membrane filtration	Particle interception		1.0–5.0	Not affected by pH Easy to use No formation of disinfection by-products	Filter blockage

are reported to be surface water (54%) and groundwater (41%), with the remainder coming from public mains water, rainwater and other sources [32,33]. However, in arid and semi-arid parts of Europe, such as the South of

Spain, Italy and Greece, which are confronting increasing water shortages, treated municipal wastewater is a valuable water source for recycling and reuse in agriculture. As an example, 347 hm^3 of treated wastewater were reused in 2010 in Spain, particularly in the South East of Spain where almost 60% of the reclaimed water produced was reused [34] and 233 hm^3 in Italy [35]. Although the use of untreated wastewater for crop production is not a recommended practice, research results indicate that tertiary water treatment, including final disinfection using UV light, chlorination and/or ultrasound have been shown to be effective in removal of indicator microorganisms and pathogens to below limits of detection [5,36].

Research related to microbial quality of agricultural water in Europe is limited, although recent research studies have highlighted the microbial risks associated with different water sources (Table 1). In most cases, longitudinal microbiological surveys have been carried out to evaluate the microbiological quality of produce samples, excluding the monitoring of risk factors such as irrigation water. It should be taken into account that the performance of irrigation water-quality monitoring and management practices are highly variable within countries as well as among different countries, with large differences across Europe. In 2006, Tyrrel *et al.* [37] reported that in UK, the majority of growers do not irrigate with water that would conform the European Union Drinking Water Standard [38] and used water that was faecally contaminated, but that typical faecal coliform concentrations were ≤1000/100 mL. More recently, Holvoet *et al.* [16] monitored eight Belgian lettuce farms to establish the relationships between levels of indicator bacteria and detection of enteric zoonotic pathogens. A high prevalence (75%, $n = 120$) of *E. coli* was found in the irrigation water sources with 65% of the positive samples having *E. coli*levels ≥1 log cfu/100 mL while 26% of the samples showed *E. coli* counts ≥2 log cfu/100 mL, which is above most of the irrigation water-quality standards. Additionally, 35% of the collected samples were positive for at least one pathogen (*Salmonella*, *Campylobacter* or Shiga toxin-producing *E. coli* (STEC)). In this study, most of the monitored farms used open wells to hold collected rainfall water. Previously reported *E. coli* values for surface water collected from rivers within an agricultural landscape were within the range 1.5–3.3 log cfu/mL [39], confirming the evidence that surface water, including rivers, streams, and creeks have unpredictable water quality and activities upstream can rapidly change the levels of contaminants [9]. A study focused on Belgian strawberry production reported significant differences in the microbiological quality of irrigation water obtained from different water sources including groundwater and collected rainfall water stored in ponds [19]. They reported that water obtained from the ponds was positive for *E. coli* (40/56) with an average level of 1.6 log cfu/100 mL and almost 20% of these samples were positive for STEC. However, groundwater samples showed much lower levels of *E. coli* and these samples did not contain STEC. Water stored in reservoirs or lagoons provides both an opportunity for pathogen die-off through natural UV but also for new contamination associated with the use of reservoirs as wildlife habitats [37]. In many European countries, a significant proportion of irrigation water is abstracted and stored in farm reservoirs. Castro-Ibañez *et al.* [17] monitored several water reservoirs used to irrigate growing fields of leafy greens. Samples were analysed for presence of foodborne pathogens and only one sample was positive for *Salmonella* spp. Reported*E. coli* prevalence and concentration in irrigation water samples from water reservoirs was lower than that reported from open wells in Belgium. Observed differences between Belgium and Spain might be due to differences in weather conditions as solar radiation has been highlighted as an important mechanism for bacterial decline in environmental samples including irrigation water [18]. Urban wastewater has been used in agriculture as a way to overcome water scarcity in the South of Europe. The European Water Framework Directive [40], specifies that treated wastewater should be used in agriculture where and whenever possible. Currently, urban wastewater is mainly used for irrigation in combination with production systems that avoid direct contact between the water and the edible part of the fresh produce. In countries such as Spain, Greece and Italy, commercial production using urban wastewater as irrigation water usually involves production of tomatoes, peppers and cucumbers grown hydroponically in greenhouses. Sometimes, wastewater treatment plants, consisting of a train of individual unit processes, are located close to the greenhouse (Figure 1).

Studies focusing on the microbial quality of reclaimed water used for irrigation reported presence of faecal contamination within the range 2–4 log *E. coli* cfu/100 mL and also presence of pathogenic microorganisms such as*Salmonella* spp. [41,42] (Table 1). Production systems that minimize irrigation water contact with the edible portion of the crop seem to reduce the risk of contamination. *Codex Alimentarius* [43] reported that plants grown in hydroponic systems absorb nutrients and water at varying rates, constantly changing the composition of the re-circulated nutrient solution and because of this water used in hydroponic culture should be changed frequently, or if recycled, should be treated to minimize microbial and chemical contamination.

Figure 1. A secondary treatment wastewater plant located in a greenhouse production unit to provide irrigation water for tomato production. Reprinted with permission of Quality and Safety Lab CEBAS-CSIC.

3. REGULATIONS, GUIDELINES AND MICROBIAL QUALITY STANDARDS FOR WATER USED IN PRIMARY PRODUCTION

The microbial quality of irrigation water has been related with food safety for more than 25 years. In Europe, specific microbial criteria are only currently established in guidelines and quality assurance standards (QAS), but specific microbial criteria have been introduced in U.S. legislation and these are starting to have an impact on European standards.

3.1. Legislation—The European Commission

Regulation (EC) No 852/2004 on the hygiene of foodstuffs defines potable water as *"meeting the minimum requirements laid down in Council Directive 98/83/EC* [38] *on the quality of water intended for human consumption"*, clean water is defined in Regulation (EC) No. 852/2004 [44], as *"clean seawater and fresh water of a similar quality"* and clean seawater is defined as *"natural, artificial or purified seawater or brackish water that does not contain micro-organisms, harmful substances or toxic marine plankton in quantities capable of directly or indirectly affecting the health quality of food"* [44]. Specific microbiological criteria are not defined but growers must be able to demonstrate that their operations are managed in a way that controls food safety risks, including those associated with the use of water.

3.2. Guidelines

3.2.1. World Health Organization (WHO)

Whilst it was clear that care was needed in the safety of water used to irrigate crops that are eaten raw, such as leafy salads; prior to 1973 there was no generally accepted standard of microbial water quality. In 1973 the World Health Organization (WHO) addressed the issue that drinking water microbial quality standards (*i.e.*, ≤2.2 cfu/100 mL coliforms) were not a realistic target and produced a guideline value of ≤100 cfu/100 mL coliforms for unrestricted irrigation water (*i.e.*, water used to irrigate crops that will be eaten uncooked) derived from wastewater [45]. Following consideration of what was achievable by wastewater treatment processes and associated epidemiological studies of wastewater use, these guidelines were revised in 1989 to a geometric mean of ≤1000 cfu/100 mL faecal coliforms and an arithmetic mean of ≤1/L intestinal nematode during the irrigation period [46]. These standards have been criticised, particularly from those developing guidelines with the aim for "zero-risk" irrigation water [47]. The WHO has more recently changed the approach and there are now no definitive values for microbiological guidelines for irrigation water. Instead, irrigation water safety should be based upon risk assessment as recommended in WHO documents and water guidelines in advanced economies should rely on in-country standards [48,49].

3.2.2. Codex Alimentarius Commission

Baseline guidance on safety requirements of irrigation water is provided through *Codex Alimentarius:* both the General Principles of Food Hygiene-CAC/RCP 1–1969 [50] and the Code of hygienic practice for fresh fruit and vegetables-CAC/RCP 53-2003 [42] address the issue. However this only provides a *"general framework of recommendations [...] rather than providing detailed recommendations for specific agricultural practices..."* [43] *i.e.*, the water used in primary production must be safe. However the microbial requirements are not defined. For example, in the Leafy Green annex of the Code of hygienic practice for fresh fruit and vegetables-CAC/RCP 53–2003 growers are required to *"seek appropriate guidance on water quality and delivery methods to minimize the potential for contamination with microbial pathogens"* [43]. And where water comes in to *"substantial contact with the edible portion of the leafy vegetable should meet the standards for potable or clean water"* [43] where these are defined as: *Potable water*–water which meets the quality standards of drinking water such as described in the WHO Guidelines for Drinking Water Quality; *Clean water*–water that does not compromise food safety in the circumstances of its use [43].

3.3. Microbial Quality Standards for Water Used in Primary Production

3.3.1. QAS—1st Generation

In the early 1990s the UK fresh produce supply chain started to develop Quality Assurance Schemes (QAS) also known as Retailer Codes of Practice partly in response to the 1990 UK Food Safety Act that required growers to be treated as running food businesses within a consolidated supply and as a consequence growers were exposed to due diligence imposed from the retail end of the supply chain [51,52]. Amongst these QAS were Assured Produce–1991; Tesco Natures Choice–1992 and, more widely relevant, EurepGAP–1997 [53]. In essence these QAS followed Codex guidelines and required systems to be in place to ensure that the water used in production was safe. The approach followed Hazard Analysis and Critical Control Point (HACCP) principles such as to risk assess water sources and use but did not define water quality criteria. In 2004 Marks and Spencer (M&S) released a new QAS "Field to Fork" that required growers to test their irrigation water for *E. coli*, but again no criteria were stipulated as to what an acceptable level of indicator organism would be [53].

3.3.2. QAS—2nd Generation

The QAS increased their focus on microbial risks associated with irrigation water and a new generation of schemes (e.g., Red Tractor Fresh Produce Scheme, GlobalGAP, Tesco Nurture, M&S Field to Fork version 2) was developed in the 2000s [51]. As with the first generation of schemes, growers were still required to develop and implement management systems along HACCP principles and to risk assess all water sources. There was also more guidance on risk assessment methodology but critical levels of indicator organisms (with some confusion between *E. coli* and faecal coliforms) in water used in primary production were now defined. Generally, the criteria were based on the WHO guidance for water for irrigation of produce that can be consumed uncooked of ≤1000 faecal coliforms cfu/100 mL with an additional reporting level of ≥400 cfu/100 mL in Tesco Nurture and M&S Field to Fork version 2, which has become more stringent in later revisions for leafy crops.

3.3.3. QAS—3rd Generation

In 2011 the Food Safety Modernization Act (FSMA) became US law [54]. The FSMA proposed microbial quality standards for irrigation water and wash water that came into contact with the edible portion of fresh produce crops. Although the original microbial quality standards are being revised to include rules on produce safety that are more flexible and less burdensome in key areas [55], these standards have been applied in Good Agricultural Practices (GAP) used in the USA and also more widely. One example is the McDonalds GAP [28] where the grower is required to implement systems to ensure food safety, based on HACCP principles and risk assessment as with previous generations of QAS, but clear metrics and microbial criteria are now defined. Irrigation water that may contact the edible portion of the crop must have no more than 235 cfu (or most probable number (MPN), as appropriate) generic *E. coli* per 100 mL for any single sample, or a rolling geometric mean (n = 5) of more than 126 cfu (or MPN, as appropriate) generic *E. coli* per 100 mL of water [28]. A pre-planting water sampling programme is needed for any irrigation water used (and stored rain water). Each water source must have a minimum of one sample analyzed from the point of application closest to the crop to establish that water quality is acceptable before being used for irrigating the crop (*i.e.*, ≤235 cfu generic *E. coli* per 100 mL). Subsequently, during crop growth, a set of at least five irrigation water samples must be collected prior to harvest to establish the rolling geometric mean. The sampling frequency is not stipulated but five samples must be taken within the crop growth period or 30 days, whichever is the shorter [28].

More recently the FDA has revised the upper limit of the irrigation water quality criteria to a statistical threshold value (*i.e.*, a value that should not be exceeded by more than 10 percent of the samples taken) ≤410 CFU of generic *E. coli* in 100 mL of water [55] and it would be anticipated that this standard would start to be used in some QAS.

It has become clear that the industry is being pushed towards zero-risk irrigation water sources in response to supply chain concerns over food safety and that there is a greater requirement for frequent water testing regimes with complex calculations of critical values. These high standards will be difficult to achieve with open water sources such as rivers or stored water that are regularly tested, and for some producers their water sources will not comply with the requirements. This may lead to alternative strategies being required.

4. SOLUTIONS

4.1. Monitoring Pathogen and Indicator Species in Water Sources

Testing water can be used to establish a history of microbiological quality and inform a risk ranking for a water source [9]. The prohibitive cost and time requirement of pathogen detection make microbial indicators a good strategy to characterize microbial contamination in agricultural water [56]. Several studies report that indicator bacteria, and particularly generic *E. coli* concentrations, are not correlated with the presence of pathogens such as *E. coli* O157 or*Salmonella* spp. in water samples, suggesting that *E. coli* data might not be suitable to predict the risk of exposure to pathogenic strains [10,57,58]. However, a logistic regression analysis of samples taken monthly for 12 months from 18 locations throughout Central Florida showed that *E. coli* concentration can predict the probability of enumerating selected*Salmonella* levels, indicating that *E. coli* provides a reasonable way to predict *Salmonella* levels in surface water. According to this, in recent longitudinal microbial surveys carried out in Europe, presence of elevated levels of *E. coli* increased the probability of presence of pathogens (STEC and *Salmonella* spp.) [16,18]. Additionally, Wilkes *et al.* [59] reported that faecal indicators such as *E. coli* spp. were conservative surrogates for a variety of pathogenic microorganisms in surface waters within an agricultural landscape. Based on these reports, *E. coli* spp. could be identified as suitable for a hygiene criterion at primary production of leafy greens and can be applied for validation and verification studies of GAP.

However, factors such as the dynamic nature of agricultural water microbial quality, the time lag between obtaining agricultural water testing results and water use, and the fact that current water sampling strategies are based on an assumption that bacteria are floating as single cells in water, make water monitoring limited in its ability to monitor microbial risks [10]. In addition, Won *et al.* [60] reported that $n > 5$ canal and $n > 14$ reservoir samples were needed to calculate *E. coli* concentrations at a precision level of 85% with 95% confidence interval under the same environmental conditions during the testing period; a frequency much greater than required by the most stringent QAS. The frequency of testing will vary depending on the water source and the risk of environmental contamination [3]. For example, Holvoet *et al.* [61] observed in a study of leafy vegetable farms in Belgium that contamination of irrigation water was more prevalent in open-field production compared to greenhouse production. An additional limitation of the testing approach is the costs of an effective monitoring program which are usually too expensive for most small to medium growers [9]. Hence, alternatives to monitoring sampling plans such as water treatments and the selection of less risky production systems, which avoid contact between the edible part of the plant and irrigation water, may be an alternative option in reducing microbial risks to leafy vegetables.

4.2. Water Treatments

Physical and chemical disinfection systems have been explored as methods to remove human pathogens from agricultural water sources (Table 2), although disinfection treatment of irrigation water is still a very limited practice [4]. Nowadays, chemical sanitizers are the most commonly used water treatments, although environmentally friendly alternatives are being demanded, particularly for organic production. In fact, concerns have risen recently regarding both the absence of water treatment and the excessive use of potentially toxic chemicals to treat irrigation water. Norton-Brandão*et al.* [35] presented a critical review of a wide range of urban reclamation technologies that could be applied to water; of these, the treatments relevant to agricultural irrigation water included coagulation, flocculation, filtration and chemical disinfection [8].

Among commercially available water treatments, chlorine-based sanitizer remains the most common water treatment used for the removal of biohazards from irrigation water [4,8]. The advantages of using water treatment are generally associated with microbial reductions of both foodborne pathogens as well as phytopathogens and the reduction of contamination from biofilm formation in the irrigation pipes (Table 2) [61]. There are many inexpensive commercial applications of chlorine-based sanitizers available for growers. When using this type of water disinfectants, a key step is the estimation of the peak chlorine demand over a range of typical operating conditions, where the peak chlorine demand is defined as the maximum amount of free chlorine in a batch of water that is "used up" by soil and organic materials added with product during washing [62]. Two of the more widespread treatments in Europe and also U.S., are calcium hypochlorite or chlorine dioxide but limitations of chlorine-based disinfectant in terms of formation of disinfection by-products and the potential negative effect on the environment have limited its use in agricultural water. Nevertheless, the goal of the dosing system is, in most cases, to reduce *E. coli* levels to be within a compliant range, for which a minimal dose of disinfectant is usually enough. This reduces the potential detrimental effects on the farm soil or environment from disinfection by-products [63] in the short-term but there are still concerns for chronic effects of large-scale use over long periods of time [4]. Additionally, the volumes of water that are commonly used in a medium-large production field are very high (50–100 m^3), and the cost associated to the treatment of irrigation water can be substantial.

Greener technologies based on physical treatments such as ultrasound (US), ultraviolet light (UV-C) and filtration have been successfully tested to reduce microbial loads of irrigation water [4,29,36]. US technology has the advantage of reducing microbial loads, including algae, without the detrimental effects of the formation of disinfection by-products [36]. Other reported advantages of US include potential simultaneous oxidation, thermolysis, shear degradation, and enhanced mass transfer processes [64]. UV-C light has also been used frequently to disinfect irrigation water and requires the installation of a UV-C treatment system where water passes through a vessel while it is illuminated by UV-C lamps located in the vessel (Figure 2). These systems are usually relatively cheap and represent a good treatment option if water turbidity is low [65]. However, a pre-treatment usually based on sand filtration and also regular maintenance of the lamp are needed to ensure proper efficacy of the system. Membrane and sand filtration have been reported as effective technologies to remove pathogens from water [35]. Different pore sizes are applied for different applications. Thus, membrane filters classified as micro- (0.1–10 µm), ultra (0.002–0.1 µm), nano (0.0005–0.002 µm) filtration or reverse osmosis (<0.0005 µm) are used for removal of microorganisms while larger pore sizes and rapid sand filters are used to reduce soil and plant material that may clog the membrane filters [66]. Membrane filtration is more effective at controlling pathogens when combined in a series of multi-stage filters and in combination with other treatments that have other modes of action such as chlorine [67]. Innovative filtration systems such filters containing sand and/or materials with reactive components have been explored as potential water treatments [10]. Slow sand filtration controls microorganisms present in the water by biological, physical and chemical reactions. In these systems, microbial removal is carried out by a complex microbial community located in the upper layer on the sand bed but also by physically entrapping the pathogens and debris [67]. Biosand filter zero-valent iron incorporated (ZVI) treatments, which have been used in permeable reactive barriers to remove a broad range of chemical contaminants in groundwater, have also been reported as potential water treatments [68]. Ingram *et al.* [69] have recently proposed ZVI treatment as a cost-effective mitigation option for irrigation water to help small farmers reduce risk of foodborne *E. coli* infections associated with contamination of leafy greens.

Whilst alternative systems are being proposed in scientific studies, the evaluation of the site-specific applicability of any of these technologies, such as maintenance costs, safety, and biological effects on crops and humans, is missing [4,5]. In general, water treatment technologies for agricultural water have not been deeply evaluated and in most of the cases, operational and maintenance costs are knowledge gaps crucial to the decision making process of farmers [4]. On the other hand, process control for water treatment technologies is mandatory to guarantee the efficacy of the treatment. A good understanding and consistent implementation of systems to monitor, control, and document water treatment performance are needed because without the process control, all water treatment systems may give a "false confidence".

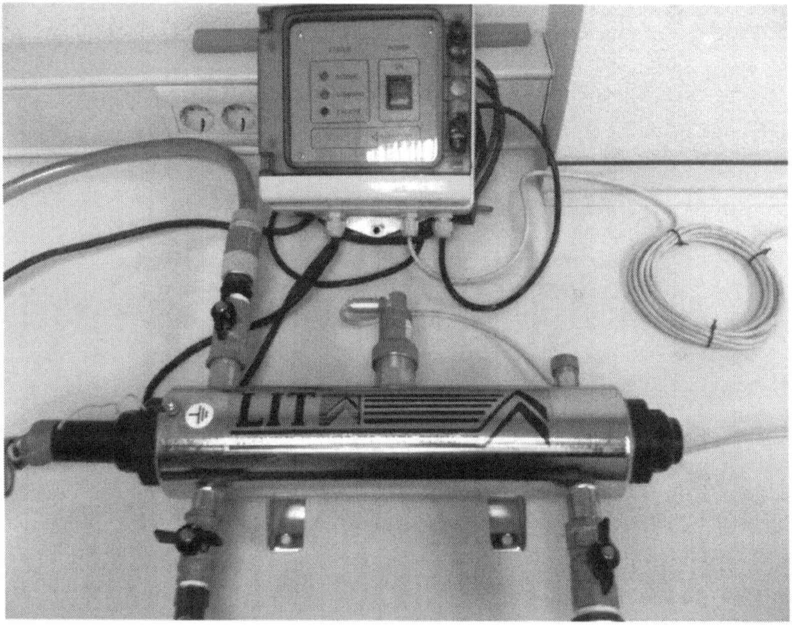

Figure 2. UV-C treatment system where water passes through a vessel while it is illuminated by UV-C lamps located in the vessel. Reprinted with permission of Quality and Safety Lab CEBAS-CSIC.

4.3. Alternative Production Systems

An alternative strategy to managing the risk of contaminating leafy vegetables through faecally contaminated irrigation water is to avoid direct contact with the edible parts of the crop. This can be achieved through selecting the irrigation method and/or the production system used to the crops. Application of microbial-contaminated irrigation water using subsurface drip irrigation has been shown to reduce contamination of crops including lettuce at harvest compared to furrow irrigation [70]. However, even if direct contact between irrigation water and the edible part of the leafy crop is avoided, irrigation water may contaminate the soil or substrate, where the bacteria can survive for some time [30] and irrigation or rainfall splash may contaminate the crop [11].

Soilless systems, such as hydroponic floating systems [71] or nutrient film techniques (NFT) [72], are being used for leafy vegetables with short production cycles allowing a better control and standardization of the cultivation process (Figure 3). Many advantages have been attributed to the use of soilless systems in greenhouses to produce leafy greens, but reductions in product quality and shelf life have been observed [73] which may limit the use of these systems. However, recent studies carried out in commercial agricultural production sites showed that the use of poor quality irrigation water combined with the use of soilless production systems considerably reduced microbial contamination risks to fresh produce [42,74].

Figure 3. Nutrient film techniques (NFT), a type of soilless systems where a thin film of nutrient solution flows through plastic channels which contain the plant roots and laid on a slope in order to grant the constant flow of nutrient solution. Source: James M. Monaghan (Harper Adams University).

This reduction can be attributed to preventing contact between the irrigation water and the edible part of the plant. In a non-commercial hydroponic system lettuce grown using diluted effluent from secondary UV treated grey-water was shown to have an acceptable level of *E. coli* spp. on the harvested leaves, even though levels on the roots were high [75]. The use of alternative production systems such as greenhouses and hydroponics are assumed to be safer than open field production from the microbiological point of view due to the minimization of some risk factors associated with sources of pre-harvest contamination, but also a greater control of water disinfection with water being recirculated and cleaned periodically [76]. The nutrient solution, and hence water, used in soilless systems is one of the most important aspects for the success of leafy greens production [77] and the quality of water can be better controlled in soilless systems [63] both from a plant pathogen and human pathogen perspective. Nevertheless, before hydroponics could be used to grow crops using lower quality water more knowledge is required on the risks posed from root contamination and internalization of bacteria (e.g., [78]) from contaminated hydroponic water sources.

5. CONCLUSIONS

There is an increasing evidence of contamination of produce from irrigation water, but scarce information on the microbial quality of agricultural water is available. Despite this, retailers selling leafy vegetables are risk averse and are targeting zero-risk production systems. As a consequence the requirements for irrigation water quality and safety in QAS have evolved and become more stringent. One approach is for growers to identify water sources that are contaminated with potential pathogens through a monitoring regime and only use water free of pathogens, but this approach is very costly and ultimately does not ensure safe water. The use of faecal indicators, particularly *E. coli*, allows growers to identify water sources where there is a route of faecal contamination, allowing risk categorization of water sources [9,30]. Where growers only have access to water sources of moderate quality, with low levels of faecal contamination, they have two main options available to produce leafy vegetables that will not pose unacceptable risks to the consumer: treat water before use, using physical or chemical disinfection systems, with chlorine the most commonly used system currently; or reduce/eliminate contact with the leaves from irrigation water through irrigation water placement *i.e.*, drip or furrow, or the use of soilless growing systems. It is likely that a combination of approaches will be needed to meet the microbial requirements of leafy vegetables and ensure safe food for consumers.

ACKNOWLEDGMENTS

The authors are thankful for the financial support from the Center for Produce Safety Grant Agreement 2015-374 and the MINECO (Project AGL2013-48529-R). Support provided by the COST ACTION FA1202 BacFoodNet is also appreciated.

AUTHOR CONTRIBUTIONS

Authors (Ana Allende and James Monaghan) have equally participated in this work via study of conception, acquisition of data, drafting of manuscript and/or critical revision.

CONFLICTS OF INTEREST

The authors declare no conflict of interest.

REFERENCES

1. Guide to minimize microbial food safety hazards for fresh fruits and vegetables. U.S. Department of Health and Human Services, Food and Drug Administration/Center for Food Safety and Applied Nutrition: Washington, DC, USA, 1998.

2. Steele, M.; Odumeru, J. Irrigation water as source of foodborne pathogens on fruits and vegetables. J. Food Prot. 2004,67, 2839–2849.

3. Gerba, C.P. The role of water and water testing in produce safety. In Microbial Safety of Fresh Produce; Fan, X., Niemira, B.A., Doona, C.J., Feeherty, F.E., Gravani, R.B., Eds.; Wiley: Indianapolis, IN, USA, 2009; pp. 129–142.

4. Suslow, T.V. Standards for irrigation and foliar contact water. Available online: http://extension.psu.edu/food/safety/farm/resources/water/standards-for-irrigation-and-foliar-contact-water/view (accessed on 28 May 2015).

5. Pachepsky, Y.; Shelton, D.R.; McLain, J.E.; Patel, J.; Mandrell, R.E. Irrigation waters as a source of pathogenic microorganisms in produce: A Review. Adv. Agron. 2011, 113, 73–138.

6. Olaimat, A.N.; Holley, R.A. Factors influencing the microbial safety of fresh produce: A review. J. Food Prot. 2012, 32, 1–19.

7. Park, S.; Szonyi, B.; Gautam, R.; Kendra, N.; Anciso, J.; Ivanek, R. Risk factors for microbial contamination in fruits and vegetables at the preharvest level: A systematic review. J. Food Prot. 2012, 75, 2055–2081.

8. Gil, M.I.; Selma, M.V.; Suslow, T.; Jacxsens, L.; Uyttendaele, M.; Allende, A. Pre- and post-harvest preventive measures and intervention strategies to control microbial food safety hazards of fresh leafy vegetables. Crit. Rev. Food Sci. Nutr. 2015, 55, 453–468.

9. Uyttendaele, M.; Jaykus, L.A.; Amoah, P.; Chiodini, A.; Cunliffe, D.; Jacxsens, L.; Holvoet, K.; Korsten, L.; Lau, M.; McClure, P.; et al. Microbial hazards in irrigation water: Standards, norms, and testing to manage use of water in fresh produce primary production. Compr. Rev. Food Sci. F 2015, 14, 336–356.

10. Agricultural Water: Five Year Research Review; Center for Produce Safety: Davis, CA, USA, 2014.

11. Monaghan, J.M.; Hutchison, M.L. Distribution and decline of human pathogenic bacteria in soil after application in irrigation water and the potential for soil-splash-mediated dispersal onto fresh produce. J. Appl. Microbiol. 2012, 112, 1007–1019.

12. Oliveira, M.; Viñas, I.; Usall, J.; Anguera, M.; Abadias, M. Presence and survival of Escherichia coli O157:H7 on lettuce leaves and in soil treated with contaminated compost and irrigation water. Int. J. Food Microbiol. 2012, 156, 133–140.

13. Kisluk, G.; Yaron, S. Presence and persistence of Salmonella enterica serotype Typhimurium in the phyllosphere and rhizosphere of spray-irrigated parsley. Appl. Environ. Microbiol. 2012, 78, 4030–4036.

14. Van der Linden, I.; Cottyn, B.; Uyttendaele, M.; Vlaemynck, G.; Heyndrickx, M.M.; Maes, M. Survival of enteric pathogens during butterhead lettuce growth: Crop stage, leaf age, and irrigation. Foodborne Pathog. Dis. 2013, 10, 485–491.

15. Ceuppens, S.; Hessel, C.T.; de Quadros Rodrigues, R.; Bartz, S.; Tondo, E.C.; Uyttendaele, M. Microbiological quality and safety assessment of lettuce production in Brazil. Int. J. Food Microbiol. 2014, 181, 67–76.

16. Holvoet, K.; Sampers, I.; Seynnaeve, M.; Uyttendaele, M. Relationships among hygiene indicators and enteric pathogens in irrigation water,

soil and lettuce and the impact of climatic conditions on contamination in the lettuce primary production. Int. J. Food Microbiol. 2014, 71, 21–31.

17. Castro-Ibáñez, I.; Gil, M.I.; Tudela, J.A.; Ivanek, R.; Allende, A. Assessment of microbial risk factors and impact of meteorological conditions during production of baby spinach in the Southeast of Spain. Food Microbiol. 2015, 49, 173–181.

18. Castro-Ibáñez, I.; Tudela, J.A.; Gil, M.I.; Allende, A. Microbial safety considerations of flooding in primary production of leafy greens. Food Res. Int. 68, 62–69. [CrossRef]

19. Delbeke, S.; Ceuppens, S.; Holvoet, K.; Samuels, E.; Sampers, I.; Uyttendaele, M. Multiplex real-time PCR and culture methods for detection of Shiga toxin-producing Escherichia coli and Salmonella Thompson in strawberries, a lettuce mix and basil. Int. J. Food Microbiol. 2015, 193, 1–7.

20. Généreux, M.; Grenier, M.; Côté, C. Persistence of Escherichia coli following irrigation of strawberry grown under four production systems: Field experiment. Food Control 2015, 47, 103–107.

21. Stine, S.W.; Song, I.; Choi, C.Y.; Gerba, C.P. Application of microbial risk assessment to the development of standards for enteric pathogens in water used to irrigate fresh produce. J. Food Prot. 2005, 68, 913–918.

22. Mota, A.; Mena, K.D.; Soto-Beltran, M.; Tarwater, P.M.; Cháidez, C. Risk assessment of Cryptosporidium and Giardia in water irrigating fresh produce in Mexico. J. Food Prot. 2009, 72, 2184–2188.

23. Seidu, R.; Sjølander, I.; Abubakari, A.; Amoah, D.; Larbi, J.A.; Stenström, T.A. Modeling the die-off of E. coli and Ascaris in wastewater-irrigated vegetables: Implications for microbial health risk reduction associated with irrigation cessation. Water Sci. Techn. 2013, 68, 1013–1021.

24. Castro-Ibáñez, I.; Jacxsens, L.; Gil, M.I.; Uyttendaele, M.; Allende, A. Quantitative exposure assessment of Escherichia coli in baby spinach primary production in Spain: Effects of weather conditions and agricultural practices. Int. J. Food Microbiol. 2015. submitted.

25. Jacxsens, L.; Delbeke, S.; Ceuppens, S.; Johannessen, G.; Allende, A.; Sampers, I.; Uyttendaele, M. Comparative quantitative exposure modelling of enteric pathogens Salmonella and Shiga toxin producing Escherichia coli on fresh basil, strawberries and butterhead lettuce. Food Microbiol. 2015. submitted.

26. Parker, J.S.; Wilson, R.S.; LeJeune, J.T.; Rivers, L.; Doohan, D. An expert guide to understanding grower decisions related to fresh fruit and vegetable contamination prevention and control. Food Control 2012, 26, 107–116.

27. Fulponi, L. Private voluntary standards in the food system: The perspective of major food retailers in OECD countries. Food Policy 2006, 31, 1–13.

28. McDonald. McDonald's Good Agricultural Practices Food Safety Standards, Food Safety Checklist & Produce Processing Guidelines–August 2012.

29. Jones, L.A.; Worobo, R.W.; Smart, C.D. UV light inactivation of human and plant pathogens in unfiltered surface irrigation water. Appl. Environ. Microb. 2014, 80, 849–854.

30. EFSA; Panel on Biological Hazards (BIOHAZ). Scientific opinion on the risk posed by pathogens in food of non-animal origin. Part 2 (Salmonella and Norovirus in leafy greens eaten raw as salads). EFSA J. 2014, 11. Available online: www.efsa.europa.eu/efsajournal (accessed on 16 May 2015).

31. Considering Water Quality for Use in the Food Industry; ILSI Europe Report Series; International Life Sciences Institute: Brussels, Belgium, 2008.

32. Pond, K.; Charles, K.; Pedley, S. Review of the Use of Water in UK Agriculture and the Potential Risks to Food Safety; Report B17005; Food Standards Agency: London, UK, 2007.

33. Weatherhead, K. Water—Future Availability, Demand and Management. Available online: http://www.ofc.org.uk/files/ofc/papers/08weatherhead.pdf (accessed on 14 May 2015).

34. Pedrero, F.; Kalavrouziotis, I.; Alarcón, J.J.; Koukoulakis, P.; Asano, T. Use of treated municipal wastewater in irrigated agriculture—Review of some practices in Spain and Greece. Agric. Water Manag. 2010, 97, 1233–1241.

35. Norton-Brandão, D.; Scherrenberg, S.M.; van Lier, J.B. Reclamation of used urban waters for irrigation purposes - A review of treatment technologies. J. Environ. Manag. 2013, 122, 85–98.

36. Villanueva, M.V.; Luna, M.C.; Gil, M.I.; Allende, A. Ultrasound treatments improve the microbiological quality of water reservoirs

used for the irrigation of fresh produce. Food Res. Int. 2015, 75, 140–147.

37. Tyrrel, S.F.; Knox, J.W.; Weatherhead, E.K. Microbiological water quality requirements for salad irrigation in the United Kingdom. J. Food Prot. 2006, 69, 2029–2035.

38. EC, European Commission. Council Directive 98/83/EC of 3 November 1998 on the quality of water intended for human consumption. Off. J. Eur. Commun. 1998, 5, 0032–0054.

39. Economou, V.; Gousia, P.; Kansouzidou, A.; Sakkas, H.; Karanis, P.; Papadopoulou, C. Prevalence, antimicrobial resistance and relation to indicator and pathogenic micro-organisms of Salmonella enterica isolated from surface waters within an agricultural landscape. Int. J. Hyg. Environ. Health. 2012, 216, 435–444.

40. EC, European Commission. European Water Framework Directive. Council Directive 2000/60/EC of 2nd December 2000. Available online: http://ec.europa.eu/environment/water/water-framework/index_en.html (accessed on 1 June 2015).

41. Forslund, A.; Ensink, J.H.H.; Markussen, B.; Battilani, A.; Psarras, G.; Gola, S.; Sandei, L.; Fletcher, T.; Dalsgaard, A.Escherichia coli contamination and health aspects of soil and tomatoes (Solanum lycopersicum L.) subsurface drip irrigated with on-site treated domestic wastewater. Water Res. 2012, 46, 5917–5934.

42. Lopez-Galvez, F.; Allende, A.; Pedrero-Salcedo, F.; Alarcon, J.J.; Gil, M.I. Safety assessment of greenhouse hydroponic tomatoes irrigated with reclaimed and surface water. Int. J. Food Microbiol. 2014, 191, 97–102.

43. CAC (Codex Alimentarius Commission). Code of Hygienic Practice for Fresh Fruits and Vegetables. Available online:ftp://ftp.fao.org/codex/publications/Booklets/FreshFruitsVeg/FFV_2007_EN.pdf (accessed on 16 May 2015).

44. European Commission. Regulation (EC) No 852/2004 of the European Parliament and of the Council of 29 April 2004 on the hygiene of foodstuffs. 2004. Available online: http://eur-lex.europa.eu/LexUriServ/LexUriServ.do?uri=OJ:L:2004:139:0001:0054:en:PDF (accessed on 1 July 2015).

45. Reuse of effluents: Methods of wastewater treatment and public health safeguards. In Report of a WHO Meeting of Experts, Technical Report Series No. 517; WHO: Geneva, Switzerland, 1973.

46. Health guidelines for the use of wastewater in agriculture and aquaculture. In Report of a WHO Scientific Group, Technical Report Series No. 778; WHO: Geneva, Switzerland, 1989.

47. Havelaar, A.; Blumenthal, U.J.; Strauss, M.; Kay, D.; Bartram, J. Chapter 2—Guidelines: The current position. In Water Quality: Guidelines, Standards and Health; Fewtrell, L., Bartram, J., Eds.; IWA Publishing: London, UK, 2001; pp. 17–42.

48. WHO. Third Edition of the WHO Guidelines for the Safe Use of Wastewater, Excreta and Greywater in Agriculture and Aquaculture: Guidance note for National Programme Managers and Engineers-Applying the Guidelines along the Sanitation Ladder. 2010. Available online: http://www.who.int/water_sanitation_health/wastewater/FLASH_OMS_WSHH_Guidance_note1_20100729_17092010.pdf (accessed on 18 May 2015).

49. WHO. Third edition of the WHO Guidelines for the Safe Use of Wastewater, Excreta and Greywater in Agriculture and Aquaculture: Guidance note for National Programme Managers-Guidelines for Drinking-Water Quality, 3rd ed. In Health-based targets; 2010; Available online: http://www.unwater.org/wwd10/downloads/WHO_IWA/Health-Based_Targets_3_17_10.pdf (accessed on 18 May 2015).

50. Recommended International Code of Practice: General Principles of Food Hygiene. Available online: http://www.codexalimentarius.org/standards/list-of-standards/en/?provide=standards&orderField=fullReference&sort=asc&num1=CAC/RCP (accessed on 16 May 2015).

51. Soon, J.M.; Baines, R.N. Public and private food safety standards: Facilitating or frustrating fresh produce growers?Laws 2013, 2, 1–19.

52. The Food Safety Act—A guide for businesses. 2015. Available online: https://www.food.gov.uk/sites/default/files/multimedia/pdfs/fsactguidefoodbusiness.pdf (accessed on 14 May 2015).

53. Monaghan, J.M. United Kingdom and European approach to fresh produce food safety and security. HortTechnics2006, 16, 559–562.

54. FDA, Food and Drug Administration. Fact Sheets on the Subparts of the Original FSMA Proposed Rule for Produce Safety Standards for the Growing, Harvesting, Packing, and Holding of Produce for Human Consumption. Available online:

http://www.fda.gov/Food/GuidanceRegulation/FSMA/ucm334552.htm (accessed on 14 May 2015).

55. FDA, Food and Drug Administration. FSMA Proposed Rule for Produce Safety: Information on Specific Provisions. 2014. Available online: http://www.fda.gov/food/guidanceregulation/fsma/ucm415226.htm#water (accessed on 14 May 2015).

56. Park, S.; Navratil, S.; Gregory, A.; Bauer, A.; Srinath, I.; Jun, M.; Szonyi, B.; Nightingale, K.; Anciso, J.; Ivanek, R. Generic Escherichia coli contamination of spinach at the preharvest level: The role of farm management and environmental factors. Appl. Environ. Microbiol. 2013, 79, 4347–4358.

57. Shelton, D.R.; Karns, J.S.; Coppock, C.; Patel, J.; Sharma, M.; Pachepsky, Y.A. Relationship between eae and stx virulence genes and E. coli in an agricultural watershed: Implications for irrigation water standards and leafy green commodities. J. Food Prot. 2011, 74, 18–23.

58. Benjamin, L.; Atwill, E.R.; Jay-Russell, M.; Cooley, M.; Carychao, D.; Gorski, L.; Mandrell, R.E. Occurrence of genericEscherichia coli, E. coli O157 and Salmonella spp. in water and sediment from leafy green produce farms and streams on the Central California coast. Int. J. Food Microbiol. 2013, 165, 65–76.

59. Wilkes, G.; Edge, T.; Gannon, V.; Jokinen, C.; Lyautey, E.; Medeiros, D.; Neumann, N.; Ruecker, N.; Topp, E.; Lapen, D.R. Seasonal relationships among indicator bacteria, pathogenic bacteria, Cryptosporidium oocysts, Giardia cysts, and hydrological indices for surface waters within an agricultural landscape. Water Res. 2009, 43, 2209–2223.

60. Won, G.; Schlegel, P.J.; Schrock, J.M.; Lejeune, J.T. Absence of direct association between coliforms and Escherichia coliin irrigation water and on produce. J. Food Prot. 2013, 76, 959–966.

61. Holvoet, K.; Sampers, I.; Seynnaeve, M.; Jacxsens, L.; Uyttendaele, M. Agricultural and management practices and bacterial contamination in greenhouse vs. open field lettuce production. Int. J. Environ. Res. Public Health 2015, 12, 32–63.

62. Suslow, T. It's Time to "Farmer-Up". Available online: http://ucanr.edu/datastoreFiles/234-2605.pdf (accessed on 14 May 2015).

63. Benefits and Risks of the Use of Chlorine-Containing Disinfectants in Food Production and Food Processing: Report of a Joint FAO/WHO Expert Meeting; FAO/WHO: Ann Arbor, MI, USA, 2009; pp. 27–30.

64. Sharma, S.K.; Sanghi, R.; Mudhoo, A. Chapter 1—Green Practices to Save Our Precious "Water Resource". InAdvances in Water Treatment and Pollution Prevention; Sharma, S.K., Sanghi, R., Eds.; Springer: Dordrecht, The Netherlands, 2012; pp. 1–36.

65. Newman, S.E. Disinfecting Irrigation Water for Disease Management. In Proceedings of 20th Annual Conference on Pest Management on Ornamentals, San Jose, CA, USA, 20–22 February 2004; Available online: http://ghex.colostate.edu/pdf_files/DisinfectingWater.pdf (accessed on 14 May 2015).

66. Van der Bruggen, B.; Vandecasteele, C.; van Gestel, T.; Doyen, W.; Leysen, R. A review of pressure-driven membrane processes in wastewater treatment and drinking water production. Environ. Prog. 2003, 22, 46–56.

67. Raudales, R.E.; Parke, J.L.; Guy, C.L.; Fisher, P.R. Control of waterborne microbes in irrigation: A review. Agric. Water Manag. 2014, 143, 9–28.

68. Meggyes, T.; Simon, G. Removal of organic and inorganic pollutants from groundwater using permeable reactive barriers Part2. Engineering of permeable reactive barriers. Land Contam. Reclam. 2000, 8, 175–187.

69. Ingram, D.T.; Callahan, M.T.; Ferguson, S.; Hoover, D.G.; Chiu, P.C.; Shelton, D.R.; Millner, P.D.; Camp, M.J.; Patel, J.R.; Kniel, K.E.; et al. Use of zero-valent iron biosand filters to reduce Escherichia coli O157: H12 in irrigation water applied to spinach plants in a field setting. J. Appl. Microbiol. 2012, 112, 551–560.

70. Song, I.; Stine, S.; Choi, C.; Gerba, C. Comparison of crop contamination by microorganisms during subsurface drip and furrow irrigation. J. Environ. Eng. 2006, 132, 1243–1248.

71. Nicola, S.; Hoeberechts, J.; Fontana, E. Comparison between traditional and soilless culture system to produce rocket (Eruca sativa) with low nitrate content. Acta Hortic 2005, 697, 549–555.

72. Benoit, F.; Ceustermans, N. Consequences of closed soilless growing systems for the recirculating nutrient solution and the production techniques. In Proceedings of the XXVI International Horticultural Congress: Protected Cultivation 2002: In Search of Structures, Systems and Plant Materials, Toronto, Canada, 2004; pp. 331–340.

73. Manzocco, L.; Foschia, M.; Tomasi, N.; Maifreni, M.; Costa, L.D.; Marino, M.; Cortella, G.; Cesco, S. Influence of hydroponic and soil

cultivation on quality and shelf life of ready-to-eat lamb's lettuce (Valerianella locusta L. Laterr). J. Sci. Food Agric. 2011, 91, 1373–1380.

74. Luedtke, A.N.; Chapman, B.; Powell, D.A. Implementation and analysis of an on-farm food safety program for the production of greenhouse vegetables. J. Food Prot. 2003, 66, 485–489.

75. Keller, R.; Perim, K.; Semionato, S.; Zandonade, E.; Cassini, S.; Goncalves, R.F. Hydroponic cultivation of lettuce (Lactuca sativa) using effluents from primary, secondary and tertiary + UV treatments. Water Sci. Technol.: Water Suppl.2005, 5, 95–100.

76. Orozco, L.; Rico-Romero, L.; Escartín, E.F. Microbiological profile of greenhouses in a farm producing hydroponic tomatoes. J. Food Prot. 2008, 71, 60–65.

77. Soundy, P.; Cantliffe, D.J.; Hochmuth, G.J.; Sttofella, P.J. Nutrient requirements for lettuce transplants using a floatation irrigation system. I. Phosphorus. HortScience 2001, 36, 1066–1070.

78. Deering, A.J.; Mauer, L.J.; Pruitt, R.E. Internalization of E. coli O157: H7 and Salmonella spp. in plants: A review. Food Res. Int. 2012, 45, 567–575.

Chapter 8

Challenges of Reducing Fresh Produce Waste in Europe—From Farm to Fork

Michael Blanke

INRES- Horticultural Science, University of Bonn, Auf dem Hügel 6, D-53121 Bonn, Germany

ABSTRACT

This concept paper summarizes key "hotspots" for waste generation along the food supply chain and identifies a range of existing solutions/measures that can help producers, retailers and consumers reduce the amount of food that is wasted. The majority of food waste of 71–92 kg/head/year in Western Europe was found to originate from private households (61%), followed by restaurants and canteens (17%) and then supermarkets (5%); 59%–65% (of this food waste (71–92 kg) can be avoided and 54% thereof are fruit and vegetables. Since ethylene accelerates fruit ripening and its accumulation can lead to fruit decay and waste and new portable instruments now enable continuous *in-situ*determination of ethylene along the food chain, there is a possible key to reducing food waste of perishable, fresh produce. Hence, suggested countermeasures at the field level are use of ethylene inhibitors (AVG as "Retain" or MCP as "Harvista"), the former prevents pre-mature fruit drop in pome fruit, incentives for processing fruit of industrial grade and whole crop purchase ("WCP"). Along the supply chain, applications of ethylene inhibitors (e.g., 1-MCP as "SmartFresh") absorber strips (e.g., "It's Fresh", Sensitech), bags (e.g., "Peakfresh") as well as simply cooling and venting, and shading to avoid sun exposure. Countermeasures also include superstores no longer promoting multi-packs, e.g., "two strawberry punnets for the price of one", abandon the "Display until" or "Sell by" date, conservative consumer shopping behavior, and sale of class II produce ("Wunderlinge" in Billa or "Kleine Äpfel" in REWE, "Ünique" in Coop), collection (rather than wasting) of perishable food by volunteers ("Die Tafel"), or "Food Sharing" of private household left-over perishable on social media, or any combination of the above to aid reducing fresh produce waste.

Keywords: apple; ethylene scrubber; food bank; food sharing; greenhouse gas emissions (GHG); resource conservation, Russian trade embargo; strawberry, sustainability; waste, whole crop purchase

1. INTRODUCTION—FRESH PRODUCE LEFT IN THE FIELD

Horticulture can be distinguished from agriculture and characterized by producing predominantly fresh, perishable produce, fuelling criticism of waste of perishable food. Waste assessment and reduction have hence long been essential prerequisites of sustainable horticulture. The horticultural sector in Europe is driven by a combination of surplus production, e.g., 1–2 million tonnes apples/year (Blanke, 2007 [1]) and stringent EU marketing standards (VO EU No. 543/2011) even before the Russian fresh food embargo was imposed in August 2014 and the rise of the Polish apple production. The EU standards exclude "under-sized", "mal-formed" and "unevenly colored" produce, e.g., apple, banana or tomato or pepper fruit from being marketed as fresh fruit and are primarily based on cosmetic appearance rather than nutritional value. As a consequence, some fruit and vegetables are not harvested or disposed of after harvest. Fortunately, in many cases such as apples, industrial grade fruit ("cull") go into juicing and processing, In other crops like lettuce, where this is not possible, the crop is ploughed (wasted), if the wholesale market drops, e.g., well below 10 cents per lettuce head during the peak season.

The objective of this invited contribution is to provide a brief overview on fruit and vegetable waste in Western Europe, to incorporate published reports with those unavailable in English and unpublished information (congress lectures with subsequent discussion) as a basis for discussion on the subject within and outside Western Europe. This concept paper is to summarize key "hotspots" for waste generation along the food supply chain and identifies a range of existing solutions/measures that can help producers, retailers and consumers reduce the amount of food that is wasted.

2. VOLUMES OF FRESH PRODUCE WASTE AND ASSOCIATED FINANCIAL LOSSES

The volume of fresh produce waste in many Western European countries seems to fall within a narrow range with values from 71 kg for the UK (Langley *et al.*, 2010 [2]), 72 kg in the Netherlands (Westerhoven and Steenhuisen, 2010) [3] to 71–92 kg for Germany (BELV, 2012 [4]). However, a much larger value of 140 kg/head/ year is reported again for the UK by the*Waste and Resources Action Programme* (WRAP) (2009 [5]) (Figure 1). The former values are equivalent to a financial loss of*ca.* €200–260 per person per year. The surveys and statistics (Figure 1) include food waste from the food industry, private households as well as restaurants, canteens and both wholesalers and retailers; their relative contribution is shown Table 1.

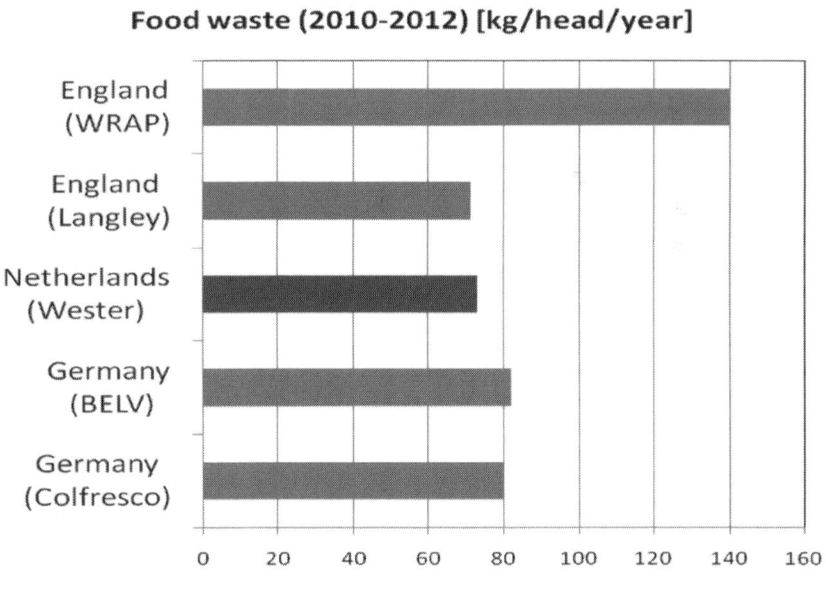

Figure 1. Food waste in Germany, England and the Netherlands (2010–2012).

Table 1. Sources of waste along the food chain in Germany (BELV, 2012 [4]).

Source	Waste (Absolute Value in t/a)	Waste (Percentage)
Private households	6,670,000 t	61%
Canteens, restaurants	1,900,000 t	17%
Industry	1,850,000 t	17%
Trade-both wholesale and retail	550,000 t	5%
Total waste	11 mil t	100%

3. SOURCES OF FOOD WASTE—A KEY TO IDENTIFYING SOLUTIONS AND AVOIDING WASTE

Marks and Spencer's, one of the major UK retailers, analyzed (Hill, 2013 [6]) the sources of their fresh produce waste: The predominant source or site of their food waste of *ca.* 2 million tonnes/year (70%) was in the supply chain compared with relatively small losses of the actual produce (25%) and 5% in the supermarket (Figure 2).

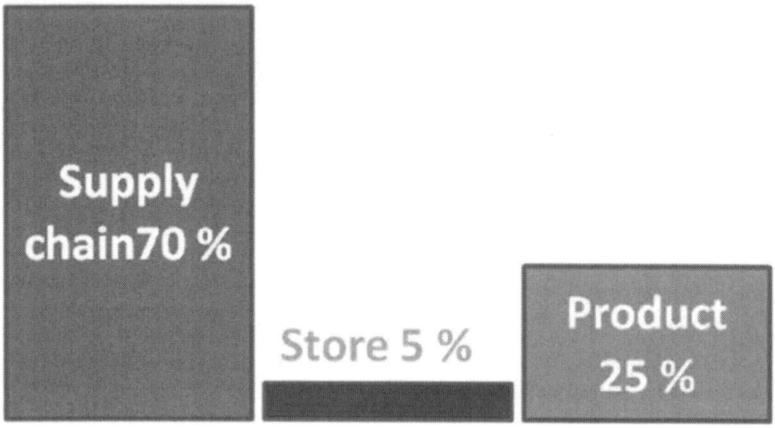

Figure 2. Sources of waste in the food supply chain (Hill, 2013 [6]).

In the US, food waste is reported to account for up to *ca.* 30% compared to 16% in Europe, where 54% of the 82 kg *versus*44 kg comprise *ca.* 26 kg vegetables (Table 2) and *ca.* 18 kg fruit (Figure 3).

In the strive for waste avoidance strategies, Cofresco (2011) [7] and BELV (2012) [4] stated that respectively 59% and 65% of this overall food waste along the supply chain may be avoided. In the UK (WRAP, 2013) [5], 51% (813,000 t) of the overall vegetable and salad waste (1.6 million tonnes) can be avoided as can 37% (340,000 t) of the fruit wasted annually in the UK (910,000 t) (Table 3). This gave rise to the "Fresh for Longer" campaign by WRAP, the film "Good food bad food" by Coline Serreau and a book and film in 2013 by Valentin Thurn entitled "Taste the waste". These three initiatives sparked public awareness and criticism of wasting fresh produce, thereby making it the challenge, task and responsibility of everyone involved in agriculture and the food supply chain.

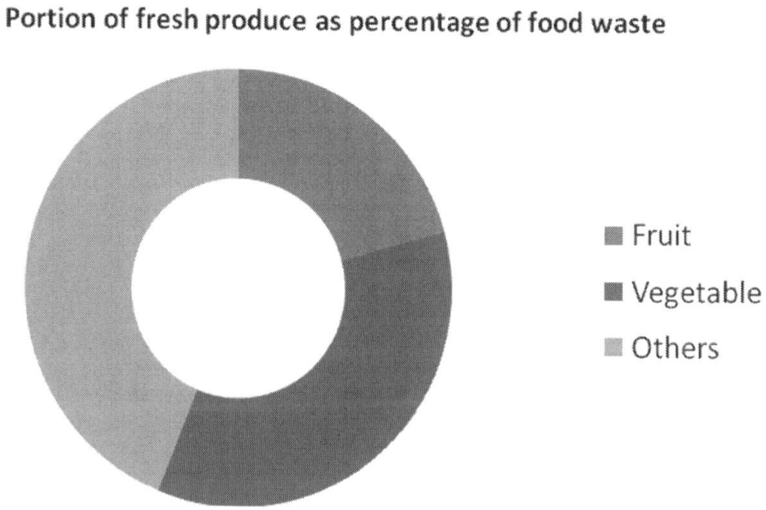

Figure 3. Proportion of perishable produce in food waste in Germany (100% = 82 kg; BELV, 2012 [4]).

Table 2. Waste of perishable fresh fruit, vegetables & salads and potatoes in the UK (WRAP, 2013 [8]).

Commodity	Waste (absolute value (tonnes/year)	Waste that can be avoided (percentage)	Waste that can be avoided (absolute in tonnes/year)
Vegetables and salads	1,600,000 t	51%	813,000 t
Fresh fruit	910,000 t	37%	340,000 t
Potato	5,900,000 t	5%	320,000 t
Total waste	2,510,000 t	n.a.	1,473,000 t

4. FIVE APPROACHES TO REDUCE WASTE IN THE FIELD

In Europe, *ca.* 1/3 of the vegetable crop remains in the field (Figure 4) due to either not meeting marketing standards or insufficient market demand, as known from waste assessment during carbon footprint studies [9]. Ethylene plays a key role; its accumulation may lead to decay or overripe fruit and vegetables. Ethylene-sensitive climacteric fruits score high in the waste reports, e.g., of WRAP (2013) of 1.4 million bananas and 1.5 million tomatoes wasted daily nationwide in the UK despite a 6% decrease in the overall food waste in the last few years.

Table 3. Countermeasures against waste of perishable fresh food at the farm level.

Countermeasure	Solution
Fruit breeding for low ethylene	Ethylene insensitive breeding lines (apple and tomato)
Blocking ethylene receptors	1-MCP treatment before/after harvest
Cooling	Retards fruit ripening and ethylene emission
Replace combustion engines	use hand or electric fork lifts, vents *etc.*
Vents	Introduce fresh ambient air
Ethylene separator	Ethylene scrubber in fruit (CA/ULO) stores
Ethylene absorber	Sachets (Ryan /Sensitech Co.USA) or strips (It's Fresh UK) in the punnet
Gas selective packaging film	"Peakfresh" packs (Peakfresh, Australia), or modified atmosphere packaging (MA)
Identify ethylene hot spots	any combination of the above
Potato and onion	Ethylene generation ("Restrain")
Class I marketing	Whole crop purchase

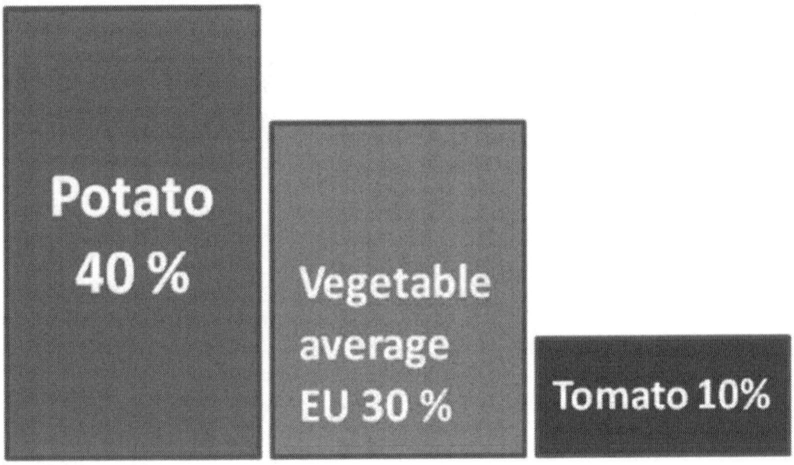

Figure 4. Waste left in the field.

4.1. Inhibitors of Ethylene Biosynthesis

In the field, AVG (aminoethoxyvinyl-glycine–"Retain®", an inhibitor of ethylene synthesis), viz. inhibition of formation of ACC from SAM via ACC synthase, can reduce pre-harvest fruit drop in affected climacteric pome fruits such as apples and pears, thereby preventing fresh produce waste. However, registration is not expected in Europe.

4.2. Blocking the Ethylene Receptors to Extend Shelf-Life

Harvista: 1-MCP (1- methylcyclopropene; "Smartfresh®"), used so far in Europe only after harvest, may be used in the future also prior to harvest ("Harvista®"), e.g., on several commodities such as apples and tomatoes to delay ripening via blocking the ethylene receptors and the ethylene-induced ripening cascade (Sisler and Serek, 1997 [10]).

4.3. Potato and Onion Storage with the Addition of Ethylene

In contrast to the majority of perishable fruit and vegetables (and flowers), onion (Bufler, 2009 [11] and potato (Rylski *et al.*, 1974 [12] can often be successfully stored with *ca.* 10 ppm ethylene (*ca.* 5–15 ppm C_2H_4 for onion, *ca.* 5–50 ppm C_2H_4 for potato) with a recent technology ("BioFresh", and "Restrain"); in this case, ethylene in the lower ppm range suppresses germination of the eyes on these tubers. Generally, the optimum 1-MCP or ethylene concentration depends on the commodity, variety, ripening stage, storage conditions (temperature, CO_2) and intended storage period.

4.4. Waste in the Field not Meeting Class I Marketing Standards

In the field, fruit and vegetables are already rejected, because they do not meet class 1 EU marketing standards, like VO EU No. 543/2011 for apple, and may go to waste, unless the rejects become useful and economic for processing, e.g., for juice, sauce, vinegar, cake or distilling, *etc.*

4.5. Whole Crop Purchase (WCP)

Instigated by the class I marketing debate, a new concept of whole crop purchase (WCP) is being developed (Table 3), but it is suitable largely for vertically integrated enterprises, who can make proper use of class II produce or even of industrial grade produce.

The described ethylene-related technologies are readily available to control ripening. There are, however, very few exceptions such as It's Fresh strips for apple or 1-MCP for cv. "Braeburn" apple, which can develop internal (Braeburn) browning (Mattheis and Rudell, 2008 [13]). If harvest was inadequately timed (too early) in apple or ripening after fruit storage lacking, 1-MCP can adversely affect ripening including firmness, color and aroma formation [13]. “BioFresh”/“Restrain” may induce off-flavor in three potato varieties [12].

5. FOUR APPROACHES TO ALLEVIATE WASTE IN THE FOOD CHAIN-TRANSPORT/WHOLESALE

Climacteric fruit are obviously the first major source of ethylene in all compartments along the food chain from wholesale market, receiving area, storage and display shelves in the supermarket. Since ethylene diffuses freely, considerable amounts of ethylene may also remain in these compartments depending on the amount, stage of ripeness and type of fruit and length of storage where climacteric fruit had been stored previously, even when they are emptied. In Australia, as much as 280 ppb ethylene has been found in an empty storage room, which previously held ripe (climacteric) avocado fruit (Wills *et al.*, 2000 [14]) long before the food waste debate in Europe. The new generation of portable instruments enables continuous*in-situ* determination of ethylene, where it is produced, during long haul of fruit, vegetables and flowers on trucks or ships or by exhaust fumes from internal combustion engines.

However, the second considerable source of ethylene may be from internal combustion engines, where substantial movements of tractors, gas or fuel powered forklifts, trucks and their cooling units, which often continue running while the truck’s engine is off, occur in the vicinity of the fruit and vegetables. Quantities of ethylene in three Australian RDCs investigated by Wills *et al.* (2000) [14] were described as “never low” as a basis to identify hot spots of ethylene accumulation in the food chain.

5.1. Cooling

In the food chain, continuous cooling controls fruit ripening and ethylene synthesis, as the simplest and most efficient approach.

5.2. Venting

Similarly, vents can prevent ethylene accumulation around the sensitive fruit and vegetables by introducing fresh ambient air with only 1–6 ppb C_2H_4 on long haul ships, or by the use of ethylene absorbers or packages with different resistance to the four relevant gases.

5.3. Inhibitor of Ethylene Receptors

1-MCP (1-methylcyclopropene; "Smartfresh®"; Table 3) can be used on several commodities such as apples or tomatoes to delay ripening via blocking the ethylene receptors and the ethylene-induced ripening cascade (Sisler and Serek 1997 [10]).

5.4. Ethylene Absorbers and Scrubbers

Ethylene absorbers sachets (Sensitech Ryan Co, Beverly, MA, USA) and clay mineral coated strips (It's Fresh, UK) are both available for bulk or retail application e.g., of kaki (Figure S3) and strawberries (Figure S4) and many other fruits. For the absorbing systems to be effective, the commodity is best packaged either in modified atmosphere (MA) or with the absorbers already at the packhouse, to be re-used by the consumer. They may lose some of their benefits once the package is opened by the consumer, unless reseal able bags are used such as, e.g., "Peakfresh", where clay mineral is integrated in the plastic pack, which then becomes opaque (Figure S5), as a sustainable strategy.

6. EIGHT APPROACHES TO ALLEVIATE WASTE AT THE RETAIL AND CONSUMER LEVEL

6.1. Supermarket

Fresh produce such as fruit and vegetables are sometimes wasted, when they remain un-sold in the store on Saturday evening, particularly in countries where supermarkets close over weekends such as Austria, Germany and Switzerland.

6.2. Canteens and Refectories

Fresh produce such as fruit and vegetables are similarly wasted in canteens at the end of the week (Friday or Saturday) before they close for the weekend, or bank holidays.

6.3. Consumer

Fresh produce often decays and is wasted, when over-bought and expected visitors did not turn up or over a weekend or bank holiday and it remains uneaten; private households are responsible for the largest portion of food waste (Figure 1: 61%).

6.4. Dedicated Recipes for Over-ripe Fruit

Optimization of fruit and vegetable storage and private shopping behavior at home; recipes utilizing ripening perishable food could be a simple but effective answer.

6.5. "Display until" or "Sell-by" Date

In the UK, some of this food waste may also be due to "Display until" and "Sell by" date on packed food items in the supermarkets (Supplement File, Figure S1), which lead to disposal of possibly still edible fruit. This practice is currently disputed and countermeasures are being discussed and enforced in both the UK and Germany (Table 4).

6.6. Sales of Class II Fruit

Since September 2013, Austrian Billa, Merkur and Adeg sell "Wunderlinge" ("odds") (Figure S6), *i.e.*, class 2 apple fruit, carrots and potatoes at a 25% lower price to reduce waste of otherwise industrial or wasted fruit at the producer level and support and encourage regional produce (Blanke and Burdick, 2005 [15]); REWE sells "Kleine Äpfel" ("small apples") in Germany (Supplemental Figure S7).

Concomitantly, Coop in Switzerland started selling class II fruit and vegetables as "Ünique". The great success of this Coop sustainability initiative led to an extension to a fair range of commodities such as bent-cucumbers, carrots, sweet pepper, asparagus, kiwi, lemon, orange, potato and in 2013, hail apricots from the Wallis region in Switzerland (Figures S8 and S9) with their own class II quality standards [16].

6.7. Food Sharing

Food sharing (https://foodsharing.de) is an advent of the social media founded by the film editor Valentin Thurn. In contrast to "Die Tafel" (Chapter 6.8), "Foodsharing" [17] provides the IT, *i.e.*, the Internet platform for private households in five districts in Germany (Hamburg, Berlin, Cologne, Munich and Frankfurt) and Austria's Capital Vienna. Private persons, once registered, can use the "Foodsharing" platform to offer their left-over perishables, e.g., before their holidays or after a weekend, and arrange for regional collection, which sometimes ends up in joint cooking of meals as a positive social side-effect.

Table 4. Sources of food waste and corrective actions at the retail-consumer interface.

Causes	Countermeasure	Corrective Actions to be Taken
EU marketing standards-class I	Whole crops purchase (WCP)	"Purchase of the whole crop of the tree" or from the field
	Deliberate sale of class II fruit	"Wunderlinge" in Billa (Sept. 2013)
"Display until" date (UK)	Abandon this practice	Less waste at super-market level
Food waste	Recipes including (over-) ripe fruit	Recipes including very ripe fruit
Food waste in private households	Food sharing	Left-over perishables in private households are offered for collection on social media
Food Waste at wholesale/retail	Die Tafel (Germany)	Volunteers collect left-over perishables from stores, canteens, airlines, restaurants etc

6.8. 'Die Tafel' in Germany versus 'Food Banks' in the UK

"Die Tafel" (www.tafel.de) [18] successfully collects left-over perishable produce like fruit and vegetables and other food with their own volunteers in *ca*. 1000 nation-wide locations from canteens, restaurants, supermarkets, catering enterprises and airlines, *etc.,* for free distribution in Germany to those in need (1.5 mil people in the last 20 years in Germany) thereby greatly reducing waste of perishable food, whereas the "food banks" in the UK help collect un-wanted, over-bought quantities of non-perishable food.

7. SUCCESSFUL COUNTERMEASURES IN WESTERN EUROPE

The waste of perishable food discussion peaked around 2012 with focused media attention in Germany, UK, Switzerland and Austria. As a consequence, UK supermarkets started discussing the "display before" label and to withdraw multi-pack offers ("Two raspberry punnets for the price of one"), and a few German, Austrian and Swiss supermarkets started to sell class II produce on a small scale. Coop (Switzerland) had to restrict the number of outlets with "Ünique" produce, since demand exceeded offers. Coop sold 180 t class II carrots, 36 t pears in 2014 and 122 t apricots damaged in the hailstorm 2013 (CS 9) [17], which would have otherwise gone to waste in Switzerland.

Food sharing [16] is a new concept mostly for the younger generation, where social media is popular, and has gained popularity in a very short time. 'Die Tafel' [18] successfully collected left-over perishable produce in *ca.* 1000 nation-wide locations for 1.5 mil people in need in the last 20 years in Germany, thereby greatly reducing waste of perishable food. Overall, fruit and vegetable consumption in Europe currently decreases by *ca.* 2% annually, which is partly attributed to more considerate and conservative shopping, thereby reducing food waste. In the UK, WRAP (2013 [8]) similarly reported a 6% decrease in the overall food waste in the last six years and aims at a 50% reduction of the avoidable waste, as stated inTable 2, by 2025.

Compared to the 21% decrease in avoidable household waste in the UK (Table 2) over the last six years (2007–2013), this drop was surprisingly only as little as 4% in vegetables and salads (from 851,600 t to 813,600 t), but as much as 26% with fresh fruit (WRAP, 2013 [8]).

8. CONCLUSIONS

There will still be a proportion of wastage of fresh perishable produce such as fruit and vegetables, due, e.g., to mold despite all our efforts and using some of the upcoming techniques. However, our task should be to reduce fresh produce wastage at all levels from field to fork and may be acknowledged as part of an already existing sustainability label [19] or part of the QRC code on the produce or shelf. The future will show how much each of the different approaches at the three levels can contribute to actually reducing fresh produce waste, but every attempt and idea is a much appreciated and valued effort to safeguard our natural resources including labor, water and land and contribute to reducing GHG emissions, e.g., by use of less nitrogen fertilizer as a source of N_2O.

SUPPLEMENTARY MATERIALS

Figure S1: Tomatoes in a UK high street store with both the "display until" and "best before" date in 2014.

Figure S2: Unnecessary exposure of perishable commodities (lettuce) to sun, freshly delivered lettuce in a med county store left out in the sun (30°C at the time of the photo) where it is prone to wilt, which could simply be prevented by shading.

Figure S3: Kaki shipment with the Ryan ethylene sachets from Sensitech USA in a wholesale box.

Figure S4: Strawberry punnet with the ethylene absorber strip at the bottom from It's Fresh, UK; the pad is added at the pack house and remains active until the consumer's home.

Figure S5: Ethylene absorbing clay minerals integrated in the plastic bag by PeakFresh, Australia, which becomes opaque.

Figure S6: Uneven or odd-shaped fruit like apples and vegetables like carrots sold at a 25% discount price as "Wunderlinge-Vielfalt der Natur" in a Billa store, Austria (©A Schroeckl).

Figure S7: Class II apples of the small-fruited cv. "Gala" as "Kleine Äpfel" in a REWE store in Germany.

Figure S8: Sale of class II bend cucumbers in a Co-OP store in Switzerland (©Ünique, Coop, Switzerland).

Figure S9: Sale of class II "Ünique" hail-hit apricots from the Swiss Wallis region in 2013 (©Coop Switzerland).

ACKNOWLEDGEMENTS

The author is grateful to Professor Daryl Joyce, Australia for the invitation to this concept paper and editors for editorial guidance, the two anonymous referees for their valued suggestions, Andrea Schroeckl, University of Graz for the photo of the "Wunderlinge" in Billa, Austria and David Cooke, St. Austell, UK for revising the English.

CONFLICTS OF INTEREST

The author declares no conflict of interest.

REFERENCES

1. Blanke, M.M. Perspectives of fruit research and apple orchard management in Germany in a changing climate. InActa Horticulturae; Palmer, J., Ed.; ISBN: 978-90-6605-531-5. ISHS: Korbeek-Lo, Belgium, 2007; Volume 772, pp. 441–447.

2. Langley, J. Food for Thought?—A UK pilot study testing a methodology for compositional domestic food waste analysis. Waste Manag. Res. 2010, 28, 220–227.

3. Van Westerhoven, M.; Steenhuisen, F. Bepaling Voedselverliezen bij Huishoudens en Bedrijfscatering in Nederland; CREM: Amsterdam, The Netherlands, 2010. (In Dutch)

4. BELV. Ermittlung der weggeworfenen Lebensmittel und Vorschläge zur Verminderung der Wegwerfrate bei Lebensmitteln in Deutschland; Institute für Abfallwirtschaft der Universitäten Stuttgart-Hohenheim und BoKu Wien: Stuttgart, Germany, 2012; p. 42. (In German)

5. WRAP. Waste and Resources Action Programme, UK Government Watchdog on Waste Household Food and Drink Waste in the UK; Final Report 2009; WRAP: Banbury, Oxon, UK, 2009; ISBN: 1-84405-430-6.

6. Hill, R. Plan A: Product footprint. In Proceedings of the BSI Product Sustainability Conference, Drummond Gate, UK, 28 November 2013.

7. Cofresco Frischehalteprodukte Europa. Das Wegwerfen von Lebensmitteln—Einstellungen und Verhaltensmuster. Ergebnisse Deutschland. Save Food Studie, Available online: http://www.cofresco.de/pdf/Results_Save_Food_Study_Germany.pdf.011 (accessed on 10 March 2015).

8. WRAP. Waste and Resources Action Programme, UK Government Watchdog on Waste; WRAP: Banbury, Oxon, UK, 6 November 2013. Available online: http://www.wrap.org.uk/content/use-your-loaf-and-save-billions (accessed on 30 March 2015).

9. Schaefer, F.; Blanke, M.M. Farming and marketing system affect carbon and water footprint of Hokaido pumpkin. J. Clean. Prod. 2012, 28, 113–119.

10. Sisler, E.C.; Serek, M. Inhibitors of ethylene responses in plants at the receptor level-recent developments. Physiol. Plant. 1997, 100, 577–582.

11. Bufler, G. Exogenous ethylene inhibits sprout growth in onion bulbs. Ann. Bot. 2009, 103, 23–28.

12. Rylski, I.; Rappaort, L.; Pratt, H.K. Dual effects of ethylene on potato tuber dormancy and sprout growth. Plant Physiol. 1974, 53, 658–662.

13. Mattheis, J.P.; Rudell, D.R. Diphenylamine metabolism in "Braeburn" apples stored under conditions conducive to the development of internal browning. J. Agric. Food Chem. 2008, 56, 3381–3385.

14. Wills, R.B.H.; Warton, M.A.; Ku, V.V.V. Ethylene levels associated with fruit and vegetables during marketing. Aust. J. Exp. Agric. 2000, 40, 485–470.

15. Blanke, M.M.; Burdick, B. Food (miles) for thought-primary energy balance for locally-grown versus imported apple fruit. Environ. Sci. Pollut. Res.–ESPR 2005, 12, 125–127.

16. Coop vergrössert Ünique Sortiment. Coop Suisse 2015. Available online: http://www.coop.ch/pb/site/search/search/2057/Lde/index.html?qs=%C3%BCnique&fr=coop2012&search=search&backend=backend_coop2012&la=de&_sid=665f691e-1a51-45ed-9aa5-3938c44b8f68 (accessed on 25 June 2015).

17. Foodsharing, 2015. Available online: https://foodsharing.de (accessed on 2 April 2015). (In German).

18. Die Tafel. Available online: www.tafel.de (accessed on 3 February 2015). (In German).

19. Schaefer, F.; Blanke, M.M. Opportunities and challenges for carbon footprint, climate or CO2 labels for horticultural products [-Nachhaltige Kennzeichnung klimafreundlicher Lebensmittel]. Erwerbs-Obstbau 2014, 56, 73–80.

Chapter 9

Emerging Perspectives on the Natural Microbiome of Fresh Produce Vegetables

Colin R. Jackson [1,*], Bram W. G. Stone [1] and Heather L. Tyler [2]

[1]*Department of Biology, the University of Mississippi, University, MS 38677, USA*

[2]*Crop Production Systems Research Unit, USDA Agricultural Research Service, Stoneville, MS 38776, USA*

ABSTRACT

Plants harbor a diverse microbiome existing as bacterial populations on the leaf surface (the phyllosphere) and within plant tissues (endophytes). The composition of this microbiome has been largely unexplored in fresh produce vegetables, where studies have tended to focus on pathogen detection and survival. However, the application of next-generation 16S rRNA gene sequencing approaches is beginning to reveal the diversity of this produce-associated bacterial community. In this article we review what is known about the composition of the microbiome of fresh produce vegetables, placing it in the context of general phyllosphere research. We also demonstrate how next-generation sequencing can be used to assess the bacterial assemblages present on fresh produce, using fresh herbs as an example. That data shows how the use of such culture-independent approaches can detect groups of taxa (anaerobes, psychrophiles) that may be missed by traditional culture-based techniques. Other issues discussed include questions as to whether to determine the microbiome during plant growth or at point of purchase or consumption, and the potential role of the natural bacterial community in mitigating pathogen survival.

Keywords: microbiome; bacteria; salad produce; herbs; phyllosphere; endophytes; 16S rRNA

1. INTRODUCTION

The last few decades have seen rapid growth in the fields of microbial ecology and environmental microbiology, such that natural microbial communities are now recognized as being much more diverse than previously thought. While much of the early research in this field focused on using molecular techniques to characterize the diversity of microbial assemblages in soils and waters; from the late 1990s onwards there has been an increased awareness of the natural microbiota associated with plants. Just as there has been a research focus on determining the make-up of the complex microbiome of animals (and especially mammals) and its role in health and disease, the importance of the plant microbiome is beginning to be examined, whether in the context of its role in plant growth and disease resistance, or in the context of human consumption of plant material (e.g., [1]).

The natural bacterial populations associated with plants are present on both aboveground (stem, leaves, flowers, fruits,*etc.*) and belowground (roots, tubers, *etc.*) structures. That roots have an associated microbial community (the rhizosphere) has long been recognized, and aboveground portions of plants (primarily leaves) also have such a community (the phyllosphere). While agricultural scientists are likely aware of the potential for microorganisms to be found naturally on the plant surface, the idea that plants also contain many microbial populations living inside of them (endophytes) is less well known. However, virtually every study that has tested for the presence of endophytes inside of plant samples has found them, so it is likely that every plant species is colonized by at least one endophytic bacterial species. While most endophytes are likely to be non-pathogenic to humans, a number of pathogenic bacteria can become internalized as at least temporary endophytes within leaves. These and other endophytes present an interesting problem from the viewpoint of consumer safety as no amount of washing or vegetable preparation will remove them, and while this is not a potential issue with the consumption of cooked vegetables, it may be with the consumption of raw vegetables in salads. Here we review the state of our current knowledge of the microbial assemblages associated with produce vegetables and present perspectives on how these assemblages relate to consumer health. We also demonstrate how next-generation sequencing technologies can provide valuable insights into the composition of these assemblages, paying particular attention to an understudied vegetable type, fresh herbs.

2. GENERAL CHARACTERISTICS OF PHYLLOSPHERE BACTERIAL COMMUNITIES

Leaves of plants present an area that is available for potential microbial colonization and growth, whether they are on the leaf surface (the phylloplane) or within the leaf itself as endophytes. As such, microbial populations in or on leaves can be abundant, and the phyllosphere community can be taxonomically diverse [2]. The leaf surface itself can be a harsh environment for microbial growth, with microorganisms being potentially exposed to extreme fluctuations in moisture availability, ultraviolet variation, temperature, and nutrient availability [2,3]. However, phyllosphere adapted bacteria may be capable of modifying their local environment, increasing the leakage of nutrients from the host plant and producing extracellular polysaccharides to resist desiccation and help attachment [4]. Studies of the phyllosphere using traditional culture-based approaches have been a part of agricultural and environmental microbiology since at least the 1950s, although for obvious reasons many studies have tended to focus on the detection of known plant pathogens (e.g., species of*Xanthomonas*, *Erwinia* and *Pseudomonas*) or human pathogens from a food safety standpoint. However, the development and application of molecular (largely 16S ribosomal RNA) approaches in microbial ecology, has led to a more thorough understanding of the diversity and complexity of leaf-associated bacterial communities.

Building upon Carl Woese's work demonstrating that ribosomal RNA (rRNA) gene sequences could be used to determine bacterial phylogeny [5], microbial ecology underwent a dramatic shift in the 1980s and 1990s in that it became feasible to directly obtain bacterial DNA from environmental samples, polymerase chain reaction (PCR)-amplify bacterial 16S rRNA genes, and then to use those genes to examine bacterial diversity either through DNA sequencing or through techniques such as denaturing gradient gel electrophoresis (DGGE) or terminal restriction fragment length polymorphism (T-RFLP) analysis. Such analyses revealed that microbial communities were much more complex than previously believed and that perhaps as few as 1% of bacteria had been identified using traditional culture techniques [6,7]. Initially, these molecular microbial ecology studies focused on marine systems (likely because of easier analysis), with a greater emphasis on soils from the mid-1990s onwards [8]. However, despite an increased interest in the microbial diversity of soils from an agricultural standpoint, even including the plant rhizosphere [9,10,11,12,13]; initially, only a few studies applied this concept to examine the natural microbial diversity on the aboveground structures of plants.

Yang *et al.* [14] used DGGE to examine the natural bacterial communities in the phyllosphere of seven agriculturally important plant species: cotton (*Gossypium hirsutum*), corn (*Zea mays*), sugar beet (*Beta vulgaris*), green bean (*Phaseolus vulgaris*), Valencia and navel oranges (both *Citrus sinensis*), and the grapefruit-pomelo hybrid OroBlanco (*C. grandis* × *C. paradisi*). DGGE banding patterns indicated that leaves sampled from different individuals of the same plant species tended to show similar bacterial community profiles, but the profiles for the different species were generally unique. While the phyllosphere communities on the two orange varieties examined (Valencia, navel) were similar, they were very different from those on the other citrus variety (OroBlanco) [14]. Sequence analysis of 17 of the dominant DGGE bands obtained from the Valencia orange samples showed that only four (identified as *Enterobacter agglomerans*, *Bacillus pumilus*,*Acinetobacter* sp., and a member of the Cytophagales) represented established phyllosphere taxa. The other sequences included representatives of the Alpha-, Gamma-, and Deltaproteobacteria that had not previously been reported to inhabit the phyllosphere. This was in contrast to the results of the same analysis conducted on bacteria growing in mixed culture in BIOLOG plates, which were predominantly members of *Pseudomonas*, *Erwinia*, *Acinetobacter*, and *Enterobacter* [14]; all genera that are recognized phyllosphere inhabitants based on traditional culture techniques. Thus, using culture-independent 16S rRNA based methods revealed a phyllosphere community that was very different, and presumably more diverse, than the one that would have been determined through culture-dependent methodologies.

The last 10–15 years has seen a rapid increase in the application of rRNA-based methods to describe leaf-associated assemblages. DGGE and other approaches have been used to characterize the bacterial community in the phyllosphere of food crops, such as corn (*Zea mays*) [15], pepper (*Capsicum annuum*) [16], cucumber (*Cucumis sativus*) [17], and spinach (*Spinacia oleracea*), celery (*Apium graveolens*), broccoli (*Brassica oleracea var. italica*), and cauliflower (*Brassica oleracea var. botrytis*) [18]. Similarly, the diversity and structure of the bacterial community in the phyllosphere of a number of tree species has been described [19,20,21,22]. More than basic descriptions of community structure, 16S rRNA techniques have also been used to examine how the phyllosphere community is influenced by environmental factors, such as ultraviolet light [15] and rainfall [23], as well as pesticide [24,25] or biological control agent [16,26] application. At an ecological level, both spatial [20] and temporal [21] patterns in the composition of the phyllosphere have been examined. Thus, while our understanding of the phyllosphere initially lagged behind our knowledge of soil or rhizosphere microbiology, these, and other, culture-independent studies are leading us to insights into both the composition of leaf-associated microbial communities, and how these communities interact with the host plant [27,28].

3. BACTERIAL COMMUNITIES ASSOCIATED WITH LEAVES OF SALAD PRODUCE VEGETABLES

Fresh produce vegetables such as lettuce, spinach, *etc.* present an example of leaves that may be consumed by humans without the use of preparation methods that would remove or kill the associated bacterial community. As such, studies of phyllosphere or endophytic bacterial assemblages on these plants are interesting from both a basic scientific and applied aspect. From an applied viewpoint, the presence of potential human pathogens is of interest given that these are not likely to be removed prior to consumption. Most studies of bacteria in produce have used culture-dependent methods to determine the culturable pathogenic populations present, or have examined the recovery of bacteria following intentional exposure to pathogens such as *Escherichia coli* O157:H7, *Salmonella*, or *Listeria* [29,30,31,32,33]. Direct visualization of bacterial cells in the phyllosphere through microscopy has also been used to examine the microbial colonization of leafy vegetables [29,32,33,34]. Together, these studies indicate that pathogenic bacteria are capable of colonizing leaf surfaces and internal tissues (*i.e.*, they can be part of both phyllosphere and endophytic assemblages), and thus are not necessarily removed by washing. Both *Salmonella* and *E. coli* have been found to be capable of becoming endophytic and colonize the interior tissues of plants [30,33,35,36,37,38], although the ability to colonize and persist within the leaf varies between specific strains of bacteria. The ability of these human pathogens to survive and persist on the leaf surface is also less than that of the naturally occurring bacterial populations that reside there. For example, *Salmonella enterica* serovar Thompson has been shown to be capable of colonizing the phyllosphere of the herb cilantro (*Coriandrum sativum*), but was not as resilient to environmental conditions on the leaf surface as native bacterial epiphytes [32]. Similarly, plant-associated *E. coli*strains that were specifically isolated from cabbage roots were more proficient at colonizing the surface of alfalfa sprouts than an O157:H7 clinical isolate [39]. These examples highlight the importance of considering the naturally occurring microbiome on leafy produce, as it may be more persistent than, and even interact with, potential pathogens.

Most of the earlier studies on the presence and persistence of potential human pathogens on salad produce relied on the use of selective microbiological culture media and biochemical tests for the identification of bacterial isolates [29,32]. However, as with general microbial diversity studies, this has been supplemented by culture-independent molecular approaches using 16S rRNA gene amplification and subsequent community profiling by techniques, such as DGGE, T-RFLP, DNA sequencing, and/or microarrays [34,40]. More recently, metagenomic or next-generation sequencing approaches have been used [1,41,42,43].

An interesting question regarding the microorganisms present on consumable produce is when to sample, or even broader, at what time point do we say that the leaf-associated bacteria that are present are the representative bacterial community? The naturally occurring microbiome that is associated with the plant (*i.e.*, the true phyllosphere and endophyte community) should probably be defined as the microbial populations that are present on or in vegetables growing in the field. However, from the standpoint of the consumer, the microbial populations present at the point of sale, or even the time of consumption, may be more relevant, and both phyllosphere and endophytic microorganisms may differ at these different time points. DGGE of 16S rRNA genes showed reduced bacterial diversity in the phyllosphere of lettuce (*Lacutuca sativa var. capitata*) samples collected from grocery stores compared to samples taken directly from a farm field site, suggesting a simplification of the phyllosphere community during processing of salad vegetables [44]. Furthermore, commercially pre-bagged, refrigerated versions of the same lettuce samples showed evidence of the presence of additional bacterial populations (notably *Pseudomonas libaniensis*) [44] indicating the potential for either contamination or the selective growth of specific bacterial populations during processing and storage. A similar observation has been made for spinach (*Spinacia oleracea*), for which numbers of culturable Pseudomonadaceae and Enterobacteriaceae were found to increase after 12 days of cold storage [29]. The latter study is probably a more reliable assessment of changes that occur post-harvest and following storage, as changes in DGGE profiles or in the composition of 16S rRNA gene clone libraries represent a proportional change in community composition not a change in the absolute number of organisms. Thus, for example, while the presence of *Pseudomonas libaniensis* in clone libraries generated from lettuce samples following refrigerated storage [44] could indicate the growth of that species, it could also indicate a reduction in overall bacterial numbers following refrigeration, especially of more dominant taxa, so that *P. libaniensis* could only then be detected. Regardless, it seems reasonable to assume that the microbiota on vegetables are likely to change following harvest, and while the microbiome present in fresh produce vegetables at the time of purchase and consumption may not entirely be indicative of that present in the growing plant, it does represent the microbial populations to which consumers are potentially exposed, and may be more relevant to study.

Two recent studies took that viewpoint and used 16S rRNA techniques to survey the bacterial populations present in the phyllosphere [41] or in both the phyllosphere and internalized within the leaves [1] of vegetables obtained directly from grocery stores/supermarkets. Both report 16S rRNA gene sequences identified as representing members of the Gammaproteobacteria to be the most abundant sequence types detected, with the Enterobacteriaceae being particularly prevalent. Sequences identified as Enterobacteriaceae were proportionally much more abundant on fresh salad vegetables such as lettuce and spinach, than they were on apples, peaches or grapes [41]. Clustering 16S rRNA gene sequences together using a 97% similarity criterion to define operational taxonomic units (OTUs; a surrogate for species), suggested the presence of around 40–70 different OTUs or species on the lettuce or spinach leaf surface [1,41]. Thus, while not as diverse as soils or natural waters, the leaf-associated community of consumable fresh produce is quite diverse, and consumers are likely being exposed to 50 or more species of bacteria during salad consumption. While many of these populations are likely to be plant symbionts or pathogens, some of the bacterial genera and species detected in those studies include strains that can be human pathogens [1].

Taken together, these studies suggest that there are some families and genera of bacteria that are commonly found in the phyllosphere of leafy salad vegetables. Members of the Enterobacteriaceae have been consistently detected, and are often among the most abundant members of the phyllosphere community [34,40,41,42,43]. Members of the genus *Pseudomonas*have been found to be among the dominant bacterial populations in spinach [42] and lettuce [40,43], as well as in bagged salad mixes containing lettuce, red cabbage, and carrots [34]. However, species of *Pseudomonas* are not always dominant members of the phyllosphere, as shown in the survey of Leff and Fierer [41] which found that the entire family Pseudomonadaceae accounted for less than 5% of the bacterial 16S rRNA gene sequences obtained from spinach or lettuce, compared to Enterobacteriaceae which accounted for 38% (lettuce) or 58% (spinach) of the sequences obtained. This same variability is seen in the less abundant bacterial populations detected in phyllosphere studies, where certain bacterial genera can account for >1% of the community in some studies but are not detected at all in others. Thus, despite the generally widespread occurrence of the Enterobacteriaceae and, to a lesser extent, the genus *Pseudomonas*, there does not appear to be a consistent core set of taxa that are found in the phyllosphere of all produce vegetables.

While there may be no consistent core taxa, environmental conditions during produce growth and storage likely impact the leaf-associated bacterial community. Although they are almost always abundant, Enterobacteriaceae have been reported as being more prevalent in the phyllosphere of conventionally grown spinach and lettuce compared to those under organic cultivation [41]. At least that was the case for produce sampled at the point of purchase: the opposite trend was reported when lettuce was sampled after being freshly harvested, when organically grown plants showed higher numbers of Enterobacteriaceae than conventionally grown ones [45]. However, these two studies may not be entirely comparable as they differed in the style of analysis used (culture-dependent *vs.* molecular) as well as in the time of sampling. A comparison of the plant-associated bacterial communities on six varieties of organically and conventionally grown salad produce sampled at point of purchase revealed no consistent differences in community composition between the two growth approaches [1]. Regardless of the influence of cultivation style, other environmental variables during growth such as soil moisture, organic content, or nutrient (fertilizer) availability could also affect microbial community composition.

As previously mentioned, storage undoubtedly has an influence on the leaf-associated microbial community. When fresh-cut spinach was stored at 10 °C, numbers of both Enterobacteriaceae and Pseudomonadaceae increased at least 1000-fold over 12 days [29]. Storage under refrigeration also lowered the diversity and richness of the phyllosphere community on spinach [42], and the temperature of storage influenced how much the community changes over the storage period. After 15 days of storage, spinach samples that were held at 15 °C harbored bacterial communities that were more similar to the original microbiome than samples that were stored at 10 °C [42]. Changes in the bacterial community associated with bagged lettuce mixes have also been reported following storage at 10 °C, with an increase in the relative abundance of Enterobacteriaceae and a decrease in the relative abundance of *Pseudomonas* [34]. Interestingly, when the bagged lettuce mixes were stored at refrigerator temperature (4 °C), the decrease in *Pseudomonas* was less pronounced and this genus was still the dominant taxonomic group, as it was prior to storage [34]. Thus, refrigerated storage might help retain the natural microbiome, while extended storage at cool, but not cold, temperatures might be more likely to promote shifts in the phyllosphere community, and potentially favor pathogenic strains.

4. BACTERIAL COMMUNITIES ASSOCIATED WITH CONSUMABLE HERBS: A DEMONSTRATION OF NEXT-GENERATION SEQUENCING APPROACHES

Studies of microbial diversity, even those using culture-independent molecular techniques, are limited by the ability of the technique used to adequately sample the microbiome in question. With thousands to millions of bacteria cells present on a leaf surface, even 16S rRNA gene based methods such as DGGE or cloning and Sanger sequencing of amplified 16S rRNA gene fragments only reveal a miniscule subset of the bacteria that are present. The last few years, however, have seen the emergence of next-generation sequencing technologies that allow a far more in-depth profiling of bacterial assemblages to occur. Such approaches can readily yield thousands or tens of thousands of partial 16S rRNA gene sequences from a single sample of plant material (often from just 0.1–0.5 g of material), providing much more coverage of the community present, and increasing the ability to detect less common taxa. Such approaches were used by the previously mentioned studies of fresh produce at the point of consumer purchase [1,41] suggesting that those studies may well be better assessments of the bacterial component present on fresh vegetables than those before. That said, even those studies only assessed around 2000 [1] or 200 (this lower amount because more vegetables were sampled) [41] sequences, largely because they were based around the Roche 454 next-generation sequencing platform. Illumina next-generation sequencing has replaced 454 as the procedure of choice, and emerging protocols on that platform now facilitate more in-depth sampling of microbiome structure [46,47]. This approach is beginning to be used to characterize various plant-associated bacterial communities, with studies being reported for the phyllosphere of neo-tropical rainforest trees [48] and wetland plants [49]. However, other than a recent study examining how pesticide application can change the phyllosphere of tomatoes [50], there have been few, if any, studies that have used Illumina next-generation sequencing to characterize the bacterial communities on fresh fruit or vegetables. Here, we demonstrate the use of this platform to describe the natural bacterial assemblages present on a generally ignored subset of fresh vegetables, fresh herbs, and suggest some guidelines for using next-generation sequencing approaches in this context.

4.1. Sample Collection, Processing, and 16S rRNA Gene Illumina Sequencing

Samples of six fresh herbs were purchased from an Oxford, Mississippi, USA, grocery store in September 2014. Herbs were basil (*Ocimum basilicum*), chives (*Allium schoenoprasum*), dill (*Anethum graveolens*), mint (*Mentha spicata*), rosemary (*Rosmarinus officinalis*), and thyme (*Thymus vulgaris*). All were packaged in polymer film-sealed plastic containers, labeled as products of the USA. For each herb type, three representative leaves/fronds were sampled aseptically and 0.1 g weighed and used for DNA extraction. DNA was extracted from the leaf sample using a PowerPlant DNA Isolation kit (Mo Bio Laboratories, Carlsbad, CA, USA). Thus, 16S rRNA gene sequences obtained reflect those of bacterial populations in the phyllosphere and potential endophytes. An alternative approach would have been to remove bacterial cells from the leaf surface prior to extracting DNA from them, and this has been done for some studies [41,42,43,44]. While that approach has the advantage of minimizing the co-amplification of chloroplast and mitochondrial sequences (which typically can amplify with bacterial specific primers), it prevents the sampling of endophyte populations, which may be as relevant as phyllosphere populations in regards to food safety. Furthermore, for this specific study, removing bacteria from the leaf surface would have been difficult given the small size of the particular leaf samples used. Given the large number of sequences per sample that can be generated using next-generation sequencing, we recommend extracting DNA from the intact plant material rather than leaf washes, as the ability to detect the entirety of the bacterial community associated with the plant (*i.e.*, both phyllosphere and endophyte populations) likely outweighs the potential loss of some data to chloroplast or mitochondrial sequence reads.

A dual-index barcoding approach was used for Illumina next-generation sequencing whereby each sample was amplified with bacterial specific 16S rRNA gene primers, each tagged with a specific 8-nucleotide barcode [47]. These primers amplify a 250 nucleotide long region of the V4 variable region of the 16S rRNA gene. The specific amplification approach used involves a single round of PCR, thereby limiting the risk of amplification artifacts [47]. Following amplification, amplicons from all samples were pooled, and the assembled library spiked with 5% PhiX to increase nucleotide base diversity prior to sequencing (a necessary step for bacterial community analysis by this method). The final library was sequenced on an Illumina MiSeq instrument, via two index sequencing reads, at the University of Mississippi Medical Center (UMMC) Molecular and Genomics Core Facility. Illumina MiSeq-based sequencing was chosen as it reflects a suitable balance of a large number of sequence reads per run (and therefore per unit cost), the ability to obtain high quality data, and the ability to obtain reads in excess of 200 nucleotides long, thereby facilitating more accurate sequence classification. While the Illumina HiSeq or NextSeq systems can provide higher sample throughput, the MiSeq system is currently more commonly used for 16S rRNA gene studies, and at a cost at or below U.S. $100,000 is more suitable for individual investigators or testing laboratories [46,47].

4.2. Data Processing and Bioinformatics Pipeline

Raw sequence data (fastq files) were downloaded and accessed via the 16S rRNA bioinformatics software package mothur [51,52]. Sequences were processed along a bioinformatics pipeline following the general guidelines of Kozich *et al.*[47], which are optimized for Illumina MiSeq-generated 16S rRNA gene data. Briefly, contigs were assembled using forward and reverse reads and screened to only include those with a maximum length of 275 bp and no base ambiguities (*i.e.*, sequence contigs had to have identical base calls for both the forward and reverse read). Sequences were aligned against the SILVA 16S rRNA database [53] and misaligned sequences deleted. Sequences were clustered together by roughly 1% sequence similarity to account for potential amplification and sequencing errors, and chimeras removed using UCHIME [54]. Valid sequences were classified using the Greengenes [55] 16S rRNA classification scheme, and erroneous (archaeal, eukaryotic) sequences removed. Remaining bacterial sequences were grouped into operational taxonomic units (OTUs) based on >97% sequence similarity. As stated, these procedures follow published recommendations for this type of data [47], which are also routinely updated online [56]. These procedures were developed after a series of studies that assessed issues such as the impacts of clustering level, chimera removal procedure, alignment and classification reference, and OTU similarity criterion on bacterial community analysis [56] and we have found them suitable for sequences amplified from a range of samples including plant material, soils, and waters.

4.3. Taxonomic Composition of Bacterial Communities on Fresh Herbs

The microbiome of each of the herbs examined was generally dominated by members of the Firmicutes, Bacteroidetes, and Proteobacteria (Table 1). Of the latter, Alpha- and Gamma- sub-phyla were the most prevalent. However, the proportional abundance of each of these phyla in the sequence dataset varied between herb species with members of the Firmicutes being more prevalent in/on basil, rosemary, and thyme, and the Proteobacteria (especially Gammaproteobacteria) being more prevalent in/on dill and, to a lesser extent, chives (Table 1). A substantial proportion of the sequences recovered from mint were also classified as Proteobacteria, although these were Alphaproteobacteria and their proportional abundance was highly variable between samples (Table 1). Other major bacterial groups that were detected in relatively common proportions (accounting for >1% of the sequence dataset) were the Bacteroidetes (accounting for an average of 6.1%–17.9% of sequences, depending upon the herb species) and Actinobacteria (an average of 0.3%–6.5% of sequences).

Table 1. Major phyla and subphyla of bacteria associated with prepackaged fresh herbs sampled from a grocery store in Mississippi, USA, as determined from high throughput 16S rRNA gene sequencing. Numbers represent mean (+/− SE) percentage of total sequence reads obtained from 3 samples per herb that were identified as that phylum.

Phylum	Basil	Chives	Dill	Mint	Rosemary	Thyme
Acidobacteria	0.02 (0.01)	0.07 (0.07)	0.10 (0.04)	0.04 (0.03)	0.01 (0.00)	0.01 (0.01)
Actinobacteria	0.26 (0.04)	1.00 (0.13)	6.52 (1.46)	0.74 (0.10)	0.36 (0.05)	0.67 (0.07)
Bacteroidetes	6.1 (1.26)	13.7 (7.14)	6.1 (2.73)	17.9 (3.34)	12.2 (1.28)	13.1 (1.21)
Firmicutes	84.2 (1.87)	51.6 (24.19)	13.4 (5.33)	48.8 (23.94)	82.2 (1.02)	81.5 (0.74)
Lentisphaerae	0.02 (0.01)	0.08 (0.04)	-	0.13 (0.07)	0.07 (0.01)	0.10 (0.02)
Proteobacteria	9.2 (3.08)	28.9 (26.74)	73.6 (4.32)	31.99 (27.29)	5.05 (1.05)	4.40 (0.54)
Alpha-	1.41 (0.48)	0.22 (0.17)	2.08 (0.52)	25.56 (22.82)	2.36 (0.58)	0.15 (0.05)
Beta-	0.28 (0.04)	0.78 (0.29)	1.12 (0.36)	2.65 (1.60)	0.46 (0.09)	0.51 (0.04)
Gamma-	7.29 (2.63)	27.70 (26.94)	70.33 (4.10)	3.40 (3.01)	1.89 (0.51)	3.40 (0.58)
Delta-	0.17 (0.02)	0.23 (0.12)	0.02 (0.00)	0.37 (0.14)	0.33 (0.04)	0.34 (0.07)

The total number of valid bacterial sequences obtained from each individual sample averaged 18,140, but ranged from 5129 to 47,548. Compared to prior molecular surveys of bacterial assemblages on fresh vegetables, this represents a substantial increase in sequencing depth, but raises the question as to how much sequence data is needed to accurately assess the phyllosphere and endophyte communities associated with such samples. From analyses of Roche 454 next-generation sequencing datasets, 1000 valid sequences have been found to capture 90% of the patterns in community structure between samples (beta-diversity), while 5000 sequences were necessary to accurately predict total diversity within a sample (alpha-diversity) [57]. However, those numbers were derived from analyses of sediment samples, and bacterial communities in sediments are generally more diverse than those associated with plants. Thus, while we recommend a minimum of 5000 sequences to be analyzed from a sample in order to accurately assess the bacterial community present, in practice 2000–3000 sequences is likely to be enough to obtain a reasonable determination of community composition for plant samples. Rarefaction curves of the number of distinct OTUs observed as a function of sampling effort suggested that similar sequencing depth (2000–5000 sequences per sample) was sufficient to describe the bacterial assemblages associated with fresh salad produce [1]. However, the more reads per sample the greater the ability to detect rare taxa, and if the goal is to detect specific bacterial populations (e.g., pathogens) that may account for a minor proportion of the total bacterial community, then greater sequencing depth may be needed.

The dominant specific bacterial populations (as determined from proportional abundance of each OTU) associated with the fresh herb samples included various members of the Firmicutes, Bacteroidetes, and Proteobacteria (Table 2). An unclassified member of the Bacillaceae represented by far the most numerous sequence type followed by sequences identified as *Streptococcus* sp. and *Prevotella copri*. Other major OTUs included *Pseudomonas* sp., *Bacillus cereus*, and*Bacteroides* sp. (Table 2). While some sequences could clearly be identified to the species level, others could only be identified to genus, or even less resolved (e.g., the dominant unclassified Bacillaceae, as well as unclassified members of the Enterobacteriaceae and Planococcaceae). This highlights a potential limitation of next-generation sequencing in that sequence read lengths are currently limited to a few hundred bases (in this case 250 bp). However, even with that limitation some patterns are apparent. For example, a number of well-represented OTUs were identified as taxa that are obligate anaerobic bacteria (e.g., *Prevotella copri*, *Bacteroides* sp., *Phascolarctobacterium* sp., *Catenibacterium* sp.). These taxa represent bacteria that are typically found in the human large intestine, and were initially described after isolation from fecal samples [58,59,60]. Thus, their presence in prepackaged fresh herbs suggests that those herbs show evidence of past fecal contamination. Regular microbiological culture methods would not necessarily have detected these organisms because of their anaerobic culture requirements, demonstrating that while culture-independent methods, such as next-generation sequencing may have limitations, they can be a useful tool in detecting the presence of intestinal microbial populations.

Table 2. Identities of the most prevalent bacterial populations (OTUs) associated with prepackaged fresh herbs (basil, chives, dill, mint, rosemary, thyme) sampled from a grocery store in Mississippi, USA, as determined from high throughput 16S rRNA gene sequencing. Number of reads represents the total number of sequence reads identified as that taxon, out of a total of 392,272 reads.

OTU	Phylum (Class)	Number of Reads
Unclassified Bacillaceae	Firmicutes (Bacilli)	204,483
Streptococcus sp.	Firmicutes (Bacilli)	22,984
Prevotella copri	Bacteroidetes (Bacteroidia)	22,597
Pseudomonas sp.	Proteobacteria (Gamma)	15,955
Bacillus cereus	Firmicutes (Bacilli)	12,823
Bacteroides sp.	Bacteroidetes (Bacteroidia)	10,045
Phascolarctobacterium	Firmicutes (Clostridia)	6124
Bacteroides sp.	Bacteroidetes (Bacteroidia)	5231
Psychrobacter sp.	Proteobacteria (Gamma)	4293
Unclassified Enterobacteriaceae	Proteobacteria (Gamma)	4240
Faecalibacterium prausnitzii	Firmicutes (Clostridia)	4232
Unclassified Planococcaceae	Firmicutes (Bacilli)	3485

Table 2. *Cont.*

OTU	Phylum (Class)	Number of Reads
Methylobacterium adhaesivum	Proteobacteria (Alpha)	2821
Catenibacterium sp.	Firmicutes (Erysipelotrichi)	2738
Pseudomonas veronii	Proteobacteria (Gamma)	2702
Sphingomonas sp.	Proteobacteria (Alpha)	2572
Agrobacterium sp.	Proteobacteria (Alpha)	2546
Sphingomonas sp.	Proteobacteria (Alpha)	2445
Sphingomonas sp.	Proteobacteria (Alpha)	2327
Unclassified Rikenellaceae	Bacteroidetes (Bacteroidia)	1950
Arthrobacter psychrolactophilus	Actinobacteria (Actinobacteria)	1943
Portiera aleyrodidarum	Proteobacteria (Gamma)	1868
Exiguobacterium sp.	Firmicutes (Bacilli)	1733
Sutterella sp.	Proteobacteria (Beta)	1633
Alistipes putredinis	Bacteroidetes (Bacteroidia)	1600

Three taxa (*Psychrobacter* sp., *Arthrobacter psychrolactophilus*, *Exiguobacterium* sp.) detected as important OTUs were bacterial species that are psychrophilic or psychrotolerant (*i.e.*, cold tolerant). The presence of sequences affiliated with these groups likely arises from the refrigerated storage of the produce sampled, as they are unlikely to be abundant in the natural herb phyllosphere. This confirms the accepted idea that the bacterial community associated with produce likely does change during processing and storage, with an increased prevalence of psychrotolerant groups following refrigeration. Indeed, given that the herbs sampled were packaged in plastic containers sealed with polymer file, the high prevalence of OTUs that would be classified as anaerobic likely also reflects microaerophilic storage conditions, and while those OTUs may have originated from fecal contamination, their relative abundance appears to have increased during storage. This provides some support for the argument to sample at point of sale, rather than at point of growth, if the goal is to determine the bacterial community present on produce from a consumer standpoint [1,41]. Further supporting the idea that changes in the herb phyllosphere and endophytic communities must have occurred during processing and storage is the finding that relatively few of the dominant OTUs (essentially just species of *Methylobacterium*, *Agrobacterium*, and*Sphingomonas*) would be regarded as primarily plant-associated bacteria.

4.4. Overall Summary and Conclusions of the Example Study

Overall, the results of the herb study demonstrate that next-generation sequencing technologies such as Illumina 16S rRNA gene sequencing can serve a useful function in the analysis of bacterial communities associated with fresh produce such as herbs. These approaches are rapidly becoming more affordable and easier to perform, and in terms of labor intensity tend to be more efficient than culture-based approaches. In terms of time expenditure, processing and DNA extraction from all herb samples was accomplished with 1–2 days of purchase, and could realistically be completed within a few hours by a dedicated technician. Illumina 16S rRNA gene library preparation takes another few days, with the actual sequencing run taking 24–48 h. Thus, with a dedicated laboratory, it would be possible to generate complete community data in <2 weeks. In this particular study, the bacterial assemblages associated with herbs were found to be dominated by members of the Firmicutes and Gammaproteobacteria, although there was some variation between the particular herbs sampled. Plant-associated taxa were relatively scarce, supporting previous findings [34,42] that phyllosphere and endophyte communities can change during storage, whether from refrigeration or being packaged in modified atmospheres. These changes appeared to result in a higher presence than expected of 16S rRNA sequences identified as being from psychrophilic organisms, and a substantial number of sequences classified as coming from anaerobic taxa. The latter finding is particularly alarming as many of the genera identified (e.g., *Bacteroides*, *Catenibacterium*,*Phascolarctobacterium*, *Prevotella*) are associated with the human large intestine so are typically only detected in fecal samples. Thus, the herbs likely were subject to fecal contamination at some point, and these bacterial populations increased during the storage period. Given the obligate anaerobic nature of these taxa, they would not have been detected by routine microbiological culture, demonstrating that the use of culture-independent approaches can provide useful insights into bacteria present on fresh produce.

5. CONCLUSIONS

Assessing the natural microbiome of consumable plants is important, as natural microbial assemblages may reduce the likelihood of pathogen colonization or survival. For example, reduced levels of *Salmonella enterica* colonization have been observed in lettuce that has a more diverse endophyte community [35]. The mechanism for this pathogen reduction with increased endophyte diversity is not conclusively known, but may well be an increased likelihood of antagonists to *S. enterica* being present with increased overall diversity. Additional support for this concept comes from a study examining the viability of *E. coli* O157:H7 on romaine lettuce, where the phyllosphere bacterial diversity in plants that had culturable*E. coli* O157:H7 cells differed from that on plants where the *E. coli* was no longer viable [61]. That native plant-associated microorganisms can act as competitors to potential human pathogens such as *Salmonella* species and *E. coli* O157:H7 has been shown in lettuce and alfalfa sprouts [62,63], suggesting that even in the absence of specific antagonistic interactions, natural phyllosphere and endophytic communities may limit the presence and abundance of pathogenic bacteria by simply outcompeting them in the living plant. Determining the structure of these communities might therefore provide insights into produce-borne outbreaks of disease, and even lead to the development of tools to assess the likelihood of these outbreaks occurring [1]. Intentional addition of competitive native microbiota has even been proposed as a potential method to reduce enteropathogen contamination of fresh produce [64].

Whether it is most important to determine the microbiome of fresh produce vegetables in the field or at point of sale/consumption is debatable. From the viewpoint of the potential for the natural microbial community to mitigate the growth and persistence of pathogenic bacteria, then the composition of the phyllosphere community in the field may be most important. However, a number studies have shown that the composition and diversity of this community changes during processing, handling, and storage [29,34,42,44], so it could be questioned as to whether the field community is as important if exposure to pathogens occurs post-harvesting. From the viewpoint of the consumer, the bacterial populations that are in or on the produce at point of sale, or more specifically the point of consumption, are the most relevant, as these are the bacteria to which they are being exposed. Interestingly, there have been few, if any, studies that have determined the changes that may occur in produce-associated microbial communities in the time between purchase and consumption, even though that time could well span a number of days. More in-depth tracking of the changes that can occur over the entire period from growth through harvesting, processing, storage, purchase, and final consumption is certainly merited.

The methodologies used to assess the composition of the bacterial community on produce need to be current. The use of culture-dependent methods is certainly justified when the goal is to determine the presence or viable abundance of specific bacterial populations such as pathogens or indicator species, whose culture requirements are known, and for which specific selective and differential growth media exist. However, culture-independent molecular methods allow the entire bacterial community to be examined, facilitating a more thorough examination of the microbiome present. Emerging technologies such as next-generation sequencing can be used to detect populations that may be missed by standard culture approaches (such as the anaerobic taxa detected on the herb samples described above), and as they become more affordable are likely to compete with traditional culture approaches in routine assessment and monitoring of produce. That said, when we have used culture-dependent approaches and culture-independent next-generation sequencing to analyze the same produce samples, we found that while sequencing revealed the presence of more bacterial taxa, the dominant taxa in our sequence libraries were the same ones detected by the culturing approach [1]. Thus the two approaches are best viewed as complementary and, ideally, both would be used in analyses of fresh produce vegetables.

Just as microbial ecologists have come to understand the complexity and diversity of naturally occurring bacterial assemblages in soils and waters, we are now beginning to grasp the diversity of the phyllosphere and endophytic communities in agricultural crops. Determining this diversity is especially important for crops such as fresh salad produce as they are not extensively processed or cooked prior to consumption. Understanding the complexity of these communities could lead to the development of new pathogen control mechanisms [64] and a greater understanding of pathogen survival and persistence. Regardless of potential future developments, it's becoming clear that leafy salad vegetables harbor a diverse set of phyllosphere and endophytic populations, which consumers must continually be being exposed to. The impact of this native microbiome on consumer health, whether positive or negative, has been largely unexplored.

Raw bacterial sequence data (fastq files) from the herb samples analyzed as part of this manuscript have been deposited in the NCBI Sequence Reads Archive (SRA) under Project accession number SRP052782. Individual herb samples have SRA Sample accession numbers SRS826210, SRS826225, SRS826710, SRS826711, SRS826712, SRS826713, SRS826714, SRS826715, SRS826716, SRS826717, SRS826718, SRS826719, SRS826720, SRS826721, SRS826722, SRS826723, SRS826724, SRS826726, SRS826727, SRS826728, SRS826729, SRS826730, SRS826731, and SRS826732.

ACKNOWLEDGMENTS

Colin R. Jackson's contribution was in part supported by award R01AT007042 from the U.S. National Institutes of Health and also by the Department of Biology at the University of Mississippi. The work performed through the UMMC Molecular and Genomics Facility is supported, in part, by funds from the National Institute of General Medical Sciences of the National Institutes of Health, including Mississippi INBRE (P20GM103476), Center for Psychiatric Neuroscience (CPN)-COBRE (P30GM103328) and Obesity, Cardiorenal and Metabolic Diseases-COBRE (P20GM104357). The content of the manuscript is solely the responsibility of the authors and does not necessarily represent the official views of the National Institutes of Health. Mention of trade names or commercial products is solely for the purpose of providing specific information and does not imply recommendation or endorsement by the US Department of Agriculture (USDA).

AUTHOR CONTRIBUTIONS

Colin R. Jackson reviewed the literature and wrote the initial draft of the manuscript with assistance from Heather L. Tyler; Bram W. G. Stone performed the next-generation sequencing experiments in Section 4; Colin R. Jackson performed the bioinformatics analyses on the sequencing data and generated Table 1 and Table 2. All three authors reviewed and edited the final submission.

CONFLICTS OF INTEREST

The authors declare no conflict of interest.

REFERENCES

1. Jackson, C.R.; Randolph, K.C.; Osborn, S.L.; Tyler, H.L. Culture dependent and independent analysis of bacterial communities associated with commercial salad leaf vegetables. BMC Microbiol. 2013, 13, 274.

2. Lindow, S.E.; Brandl, M.T. Microbiology of the phyllosphere. Appl. Environ. Microbiol. 2003, 69, 1875–1883.

3. Hirano, S.S.; Upper, C.D. Bacteria in the leaf ecosystem with emphasis on Pseudomonas syringiae—A pathogen, ice nucleus, and epiphyte. Microbiol. Mol. Biol. Rev. 2000, 64, 624–653.

4. Beattie, G.A.; Lindow, S.E. Bacterial colonization of leaves: A spectrum of strategies. Phytopathology 1999, 89, 353–359.

5. Woese, C.R. Bacterial evolution. Microbiol. Rev. 1987, 51, 221–271.

6. Pace, N.R. A molecular view of microbial diversity and the biosphere. Science 1997, 276, 734–740.

7. Hugenholtz, P. Exploring prokaryotic diversity in the genomic era. Genome Biol. 2002, 3.

8. Rappé, M.S.; Giovannoni, S.J. The uncultured microbial majority. Annu. Rev. Microbiol. 2003, 57, 369–394.

9. Duineveld, B.M.; Rosado, A.S.; van Elsas, J.D.; van Veen, J.A. Analysis of the dynamics of bacterial communities in the rhizosphere of the chrysanthemum via denaturing gradient gel electrophoresis and substrate utilization patterns.Appl. Environ. Microbiol. 1998, 64, 4950–4957.

10. Smalla, K.; Wachtendorf, U.; Heuer, H.; Liu, W.T.; Forney, L. Analysis of BIOLOG GN substrate utilization patterns by microbial communities. Appl. Environ. Microbiol. 1998, 64, 1220–1225.

11. Kim, J.S.; Sakai, M.; Hosoda, A.; Matsuguchi, T. Application of DGGE analysis to the study of bacterial community structure in plant roots and in nonrhizosphere soil. Soil Sci. Plant Nutr. 1999, 45, 493–497.

12. Smit, E.; Leeflang, P.; Glandorf, B.; van Elsas, J.D.; Wernars, K. Analysis of fungal diversity in the wheat rhizosphere by sequencing of cloned PCR-amplified genes encoding 18S rRNA and temperature gradient gel electrophoresis.Appl. Environ. Microbiol. 1999, 65, 2614–2621.

13. Yang, C.H.; Crowley, D.E. Rhizosphere microbial community structure in relation to root location and plant iron nutritional status. Appl. Environ. Microbiol. 2000, 66, 345–351.

14. Yang, C.H.; Crowley, D.E.; Borneman, J.; Keen, N.T. Microbial phyllosphere populations are more complex than previously realized. Proc. Nat. Acad. Sci. USA 2001, 98, 3889–3894.

15. Kadivar, H.; Stapleton, A.E. Ultraviolet radiation alters maize phyllosphere bacterial diversity. Microb. Ecol. 2003, 45, 353–361.

16. Zhang, B.; Bai, Z.; Hoefel, D.; Tang, L.; Yang, Z.; Zhuang, G.; Yang, Z.; Zhang, H. Assessing the impact of the biological control agent Bacillus thuringiensis on the indigenous microbial community within the pepper plant phyllosphere. FEMS Microbiol. Lett. 2008, 284, 102–108.

17. Suda, W.; Nagasaki, A.; Shishido, M. Powdery mildew-infection changes bacterial community composition in the phyllosphere. Microbes Environ. 2009, 24, 217–223.

18. Zhang, B.; Bai, Z.; Hoefel, D.; Wang, X.; Zhang, L.; Li, Z. Microbial diversity within the phyllosphere of different vegetable species. In Current Research, Technology and Education Topics in Applied Microbiology and Microbial Biotechnology; Méndez-Vilas, A., Ed.; Formatex: Badajoz, Spain, 2010; Volume 2, pp. 1067–1077.

19. Lambais, M.R.; Crowley, D.E.; Curry, J.C.; Büll, J.C.; Rodrigues, R.R. Bacterial diversity in tree canopies of the Atlantic forest. Science 2006, 312.

20. Redford, A.J.; Bowers, R.M.; Knight, R.; Linhart, Y.; Fierer, N. The ecology of the phyllosphere: Geographic and phylogenetic variability in the distribution of bacteria on tree leaves. Environ. Microbiol. 2010, 12, 2885–2893.

21. Jackson, C.R.; Denney, W.C. Annual and seasonal variation in the phyllosphere bacterial community associated with leaves of the Southern Magnolia (Magnolia grandiflora). Microb. Ecol. 2011, 61, 113–122.

22. Kim, M.; Singh, D.; Lai-Hoe, A.; Go, R.; Rahim, R.A.; Ainuddin, A.N.; Chun, J.; Adams, J.M. Distinctive phyllosphere bacterial communities in tropical trees. Microb. Ecol. 2012, 63, 674–681.

23. Jackson, E.F.; Echlin, H.E.; Jackson, C.R. Changes in the phyllosphere community of the resurrection fern, Polypodium polypodioides, associated with rainfall and wetting. FEMS Microbiol. Ecol. 2006, 58, 236–246.

24. Zhang, B.; Bai, Z.; Hoefel, D.; Tang, L.; Wang, X.; Li, B.; Li, Z.; Zhuang, G. The impacts of cypermethrin pesticide application on the non-target microbial community of the pepper plant phyllosphere. Sci. Total Environ. 2009, 407, 1915–1922.

25. Gu, L.; Bai, Z.; Jin, B.; Hu, Q.; Wang, H.; Zhuang, G.; Zhang, H. Assessing the impact of fungicide enostroburin application on bacterial community in wheat phyllosphere. J. Environ. Sci. 2010, 22, 134–141.

26. Alberghini, S.; Battisti, A.; Squartini, A. Monitoring a genetically modified Pseudomonas sp. released on pine leaves reveals concerted successional patterns of the bacterial phyllospheric community. Antonie van Leeuwenhoek 2008, 94, 415–422.

27. Vorholt, J.A. Microbial life in the phyllosphere. Nat. Rev. Microbiol. 2012, 10, 828–840.

28. Müller, T.; Ruppel, S. Progress in cultivation-independent phyllosphere microbiology. FEMS Microbiol. Ecol. 2014, 87, 2–17.

29. Babic, I.; Roy, S.; Watada, A.E.; Wergin, W.P. Changes in microbial populations on fresh cut spinach. Int. J. Food Microbiol. 1996, 31, 107–119.

30. Franz, E.; Visser, A.A.; van Diepeningen, A.D.; Klerks, M.M.; Termorshuizen, A.J.; van Bruggen, A.H.C. Quantification of contamination of lettuce by GFP-expressing Escherichia coli O157:H7 and Salmonella enterica serovar Typhimurium. Food Microbiol. 2007, 24, 106–112.

31. Jablasone, J.; Warriner, K.; Griffiths, M. Interactions of Escherichia coli O157:H7, Salmonella typhimurium and Listeria monocytogenes plants cultivated in a gnotobiotic system. Int. J. Food Microbiol. 2005, 99, 7–18.

32. Brandl, M.T.; Mandrell, R.E. Fitness of Salmonella enterica serovar Thompson in the cilantro phyllosphere. Appl. Environ. Microbiol. 2002, 68, 3614–3621.

33. Solomon, E.B.; Yaron, S.; Matthews, K.R. Transmission of Escherichia coli O157:H7 from contaminated manure and irrigation water to lettuce plant tissue and its subsequent internalization. Appl. Environ. Microbiol. 2002, 68, 397–400.

34. Rudi, K.; Flateland, S.L.; Hanssen, J.F.; Bengtsson, G.; Nissen, H. Development and evaluation of a 16S ribosomal DNA array-based approach for describing complex microbial communities in ready-to-eat vegetable salads packed in a modified atmosphere. Appl. Environ. Microbiol. 2002, 68, 1146–1156.

35. Klerks, M.M.; Franz, E.; van Gent-Pelzer, M.; Zijlstra, C.; van Bruggen, A.H.C. Differential interaction of Salmonella enterica serovars with lettuce cultivars and plant microbe factors influencing the colonization efficiency. ISME J. 2007,1, 620–631.

36. Guo, X.; Chen, J.; Brackett, R.E.; Beuchat, L.R. Survival of Salmonellae on and in tomato plants from the time of inoculation at flowering and early stages of fruit development through fruit ripening. Appl. Environ. Microbiol. 2001,67, 4760–4764.

37. Wachtel, M.R.; Whitehand, L.C.; Mandrell, R.E. Association of Escherichia coli O157: H7 with preharvest leaf lettuce upon exposure to contaminated irrigation water. J. Food Prot. 2002, 65, 18–25.

38. Deering, A.J.; Mauer, L.J.; Pruitt, R.E. Internalization of E. coli O157:H7 and Salmonella spp. in plants: A review. Food Res. Int. 2012, 45, 567–575.

39. Barak, J.D.; Whitehand, L.C.; Charkowski, A.O. Differences in attachment of Salmonella enterica serovars andEscherichia coli O157:H7 to alfalfa sprouts. Appl. Environ. Microbiol. 2002, 68, 4758–4763.

40. Hunter, P.J.; Hand, P.; Pink, D.; Whipps, J.M.; Bending, G.D. Both leaf properties and microbe-microbe interactions influence within-species variation in bacterial population diversity and structure in the lettuce (Lactuca species) phyllosphere. Appl. Environ. Microbiol. 2010, 76, 8117–8125.

41. Leff, J.W.; Fierer, N. Bacterial communities associated with the surfaces of fresh fruits and vegetables. PLoS ONE2013, 8, e59310.

42. Lopez-Velasco, G.; Welbaum, G.E.; Boyer, R.R.; Mane, S.P.; Ponder, M.A. Changes in spinach phylloepiphytic bacteria communities following minimal processing and refrigerated storage described using pyrosequencing of 16S rRNA amplicons. J. Appl. Microbiol. 2011, 110, 1203–1214.

43. Rastogi, G.; Sbodio, A.; Tech, J.J.; Suslow, T.V.; Coaker, G.L.; Leveau, J.H.J. Leaf microbiota in an agroecosystem: Spatiotemporal variation in bacterial community composition on field-grown lettuce. ISME J. 2012, 6, 1812–1822.

44. Handschur, M.; Pinar, G.; Gallist, B.; Lubitz, W.; Haslberger, A.G. Culture free DGGE and cloning based monitoring of changes in bacterial communities of salad due to processing. Food Chem. Toxicol. 2005, 43, 1595–1605.

45. Oliveira, M.; Usall, J.; Viñas, I.; Anguera, M.; Gatius, F.; Abadias, M. Microbiological quality of fresh lettuce from organic and conventional production. Food Microbiol. 2010, 27, 679–684.

46. Caporaso, J.G.; Lauber, C.L.; Walters, W.A.; Berg-Lyons, D.; Huntley, J.; Fierer, N.; Owens, S.M.; Betley, J.; Fraser, L.; Bauer, M.; et al. Ultra-high-throughput microbial community analysis on the Illumina HiSeq and MiSeq platforms.ISME J. 2012, 6, 1621–1624.

47. Kozich, J.J.; Westcott, S.L.; Baxter, N.T.; Highlander, S.K.; Schloss, P.D. Development of a dual-index sequencing strategy and curation pipeline for analyzing amplicon sequence data on the MiSeq Illumina sequencing platform.Appl. Environ. Microbiol. 2013, 79, 5112–5120.

48. Kembel, S.W.; O'Connor, T.K.; Arnold, H.K.; Hubbell, S.P.; Wright, S.J.; Green, J.L. Relationships between phyllosphere bacterial communities and plant functional traits in a neotropical forest. Proc. Nat. Acad. Sci. USA 2014,111, 13715–13720.

49. Xie, W.Y.; Su, J.Q.; Zhu, Y.G. Phyllosphere bacterial community of floating macrophytes in paddy soil environments as revealed by Illumina high-throughput sequencing. Appl. Environ. Microbiol. 2015.

50. Ottesen, A.R.; Gorham, S.; Pettengill, J.B.; Rideout, S.; Evans, P.; Brown, E. The impact of systemic and copper pesticide applications on the phyllosphere microflora of tomatoes. J. Sci. Food Agric. 2015.

51. Schloss, P.D.; Westcott, S.L.; Ryabin, T.; Hall, J.R.; Hartmann, M.; Hollister, E.B.; Lesniewski, R.A.; Oakley, B.B.; Parks, D.H.; Robinson, C.J.; et al. Introducing mothur: Open-source, platform-independent, community-supported software for describing and comparing microbial communities. Appl. Environ. Microbiol. 2009, 75, 7537–7541.

52. Schloss, P.D.; Gevers, D.; Westcott, S.L. Reducing the effects of PCR-amplification and sequencing artifacts on 16S rRNA-based studies. PLoS ONE 2011.

53. Quast, C.; Pruesse, E.; Yilmaz, P.; Gerken, J.; Schweer, T.; Yarza, P.; Peplies, J.; Glöckner, F.O. The SILVA ribosomal RNA gene database project: Improved data processing and web-based tools. Nucleic Acids Res. 2013, 41, 590–596.

54. Edgar, R.C.; Haas, B.J.; Clemente, J.C.; Quince, C.; Knight, R. UCHIME improves sensitivity and speed of chimera detection. Bioinformatics 2011, 27, 2194–2200.

55. DeSantis, T.Z.; Hugenholtz, P.; Larsen, N.; Rojas, M.; Brodie, E.L.; Keller, K.; Huber, T.; Dalevi, D.; Hu, P.; Andersen, G.L. Greengenes, a chimera-checked 16S rRNA gene database and workbench compatible with ARB. Appl. Environ. Microbiol. 2006, 72, 5069–5072.

56. MiSeq SOP. Available online: http://www.mothur.org/wiki/MiSeq_SOP (accessed on 3 October 2015).

57. Lundin, D.; Severin, I.; Logue, J.B.; Östman, Ö.; Andersson, A.F.; Lindström, E.S. Which sequencing depth is sufficient to describe patterns in bacterial α- and β-diversity? Environ. Microbiol. Rep. 2012, 4, 367–372.

58. Hayashi, H.; Shibata, K.; Sakamoto, M.; Tomita, S.; Benno, Y. Prevotella copri sp. nov. and Prevotella stercorea sp. nov., isolated from human faeces. Int. J. Syst. Evol. Microbiol. 2007, 57, 941–946.

59. Kageyama, A.; Benno, Y. Catenibacterium mitsuokai gen. nov., sp. nov., a gram-positive anaerobic bacterium isolated from human faeces. Int. J. Syst. Evol. Microbiol. 2000, 50, 1595–1599.

60. Watanabe, Y.; Nagai, F.; Morotomi, M. Characterization of Phascolartcobacterium succinatutens sp. nov., and asaccharolytic, succinate-utilizing bacterium isolated from human feces. Appl. Environ. Microbiol. 2012, 78, 511–518.

61. Williams, T.R.; Moyne, A.L.; Harris, L.J.; Marco, M.L. Season, irrigation, leaf age, and Escherichia coli inoculation influence the bacterial diversity in the lettuce phyllosphere. PLoS ONE 2013, 8, e68642.

62. Matos, A.; Garland, J.L. Effects of community versus single strain inoculants on the biocontrol of Salmonella and microbial community dynamics in alfalfa sprouts. J. Food Prot. 2005, 68, 40–48.

63. Cooley, M.B.; Chao, D.; Mandrell, R.E. Escherichia coli O157:H7 survival and growth on lettuce is altered by the presence of epiphytic bacteria. J. Food Prot. 2006, 69, 2329–2335.

64. Heaton, J.C.; Jones, K. Microbial contamination of fruit and vegetables and the behavior of enteropathogens in the phyllosphere: A review. J. Appl. Microbiol. 2007, 104, 613–626.

Chapter 10

Microarray-Based Screening of Differentially Expressed Genes of *E. coli*O157:H7 Sakai during Preharvest Survival on Butterhead Lettuce

Inge Van der Linden [1,2,3,*], Bart Cottyn [1], Mieke Uyttendaele [3], Geertrui Vlaemynck [2], Marc Heyndrickx [2,4], Martine Maes [1,5] and Nicola Holden [6]

[1]*Institute for Agricultural and Fisheries Research (ILVO), Plant Sciences Unit—Crop Protection B. Van Gansberghelaan 96, Merelbeke B-9820, Belgium*

[2]*Institute for Agricultural and Fisheries Research (ILVO), Food Sciences Unit—Food Safety Brusselsesteenweg 370, Melle B-9090, Belgium*

[3]*Ghent University (UGent), Department of Food Safety and Food Quality, Laboratory of Food Microbiology and Food Preservation, Coupure links 653, Ghent B-9000, Belgium*

[4]*Ghent University (UGent), Department of Pathology, Bacteriology and Poultry Diseases, Faculty of Veterinary Sciences, Salisburylaan 133, Merelbeke B-9820, Belgium*

[5]*Ghent University (UGent), Department of Biochemistry and Microbiology, Laboratory of Microbiology, K.L. Ledeganckstraat 35, Ghent B-9000, Belgium*

[6]*The James Hutton Institute, Cell and Molecular Sciences, Human and Animal Pathogens on Plants, Invergowrie, Dundee DD2 5DA, UK*

ABSTRACT

Numerous outbreaks of *Escherichia coli* O157:H7 have been linked to the consumption of leafy vegetables. However, up to the present, little has been known about *E. coli* O157:H7's adaptive responses to survival on actively growing (and thus responsive) plants. In this study, whole genome transcriptional profiles were generated from *E. coli*O157:H7 cells (isolate Sakai, *stx*-) one hour and two days after inoculation on the leaves of growing butterhead lettuce, and compared with an inoculum control. A total of 273 genes of *E. coli* O157:H7 Sakai (5.04% of the whole genome) were significantly induced or repressed by at least two-fold ($p < 0.01$) in at least one of the analyzed time points in comparison with the control. Several *E. coli* O157:H7 genes associated with oxidative stress and antimicrobial resistance were upregulated, including the iron-sulfur cluster and the multiple antibiotic resistance (*mar*) operon, whereas the Shiga toxin virulence genes were downregulated. Nearly 40% of the genes with significantly different expression were poorly characterized genes or genes with unknown functions. These genes are of special interest for future research as they may play an important role in the pathogens' adaptation to a lifestyle on plants. In conclusion, these findings suggest that the pathogen actively interacts with the plant environment by adapting its metabolism and responding to oxidative stress.

Keywords: *Escherichia coli* O157:H7; lettuce; gene expression; preharvest

1. INTRODUCTION

Leafy vegetables, such as lettuce, are considered a high-risk food since numerous outbreaks with enteric pathogens have been linked to the consumption of these products [1]. *Escherichia coli* O157:H7 (*E. coli* O157:H7) is one of the pathogens that is frequently involved and is of special interest due to the severe consequences of the illness it may cause. Infection may lead to bloody diarrhea, and, on rare occasions, kidney failure with damage to the central nervous system in severe cases. Young children, the elderly and immunocompromised persons are at higher risk of severe illness. The bacterium can enter the agricultural environment via animal feces and is able to enter our food chain from this point by, e.g., contaminating the irrigation water used for growing crops, or the use of untreated or non-sufficiently treated manure. As governments promote the consumption of a wide variety of fresh fruit and vegetables, it is important that there is no increased risk for foodborne infections. Therefore, the understanding of plant-pathogen interactions such as initial adherence, invasion and establishment is essential for the development of effective control measures.

Numerous studies have investigated the factors which may influence the preharvest survival, growth and attachment of*E. coli* O157:H7 on growing plants. The majority of investigations focused on the influence of temperature, relative humidity, irrigation treatment, UV radiation, leaf age, crop stage, crop variety, *etc*. However, less is known about the underlying genetic mechanisms that *E. coli* O157:H7 uses to survive and proliferate on plants. On the plant surface, the pathogen may encounter various stresses. For example, nutrient scarcity is likely to occur, especially within the phyllosphere compared to a mammalian intestinal environment. This may enhance resistance of the pathogen to subsequent physical or chemical challenges and is thus highly relevant to the pathogen's behavior during processing and storage of the produce, but has yet to be intensively investigated [2].

Gene expression profiling such as microarray technology is one of the techniques used to determine which adaptive and regulatory processes are involved in the pathogen response to specific environmental variables. Only a few microarray studies have been performed on the interaction between *E. coli* O157:H7 and fresh produce. Two studies mainly focused on post-harvest contamination. Kyle *et al.* (2010) [3] have investigated the short-term effects of the influence of leaf injuries and damaged leaf tissue on the gene expression of *E. coli* O157:H7 EDL933 on romaine lettuce leaf lysate. Fink *et al.* (2012) [4] have performed a study on the gene expression of pathogenic and a laboratory reference strain of *E. coli* (serotype K-12) on harvested, intact lettuce leaves. They have studied the mid-term effect (one to three days) on surface sterilized leaves. Preharvest contamination conditions were also investigated for *E. coli* K12 and O157:H7 interacting with the lettuce rhizosphere [5,6] but not with the phyllosphere, and this is important as this is the part of the crop that is consumed. Only recently was a report on the gene expression of *E. coli* O157:H7 on radish sprouts published [7].

The aim of the present study was to characterize the transcriptional response of *E. coli* O157:H7 Sakai during interaction with the phyllosphere of growing butterhead lettuce plants, which has relevance to previous food-borne outbreaks and horticultural production. Gene expression changes were assessed at two time points to represent the very initial interactions (at one hour post-inoculation) and after sufficient time to allow establishment and colony development (two days post-inoculation).

2. EXPERIMENTAL SECTION

2.1. Bacterial Strains and Culture Conditions

E. coli O157:H7 Sakai strain RIMD 0509952 (Sakai; Stx- Kanamycin-resistant) was used and was previously described by Dahan *et al.* (2004) [8]. The Sakai strain was isolated from the Sakai city outbreak in Japan in 1996 [9,10] and possesses a double Shiga toxin knockout (*stx* knockout). A kanamycin cassette was inserted into the *SmaI* site in the *stx2A* gene, and a 0.6 kb BsiWI fragment, which contained the *stx1A* gene and the upstream region, was deleted [8]. Cells were pre-cultured in 5 mL LB and grown overnight at 37 °C, 200 rpm. This culture was diluted 1:100 in pre-tempered MOPS medium [11] enriched with essential and non-essential tissue culture amino acids (M5550 and M7145, Sigma-Aldrich, S. Louis, MO, USA) and supplemented with 0.2% glycerol, and grown at 18 °C, 200 rpm for 22 h to the early stationary phase of growth. The OD_{600} was measured (Genesys 10uv, Thermo Fischer Scientific Inc., Waltham, MA, USA). In order to remove nutrients from the culture broth which could bias the survival and establishment of cells once in contact with the phyllosphere, the culture was washed with sterile pretempered 10 mM $MgSO_4$ (230291, Sigma-Aldrich, S. Louis, MO, USA) at 18 °C (10 min, 5000 × *g*, 4 °C), and re-suspended in sterile pretempered 10 mM $MgSO_4$ at 18 °C to an OD_{600} of 0.5 which corresponded with 8.7 ± 0.1 log CFU/mL as determined by plate counting on the selective medium sorbitol MacConkey Agar (SMAC, Sigma-Aldrich, Sigma-Aldrich, St. Louis, MO, USA) (24 h, 37 °C). These $MgSO_4$ suspended cells were used as inoculum for the plants.

2.2. Plant Growth Conditions

Pelletized butterhead lettuce seeds (*Lactuca sativa* L. var. *capitata* "Alexandria") were obtained from Rijk Zwaan Distribution B.V., De Lier, the Netherlands. The seeds were sown in a mixture of peat, sand, limestone, perlite, celcote, sincrostart and multicote-4. One week after sowing, the seedlings were placed in square pots of 10 cm and grown in the greenhouse at The James Hutton Institute (Dundee, UK). Four-week-old plants were moved from the greenhouse to the growth chamber (Microclima1000, Snijders, Tilburg, The Netherlands) one day before the start of each experiment. Growth chamber conditions were set at continuously ±18 °C with a relative humidity (RH) of 75% and 16/8 hour light/dark cycle.

2.3. Inoculation of the Lettuce Plants and Measurement of Total Pathogen Populations on Lettuce Leaves

Four-week-old lettuce plants (±9-leaf stage) were spray-inoculated in a biosafety cabinet. A total of 100 mL inoculum (described in Section 2.1) was used to inoculate 16 plants. The pathogen population on the leaves was determined in duplicate one hour after inoculation (1 hpi) and one, two and three days thereafter. For each sample, 1 g of leaves from three lettuce plants was used. The sample was ground aseptically with sterilized sand and 2 mL phosphate buffered saline (PBS), using a pestle and mortar. Serial dilutions were made in PBS and the appropriate dilutions were plated onto SMAC (24 h, 37 °C). Maceration does not recover all of the interacting bacteria, but it is still one of the best methods for direct plate counts [12]. For each experiment two samples were analyzed (n = 2) and the experiment was performed three times. Non-inoculated plants were analyzed as control.

2.4. RNA Extraction for Microarray Experiment

A total of 10 g lettuce leaves from three lettuce plants were removed with sterile forceps and placed in a sterile beaker with 200 mL ice-cold RNA-stop wash solution (0.5% phenol pH 4.3, 9.5% ethanol absolute and 90% PBS). The lettuce was cut into pieces of ±2 cm^2 with sterile scissors and stirred on a magnetic stirrer for 5 min at level 3 (Stuart stir UC151, Bibby Scientific Limited, Staffordshire, UK). The RNA extraction was performed on the wash water. Therefore, the 200 mL RNA-wash solution was pipetted into four 50 mL RNAse free polypropylene tubes and centrifuged for 10 min, 4 °C, 5000 rpm. The supernatant was removed and the pellets were dissolved in 1 mL ice-cold RNA-stop (5% phenol pH 4.3, 95% ethanol absolute). These solutions from the four tubes were collected in a 50 mL tube. Subsequently, the emptied tubes were rinsed with another 1 mL of ice-cold RNA-stop and this was added to the 50 mL tube as well and centrifuged as described above (5 mL in total). The supernatants were removed and the pellet was once more washed with 1 mL ice-cold RNA-stop solution. The pellet was immediately stored at −80 °C. The bacterial pellets were treated with 50 mg/mL of lysozyme (5 min) (L3790, Sigma Aldrich, St. Louis, MO, USA). Subsequently, RNA was extracted with the Qiagen RNeasy kit with an additional on-column DNase digestion performed following the instructions of the manufacturer (Qiagen, Hilden, Germany). As a reference control, a cell pellet of 1 mL of the inoculum was made one hour after resuspending the cells in the MgSO4-buffer, treated with the Qiagen RNeasy protocol as described above and immediately stored at −80 °C. Total RNA was quantified spectrophotometrically (NanoDrop ND 1000, Thermo Scientific, Wilmington, DE, USA) and the quality examined with a Bioanalyzer 2100 (Agilent technologies, Palo Alto, CA, USA).

2.5. Microarray Labeling Procedure

A volume of 17.7 µL of total RNA was combined with *Enterobacteriaceae* specific 10 × 11-mer oligos, 100 ng/µL as described by [13], on ice, incubated 10 min at 70 °C and cooled on ice again. A volume of 17.3 µL of a master mix (5× first strand buffer, 9.0 µL; 0.1 M DTT, 4.5 µL, 25× aa-dNTP labeling mix 1.8 µL Superscript RT 2.0 µL) was added and the samples were incubated for 2 h at 42 °C. A total of 15 µL of 1 M NaOH and 15 µL 0.5 M EDTA (pH 8.0) was added to hydrolyze the RNA and the samples were incubated for 15 min at 65 °C. Then 15 µL 1 M HCl was added to neutralize the samples. cDNA was purified by adding 450 µL PB buffer (Qiagen) mix and 1applied to a Qiagen MinElute column. The samples were centrifuged. Centrifugation always occurred at 10,000 × *g* for 1 min unless stated otherwise. The supernatant was removed and 750 µL filter sterilized phosphate wash buffer (0.25 mL 1 M phosphate buffer, 7.7 mL sterile distilled water (SDW), 42.2 mL absolute ethanol) was added and centrifuged. The supernatant was removed and the nucleic acid trapped in the column centrifuged for another 2 min. The column was transferred to a new amber 1.5 mL tube and 10 µL filter sterilized phosphate elution buffer (0.2 mL 1 M phosphate buffer, 49.8 mL SDW) was added to the center of the membrane, left for one minute at room temperature and subsequently centrifuged. This step was repeated with another 10 µL filter sterilized phosphate elution buffer. The subsequent reactions were performed in low light and the incubations were performed in the dark. Sodium carbonate buffer (2.0 µL, 1 M) was added to the purified cDNA and mixed. Subsequently, 1.0 µL of the appropriate Cy-dye (GE Healthcare #PA23001, PA25001) (in DMSO) was added and incubated for 1 h at room temperature in the dark. The labeled cDNA was purified by adding 3.0 µL 4.0 M hydroxylamine hydrochloride, mixed and incubated in the dark for 30 min. The volume of each reaction was made up to 100 µL with SDW. Five hundred µL of PB buffer were added, mixed and applied to a Qiagen MinElute (Qiagen,Valencia, CA, USA) column and centrifuged. The supernatant was removed and 750 µL PE buffer was added and then centrifuged. The supernatant was removed and the tube was once more centrifuged. The column was transferred to a new 1.5 mL tube and 10 µL elution buffer was added to the center of the membrane and left for 1 min at room temperature and then centrifuged. The sample was re-eluted with an additional 10 µL elution buffer into the same tube. The Cy3/Cy5 incorporation was estimated by NanoDrop ND 1000.

2.6. Preparation of Prokaryotic Hybridization Samples for Agilent 8 × 15K Arrays

The volume of each labeled cDNA was calculated to give 600 ng and pipetted into a fresh 1.5 mL amber tube. Nuclease-free water (Sigma-Aldrich, St. Louis, MO, USA) was added up to 20 µL. A volume of 5 µL of Agilent 10 × Blocking Agent was added to each tube and the samples were denatured at 98 °C for 3 min and cooled to room temperature. Twenty-five µL of 2 × GEx

Hybridization Buffer HI-RPM was added to each tube, mixed well by careful pipetting. The tubes were spun for 1 min at room temperature and immediately placed on ice and loaded onto the Agilent 8 × 15K Arrays (Agilent, Santa Clara, CA, USA). Forty microliter of each mix was added to the gasket slide and the Agilent array slide was placed on top. The slides were incubated for 17 h at 65 °C with rotation at 10 rpm in a hybridization oven. A selection of four samples with the highest quality RNA from three independent experiments were prepared for each time point (1 h and two days after inoculation) and are subsequently called four biological replicates.

2.7. Uninoculated Lettuce Control

To check whether the microbial background microbiota on the lettuce may interfere with the micro-array spots, uninoculated samples were checked for the presence of the *gadA* gene, a housekeeping gene of *E. coli*. The presence of *gadA*was checked by conventional PCR. The forward primer was 5′-ACCTGCGTTGCGTAAATA and the reverse 5′-GGGCGGGAGAAGTTGATG and were described in Kim *et al.* (2006) [14]. Each PCR reaction consisted of 5 μL GoTaq Buffer Green, 3 μL dNTPs (2.5 mM), 0.2 μL 50 μM forward primer, 0.2 μL 50 μM reverse primer, 0.5 μL cDNA, 0.2 μL polymerase and 15.9 μL SDW. The reaction mixture was processed in a thermocycler (TProfessional basic Thermocycler gradient, Biometra GmbH, Göttingen, Germany) with the following settings: 2 min at 94 °C, 30 cycles: 30 s at 95 °C, 30 s at 56.7 °C, and 1 min at 72 °C, followed by a final extension time at 72 °C for 7 min. Five μL of PCR-product was loaded on a 1% agarose gel and imaged under UV illumination.

2.8. Data Analysis

The data were analyzed using the software Genespring version 7.0 (Agilent/Stratagene, Palo Alto, CA, USA). The data from all four biological replicates were further tested by a Principal Component Analysis (PCA) on the different conditions (mean centering and scaling). The outliers were removed by filtering on flags (present or marginal). The data were normalized to the control samples (inoculum) and the replicates combined.

Genes showing a greater than two-fold upregulation or downregulation following Volcano Plot analysis ($0.01 \leq p < 0.05$ or $p < 0.01$) and a raw expression value >50 were considered to be differentially regulated. Comparison of the expression levels was done for the expression after 1 h (day 0) with the inoculum, day 2 with the inoculum and day 2 in comparison with 1 h after inoculation (day 0). A False Discovery Rate (FDR) was not conducted at these significance levels.

Subsequently, a single gene list was made as the Agilent array slide contained the genes of four different *E. coli* strains: the core genome derived from *E. coli* K-12 strain MG1655, and strain-specific genes of *E. coli* O157:H7 strains EDL933 and Sakai, and uropathogenic *E. coli* strain CFT073. Genes of uropathogenic *E. coli* CFT073 were not taken into consideration. For the single gene list, gene sequences of all significantly differentially regulated *E. coli* K-12, *E. coli* O157:H7 EDL933 and*E. coli* O157:H7 Sakai assigned with a gene name were taken into consideration. Furthermore, all significantly differentially regulated *E. coli* K-12, O157:H7 EDL933 and O157:H7 Sakai genes with only a "b", "Z" or "ECs" accession pre-fix, resp. were looked up on XBase [15]. Where duplicates were present in the database, the *E. coli* Sakai gene was kept if possible, if duplicates were present between *E. coli* K12 and *E. coli* O157:H7 EDL933, the *E. coli* K12 gene was kept. Unknown *E. coli*O157:H7 EDL933 genes were discarded from the list as well. This single gene list, therefore, includes all genes spotted on the microarray relevant to *E. coli* Sakai, and was used to link the genes with a COG-annotation [16]. Furthermore, gene lists were made based on the GenProtEC Multifun classes. Since *E. coli* K-12 is often the only *E. coli* reference, all E. *coli* K-12 and Sakai genes with a gene name similar to K-12 were linked to the following classes: class 5 cell processes, class 6.3 pilus, 6.4 flagella and 6.6 ribosomes [17]. For the virulence gene list, *E. coli* O157:H7 Sakai virulence genes as described in Hayashi *et al.* were taken into account and a keyword search was performed on the GIRC-site [18,19]. The keywords that were used were: invasion, adhesion, fimbria, effacement, toxin and type III secretion.

As an internal validation, a gene list with housekeeping genes was composed based on genes that have been suggested by Rocha *et al.* (2015) and Zhou *et al.* (2011) to be used for these purposes [20,21].

For each gene list, the relative expression levels on day 0 and day 2 were plotted against the inoculum and significantly differentially regulated genes were highlighted using Matlab. The nucleotide sequence of the significantly differentially regulated unknown genes with Sakai-specific ECs number was obtained by XBase search. This nucleotide sequence was blasted in the Basic Local Alignment Search Tool (BLAST) in order to find out if the gene was already known or described in related strains.

3. RESULTS AND DISCUSSION

In this study, the response of *E. coli* O157:H7 with intact growing young lettuce plants was investigated by the microarray technique to explore the transcriptional changes in *E. coli* O157:H7. In parallel, the fate of the bacteria was determined with the conventional plate-counting technique. The results will be discussed and compared with the existing literature, in particular with two gene expression studies on lettuce and *E. coli* [3,4]. Since there are apparent similarities, a comparative overview between these and the current study has been compiled (Table 1).

Table 1. Overview of the main differences between the different gene expression studies with *E. coli* O157:H7 and fresh produce.

	Kyle *et al.*, 2010 [3]	Fink *et al.*, 2012 [4]	This Study
Strain	*E. coli* O157:H7 EDL933	*E. coli* K12 MG1655 *E. coli* O157:H7 EDL933	*E. coli* O157:H7 Sakai
Growth medium	M9-glucose [a]	LB [b]	MOPS-enriched [c]
Growth temperature of *E. coli*	28 °C	37 °C	18 °C
Time points	15 min; 30 min	One and three days	1 h, 2 days
Lettuce type	romaine lettuce	green leaf lettuce	butterhead lettuce
Type of interaction	leaf lysate supernatant	surface sterilized leaves sodium hypochlorite	growing non-sterilized plants
Inoculum level	10^8 CFU/mL	10^7 CFU/cm^2 lettuce	10^8 CFU/g lettuce
Plant growth conditions	Not applicable	100% RH, 25 °C photoperiod 16 h for three days.	Growth chamber, 18 °C, 75% RH, photoperiod 16 h for two days
Reference control for expression analysis	Cells grown in M9-glucose at 28 °C	Cells grown in LB at 37 °C	Cells were re-suspended and kept for 1 h in MgSO4 buffer after initial growth in MOPS-enriched medium, both at 18 °C

[a] M9= Minimal 9; [b] LB = Luria Bertani; [c] MOPS = potassium morpholinopropane sulfonate.

3.1. Survival and Association of E. coli Sakai on/with Growing Butterhead Lettuce

The survival of *E. coli* O157:H7 Sakai on 4-week old butterhead lettuce during 3 days is shown in Figure 1. *E. coli* Sakai was inoculated at a level of 7.45 ± 0.37 log CFU/g lettuce. A high inoculum level was used to ensure sufficient bacteria were recovered for subsequent whole transcriptome analysis, not to mimic ecologically relevant levels. The number of bacteria recovered after 2 days was 5.02 ± 0.27 log CFU/g, and did not significantly change at day 3, suggesting stabilization at this level.

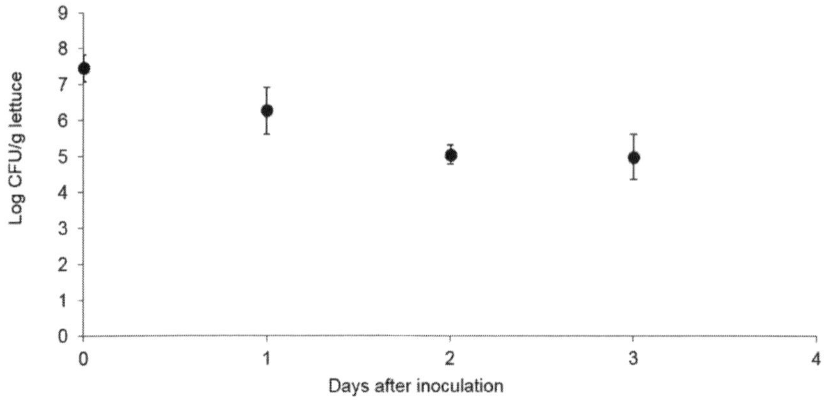

Figure 1. Survival of *E. coli* O157:H7 Sakai on butterhead lettuce. For each experiment 2 samples were analyzed ($n = 2$) and the experiment was performed three times.

3.2. Differentially Regulated Transcriptome of E. coli O157:H7 Sakai on Growing Butterhead Lettuce

A total of 273 genes (5.04%) of the Sakai genome were induced or repressed by at least two-fold ($p < 0.01$, raw value >50 for all treatments) at either 1 hpi or 2 dpi in comparison with the inoculum (cells suspended in $MgSO_4$ buffer). There was a difference between the functions of the genes that were differently regulated after being on the lettuce for one hour and after two days (Figure 2). One hour after inoculation, 164 of the selected genes were significantly differentially regulated, with the majority of the genes induced (71%). The induced genes that could be assigned to a category of orthologous genes (COG), belonging mainly to the transport and metabolism of amino acids and inorganic ions on the one hand, and to transcription, translation, ribosomal structure and biogenesis on the other hand. At day 2, 147 genes were significantly differentially regulated with the majority of the selected genes (65%) repressed. *E. coli* O157:H7 genes that were downregulated included those involved in carbohydrate transport and metabolism, cell wall/membrane, envelope biogenesis, and transcription. Only 37 of the 273 genes of the selected genes were significantly upregulated or downregulated at both time points and for only 23 genes was the expression significantly different between 1 hpi and 2 dpi. For almost 40% (39.9%) of the selected genes, a COG class could not be assigned or the genes were assigned as poorly characterized or function unknown. Furthermore, 23.8% of the selected genes were *E. coli* O157:H7 Sakai-specific genes (ECs) for which only the ECs number referred to a hypothetical or uncharacterized open reading frame.

As an internal control, the expression of 31 housekeeping genes, whose expression is expected to be consistent under a wide variety of conditions, was assessed. Only three of 35 were significantly affected in expression (*rho*, *cca* and *ftsZ*), and the majority were unchanged (see supplemental information). Furthermore, the *gadA* gene was not detected on uninoculated lettuce and generic *E. coli* was not detected on the SMAC-plates.

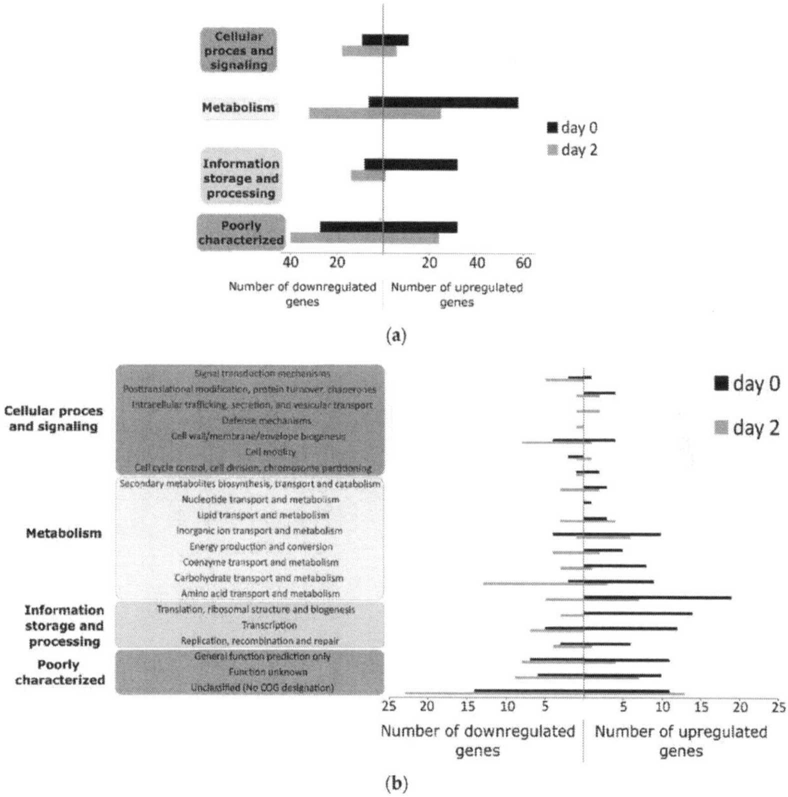

Figure 2. Overview of the significantly differentially regulated genes ($p < 0.01$, at least two-fold, raw value > 50) at 1 hpi (day 0) or 2 dpi (day 2), (**a**) classified by COG group and (**b**) COG functional category.

3.3. Transcription, Translation

One hour after inoculation, genes associated with transcription and translational processes were induced (Figure 3 andFigure 4), for example the 30S (e.g., *rpsU* at $p < 0.01$ and *rpsJT* at $p < 0.05$), 50S ribosomal subunit proteins (*rplCM* at $p < 0.05$), and cell division-related genes such as *dacA*, *groS* ($p < 0.01$) and *mnmG* ($p < 0.05$). Furthermore, genes related to transcription and translation such as *fis*, transcripts coding for, among others, RNA-helicases (*rhlE*), translation initiation factor IF-1 (*infA*), transcription termination (*rho*), and genes responsible for *r*RNA modification such as methylation of nucleotides (*mnmA*, *rumB*, *ECs4154*) were significantly upregulated. In contrast, *ftsN*, an essential cell division protein, was significantly downregulated.

The transcriptional profiles at 2 dpi were quite different. All the ribosomal subunit proteins were significantly repressed compared to their expression level 1 hpi (*rpsUJT*, *rplCM* at $p < 0.05$), as were other ribosomal subunit proteins (*rpsFP* at $p < 0.01$, *rpsDIKT*, *rplKUXY* at $p < 0.05$), and *ftsN*. This 2 dpi data agrees with the cell die-off after one day (see Figure 1) and with the data of Fink *et al.* (2012) [4], who observed a decrease in the expression of the *E. coli* O157:H7 ribosomal-related genes 1 and 3 dpi onto detached lettuce leaves.

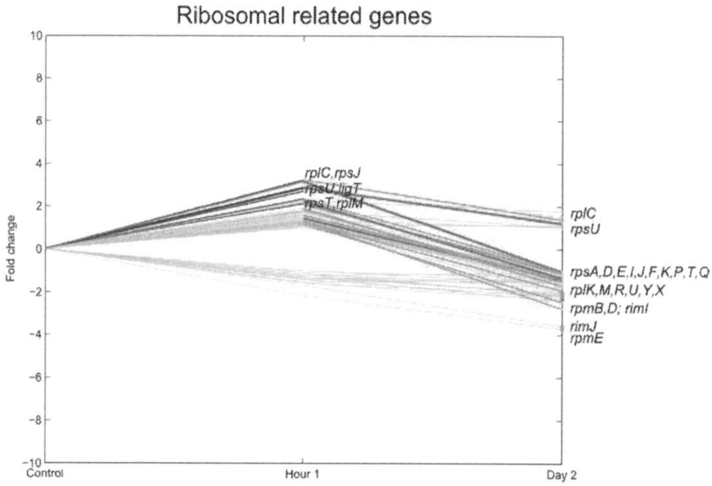

Figure 3. Relative expression of *E. coli* O157:H7 Sakai ribosomal-related genes one hour and two days after inoculation on lettuce plants. A colored line between the control and hour 1 shows that a particular gene was significantly differently expressed one hour after inoculation on the lettuce in comparison with the inoculum. A colored line between hour 1 and day 2 shows that the gene expression of a particular gene was significantly different between these two time points. Colored dots at day 2 show the genes which were significantly differently expressed on day 2 in comparison with the control. Light green: downregulation at $0.01 \le p < 0.05$, dark green: downregulation at $p < 0.01$, light red: upregulation at $0.01 \le p < 0.05$, dark red: upregulation at $p < 0.01$. Grey: no significant difference in expression. Data represent the mean of four biological replicates.

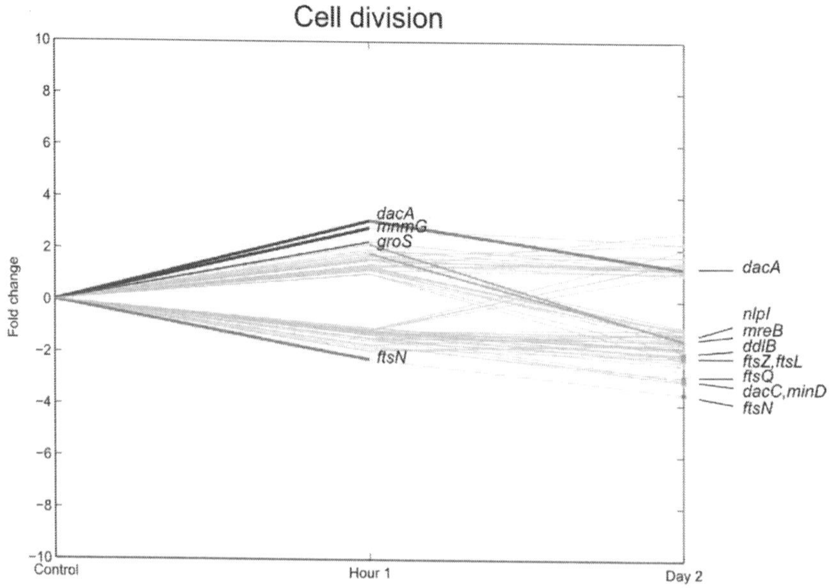

Figure 4. Relative expression of *E. coli* O157:H7 Sakai genes related with cell division one hour and two days after inoculation on lettuce plants. A colored line between the control and hour 1 shows that a particular gene was significantly differently expressed one hour after inoculation on the lettuce in comparison with the inoculum. A colored line between hour 1 and day 2 shows that the gene expression of a particular gene was significantly different between these two time points. Colored dots at day 2 show the genes which were significantly differently expressed on day 2 in comparison with the control. Light green: downregulation at $0.01 \leq p < 0.05$, dark green: downregulation at $p < 0.01$, light red: upregulation at $0.01 \leq p < 0.05$, dark red: upregulation at $p < 0.01$. Grey: no significant difference in expression. Data represent the mean of four biological replicates.

3.4. Carbohydrate Transport

One hour after inoculation, genes essential for the utilization of fructose as a carbon source (*fruB* and *fruK*) were significantly upregulated (Figure 5). The carbon starvation protein (*cstA*), which is induced by carbon starvation, was downregulated 1 hpi and 2 dpi.

The uptake of these sugars was only a short-term effect as, on day two, the expression levels did not change significantly. Further upregulation could only be seen for *alsC*, responsible for the uptake of D-allose. Several other carbohydrate transport-related genes were downregulated 2 dpi together with other genes involved in (or predicted to be involved in) sugar transport and metabolism (*araGF, dhaLM, mak, manX, sfsA, xylH, ytfQ, ytfT, yjfF,* ECs5207).

Temporary upregulation of the carbohydrate system may be explained by the fact that more nutrients are available on the lettuce leaves in comparison with the control. Sucrose is a major trans-locatable product of photosynthesis and the main soluble component of the phloem sap; its monosaccharides are glucose and fructose. It was already shown that near the guard cells of leaves, up to 150 mM [22,23,24] sucrose can be present. Moreover, for *Salmonella* it was shown that the pathogen is attracted to all three sugars and actively moves towards it on lettuce leaves [24]. In addition, Leveau *et al.*suggested that fructose is highly heterogeneously

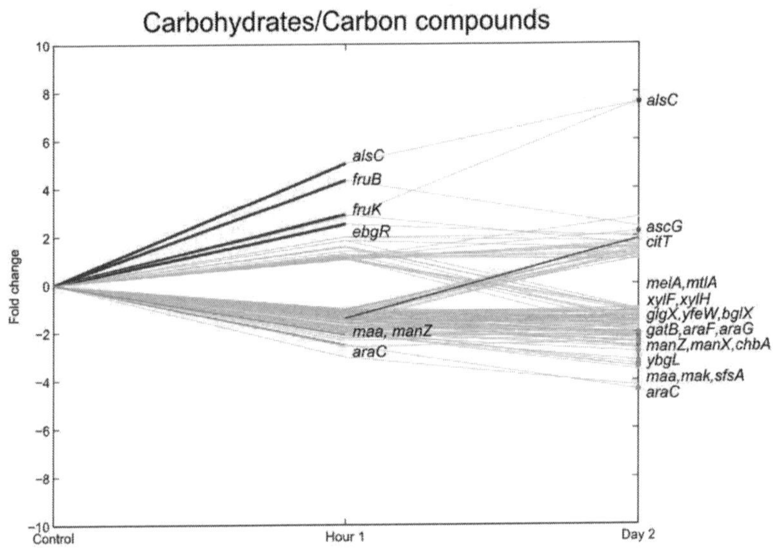

Figure 5. Relative expression of *E. coli* O157:H7 Sakai genes related with carbohydrate uptake one hour and two days after inoculation on lettuce plants. A colored line between the control and hour 1 shows that a particular gene was significantly differently expressed one hour after inoculation on the lettuce in comparison with the inoculum. A colored line between hour 1 and day 2 shows that the gene expression of a particular gene was significantly different between these two time points. Colored dots at day 2 show the genes which were significantly differently expressed on day 2 in comparison with the control. Light green: downregulation at $0.01 \leq p < 0.05$, dark green: downregulation at $p < 0.01$, light red: upregulation at $0.01 \leq p < 0.05$, dark red: upregulation at $p < 0.01$. Grey: no significant difference in expression. Data represent the mean of four biological replicates.

available in the bean phyllosphere. In their experiments with *Erwinia herbicola* harboring a fructose bioreporter, nearly all cells appeared to be actively engaged in fructose consumption 1 h after inoculation, whereas this fraction dropped to approximately 11% after 7 h and to <1% one day after inoculation [25].

3.5. Stress Responsive Genes

The expression of the stress responsive genes and drug resistance genes is shown in Figure 6 and Figure 7. A stress responsive gene set that was significantly induced at 1 hpi was the iron-sulfur cluster (*iscSRUA-hscBA-fdx*). These genes are essential for all proteins containing Fe-S clusters which participate in many vital cellular functions such as DNA repair, transcriptional regulation, nucleotide and amine acid biosynthesis and energy metabolism [26]. Two other systems with a similar function, the *suf*-system (sulfur assimilation) and the *csd*-system (cysteine sulfinate desulfinase), were not differentially regulated. Expression of the three Fe-S cluster assembly operons is regulated by oxidative stress as Fe-S clusters are very sensitive for active oxygen species and iron limitation [26,27].

Based on the data, there was no indication for iron limitation, as the *feb* operon (*febA*) related to iron transport was not differentially upregulated. However, sulfur depletion may have occurred as the cysteine operon (*cys*) was upregulated at both time points. Cysteine is, together with methionine, the only amino acid containing sulfur; therefore, it serves as a precursor for many forms of reduced organic sulfur which is present in cofactors such as Fe-S clusters [27]. Also, some genes of the *tau* operon, which are related to sulfur acquisition, were significantly upregulated at either day 0 (*tauB* at day 0 and day 2, $p < 0.05$) or day 2 (*tauA*, $p < 0.01$). Similar results were observed by Fink *et al.* (2012) [4] where genes involved in Fe-S cluster assembly and repair (*isc*) were upregulated. In addition, they also found that the two other Fe-S cluster-related operons were not differentially expressed. Brankatschk *et al.* (2013) who investigated the transcriptional profile of*Salmonella* Weltevreden during alfalfa sprout colonization also reported upregulation of the *cys* operon [28].

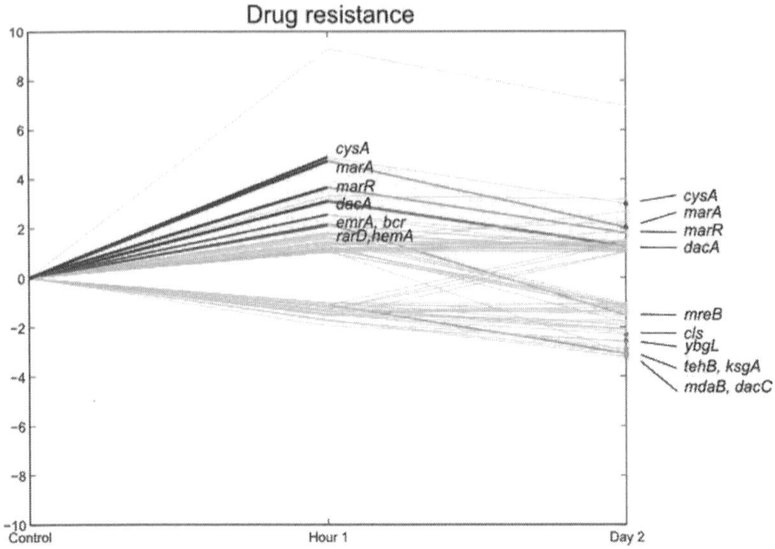

Figure 6. Relative expression of *E. coli* O157:H7 Sakai genes related with drug resistance one hour and two days after inoculation on lettuce plants. A colored line between the control and hour 1 shows that a particular gene was significantly differently expressed one hour after inoculation on the lettuce in comparison with the inoculum. A colored line between hour 1 and day 2 shows that the gene expression of a particular gene was significantly different between these two time points. Colored dots at day 2 show the genes which were significantly differently expressed on day 2 in comparison with the control. Light green: downregulation at $0.01 \leq p < 0.05$, dark green: downregulation at $p < 0.01$, light red: upregulation at $0.01 \leq p < 0.05$, dark red: upregulation at $p < 0.01$. Grey: no significant difference in expression. Data represent the mean of four biological replicates.

Several genes involved in multi-drug resistance (MDR) and response to inhibitory compounds were induced in expression. The *marRAB* operon was consistently upregulated in *E. coli* Sakai, with a higher and significant upregulation at 1 hpi. *marA*, multiple antibiotic resistance gene A, is one of the transcription factors which coordinates the multiple mechanisms that bacteria possess to survive exposure to various chemical stresses and antimicrobial compounds. It is an MDR efflux pump but different observations suggest a role of MDR pumps in plant/bacteria interactions as well [29,30].

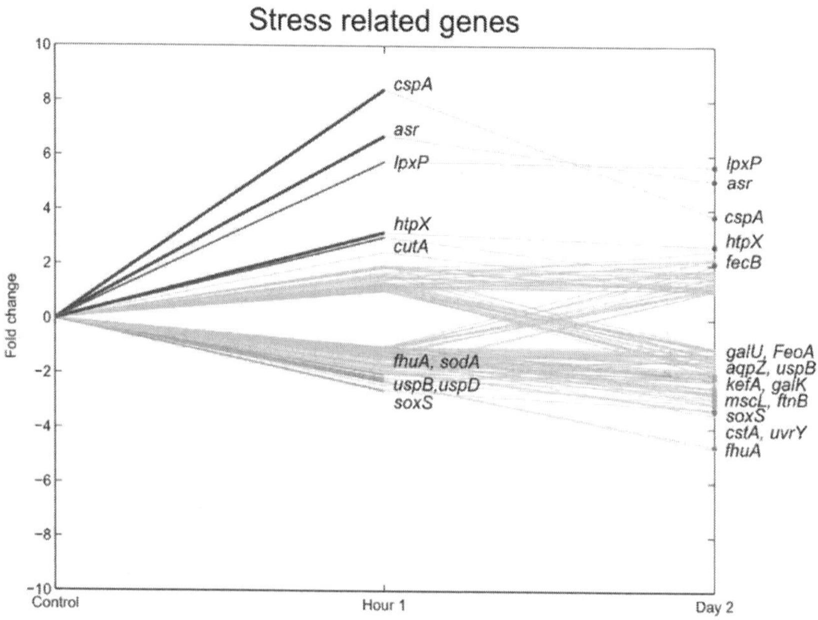

Figure 7. Relative expression of *E. coli* O157:H7 Sakai genes related with stress one hour and two days after inoculation on lettuce plants. A colored line between the control and hour 1 shows that a particular gene was significantly differently expressed one hour after inoculation on the lettuce in comparison with the inoculum. A colored line between hour 1 and day 2 shows that the gene expression of a particular gene was significantly different between these two time points. Colored dots at day 2 show the genes which were significantly differently expressed on day 2 in comparison with the control. Light green: downregulation at $0.01 \leq p < 0.05$, dark green: downregulation at $p < 0.01$, light red: upregulation at $0.01 \leq p < 0.05$, dark red: upregulation at $p < 0.01$. Grey: no significant difference in expression. Data represent the mean of four biological replicates.

Also, other genes, involved in the degradation of inhibitory phenolic compounds such as azoreductases *azoR* and *acpD*, were highly upregulated at 1 hpi and 2 dpi [31,32]. *azoR* was also found to be upregulated in *Salmonella enterica* serovar Typhimurium SL1344 on contact with macerated *Dickeya dadantii* cilantro leaves compared to healthy tissue [32] and was recently reported to be upregulated on radish sprouts as well [7]. Landstorfer *et al.* (2015) speculate on a role of this enzyme in detoxification of secondary plant metabolites such as salicylic acid (SA) and jasmonic acid (JA), produced by the plant defense system directed against, or modulating, the bacterial microbiome.

Upregulation of the *mar* operon was also described by both Fink *et al.* (2012) [4] when *E. coli* was inoculated on detached lettuce leaves and by Kyle *et al.* (2010) [3] when *E. coli* O157:H7 was inoculated into lettuce lysate. However, Kyle *et al.*(2010) [3] observed stronger upregulation compared to this study, probably in response to the presence of H_2O_2 released in the damaged lettuce tissue.

In general, many other oxidative stress genes were upregulated as well (*grxB, yeeE, yjgH, yjgI*), although the transcripts of the stress-related gene regulators *rpoS* and *relA* (stringent response) were not significantly induced. Oxidative stress may have been caused by active oxygen species, a byproduct of normal aerobic metabolism of the lettuce plant.

cspA was significantly upregulated at 1 hpi. This gene is the major cold shock protein of *E. coli* and belongs to a family of structurally analogous proteins. It is an RNA chaperone to prevent the misfolding of single-stranded RNAs due to the formation of secondary structures at cold temperatures [33,34]. Cold shock, however, was prevented as much as possible in the present study, as the MOPS preculture, the $MgSO_4$ inoculum and the temperature of the growth chamber were incubated or set at 18 °C. A plausible explanation for the increase in cold shock protein expression was suggested by Lesley*et al.* (2002), who have shown that upregulation of cold shock proteins can also be an indication of paused translation as a response to the presence of misfolded proteins, a response that is independent of any temperature shifts [35].

3.6. Flagella, Motility, Fimbrial Expression, Biofilm

Several authors have proposed that the presence of flagella and curli and the possibility of the bacteria to produce a biofilm may be involved in the attachment of enteric pathogens to plant tissue.

Flagella are cellular appendages involved in the motility of the pathogen and were already related to being involved in the attachment mechanism of pathogenic *E. coli* to fresh produce [36]. Also, for *Salmonella,* it was shown that the presence of flagella could increase the attachment capacity of the bacterium to plant tissue for some strains [24,37]. Furthermore, it was also shown that a mutant flagella-deficient *Salmonella* strain showed reduced internalization into lettuce leaf tissue compared to its wild-type strain [24]. In the present study, only a few genes related to flagella chemotaxis were significantly differentially regulated, with the majority of these genes downregulated (Figure 8 and Figure 9). *tar, cheA, cheW,* and b0374, were significantly downregulated at 2 dpi. Only *fliN,* which is responsible for flagellar export, was significantly upregulated at 1 hpi. Taking into account the reduction of cell counts at 2 dpi, these findings are in accordance with the results of Jozefzuc *et al.* (2010) and Lemuth *et al.* (2008) who observed downregulation of genes assigned to flagella and motility in response to glucose deprivation and a variety of stress conditions [38,39]. Fink *et al.* (2012) [4], who investigated the gene expression of *E. coli* and, did not observe an induction of the flagellar genes (*fli* operon) for *E. coli* O157:H7 on detached lettuce leaves. In contrast, Kyle *et al.* (2010) [3] observed an increase in *E. coli* O157:H7 *fli* genes in response to lettuce lysates as compared to growth in M9 medium. In their experiment, however, nutrient replenishment occurred (macerated leaves). This indicates that the influence of postharvest practices (e.g., presence of lettuce lysate due to wounding, e.g., in fresh-cut produce) can have a totally different influence on the gene expression of the pathogen.

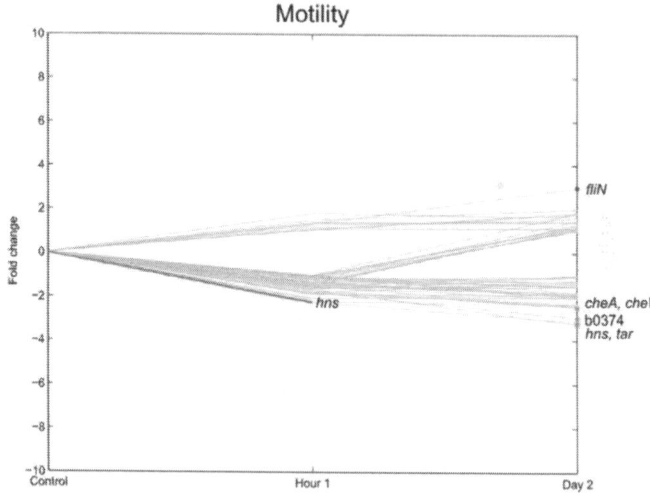

Figure 8. Relative expression of *E. coli* O157:H7 Sakai genes related with motility and flagella one hour and two days after inoculation on lettuce plants. A colored line between the control and hour 1 shows that a particular gene was significantly differently expressed one hour after inoculation on the lettuce in comparison with the inoculum. A colored line between hour 1 and day 2 shows that the gene expression of a particular gene was significantly different between these two time points. Colored dots at day 2 show the genes which were significantly differently expressed on day 2 in comparison with the control. Light green: downregulation at $0.01 \leq p < 0.05$, dark green: downregulation at $p < 0.01$, light red: upregulation at $0.01 \leq p < 0.05$, dark red: upregulation at $p < 0.01$. Grey: no significant difference in expression. Data represent the mean of four biological replicates.

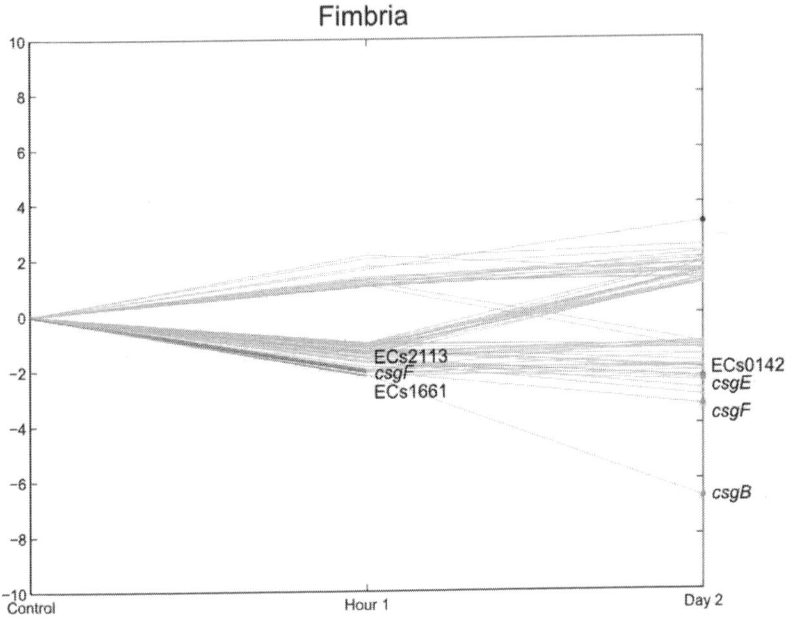

Figure 9. Relative expression of E. coli O157:H7 Sakai genes related with fimbria one hour and two days after inoculation on lettuce plants. A colored line between the control and hour 1 shows that a particular gene was significantly differently expressed one hour after inoculation on the lettuce in comparison with the inoculum. A colored line between hour 1 and day 2 shows that the gene expression of a particular gene was significantly different between these two time points. Colored dots at day 2 show the genes which were significantly differently expressed on day 2 in comparison with the control. Light green: downregulation at $0.01 \leq p < 0.05$, dark green: downregulation at$p < 0.01$, light red: upregulation at $0.01 \leq p < 0.05$, dark red: upregulation at $p < 0.01$. Grey: no significant differences in expression. Data represent the mean of four biological replicates.

Furthermore, flagellin, the protein that is the basic component of the filament of the flagella, is also known to be a major PAMP, recognized by plants a.o. *Arabidopsis thaliana* [40]. Downregulation of flagella genes may, therefore, be a reaction to the PAMP-triggered immunity response of the plant.

Curli, a specific type of fimbria, are known to mediate binding to proteins and abiotic surfaces, and are also related to biofilm formation [41,42]. In the present study, similarly to flagella-related genes, the expression of genes related to curli (operon *csgDEFG* and *csgBAC*) were either not significantly changed or were suppressed, e.g., *csgA* (major subunit curli),*csgB* (nucleator), *csgE* and *csgF* (involved in the initiation of curli subunit polymerization [43,44]). These results represent only the gene expression of loosely associated cells and confirm the results of a study performed by Macarasin *et al.* (2012). In their study, curli were only shown to be critical in strong attachment to spinach leaves, whereas loose attachment to spinach was not affected by curli expression [45]. These results differ from the observations of Fink *et al.* (2012) [4] who saw stark upregulation of *csgA* and *csgB* on detached lettuce leaves at 1 dpi. The expression levels in their study decreased at day 3 but were up to seven times higher in comparison with the control. In their experiments, the bacteria were, however, grown at 37 °C (18 °C in the present study), and it is known that curli are induced at colder temperatures. Therefore, the temperature shift that occurred in the experiment of Fink *et al.*, (2012) [4] (the lettuce was kept at 25 °C) may explain the upregulation of the curli genes. In the present experiment no temperature shift occurred and no upregulation could be seen, but curli were initially expressed as the raw value of, e.g., the *csgA* gene, which was >1000. Fink *et al.* (2012) [4] however, found no significant regulation of the generic *E. coli* and *E. coli* O157:H7 *csgAB* genes on lettuce roots, although the curli regulator *crl* was induced in generic *E. coli*. However, they could demonstrate that the deletion of the *csgA* genes (but not of the *csgB* gene), resulting in a reduced capacity of generic *E. coli* to attach to roots [6]. Curli-related genes were also expressed during the exponential growth of *E. coli* O157 EDL933 in radish seedlings [7]. In summary, it appears that the role of curli has yet to be fully elucidated.

The ability to produce curli seems to be related to the ability of *E. coli* to produce a biofilm [46]. The gene encoding *bhsA*(formerly known as *ycfR*), biofilm formation through hydrophobicity and stress response were significantly upregulated. Upregulation of *bhsA* induces repression of biofilm formation by increasing indole synthesis. *bhsA* is upregulated under a variety of stresses such as heavy metals, drastic pH changes, temperature shock, sodium, hydrogen peroxide and in the presence of sodium salicylate [47,48]. Similar to other stress-related genes, upregulation of this gene confers resistance to these stresses [48]. Upregulation was also observed by Kyle *et al.* (2010) [4] and Fink *et al.* (2012) [4] on lettuce leaves and Hou *et al.* (2013) [6] on lettuce roots. The last two authors also showed that a deletion mutant showed reduced attachment to both lettuce roots and lettuce leaves.

3.7. E. coli O157:H7 Sakai Virulence Genes

The Sakai O157:H7 chromosome encodes at least 131 proteins that are assumed to have virulence-related functions and that are not present in *E. coli* K12 [19]. A total of 31 genes are related to fimbria. Other virulence genes are encoding other adhesins/invasion-like proteins (at least 14 genes), type III secretions system-related proteins and toxin or toxin-like proteins [19].

The type III secretion system (T3SS) encoded by the LEE locus is responsible for the formation of attaching and effacing (A/E) lesions. *ler* (LEE encoded regulator, ECs4588), a central regulator for the genes encoded on the LEE locus [49], was repressed during the experiments and, consequently, most T3SS-related genes were downregulated (Figure 10). Some T3SS genes, however, were significantly upregulated, including ECs3732 (*eivE*), ECs4558 (*escD*) and ECs4580 (*escU*).

In general, the majority of virulence genes were downregulated. The best known virulence factors are the Shiga toxins (*stx1* and *stx2*) and, to a lesser extent, enterohemolysins. Here we used a strain that has an insertional inactivation of the*stx2* gene and is lacking *stx1*ORF; therefore, it was not possible to accurately assess the expression of these genes. However, the expression of Shiga toxin I subunit B precursor (ECs2973) and EHEC-*hlyA* (hemolysin toxin protein, located on the pO157 plasmid) were significantly downregulated two days after inoculation ($p < 0.01$). ECs1283, a homologue of the major virulence factor in *Clostridium difficile*, was not differentially regulated. This contrasts with the findings from Carey *et al.*(2009) who showed that the expression of *stx1* and *stx2* increased on postharvest romaine lettuce at 4 °C and 15 °C, with higher expression values at 4 °C [50]. Sharma *et al.* (2011) also reported increased expression of virulence genes at postharvest at 15 °C, although not at 4 °C. Their experiment was relevant for near-ambient conditions compared to modified atmosphere packaging [51]. Small amounts Shiga toxins were also detected in an *E. coli* O157 population supposed to be in the viable but non culturable (VBNC) state after inoculation on lettuce leaves [52]. In these cases it is possible that Shiga toxin was expressed and detected as a result of a stress response in the bacteria (e.g., from temperature or iron depletion), which may have resulted in expression and even lysogeny of the Shiga toxin phage, via the SOS response [53,54]. *eae* was also found not to be differentially regulated, which is in accordance with Xicohtencatl-cortes *et al.*(2009), who showed that the *eae* gene was not involved in attachment to fresh produce [36]. Finally, Noel *et al.* (2010) found that *Salmonella* motility and animal virulence genes did not contribute significantly to fitness of the bacteria inside tomatoes [55].

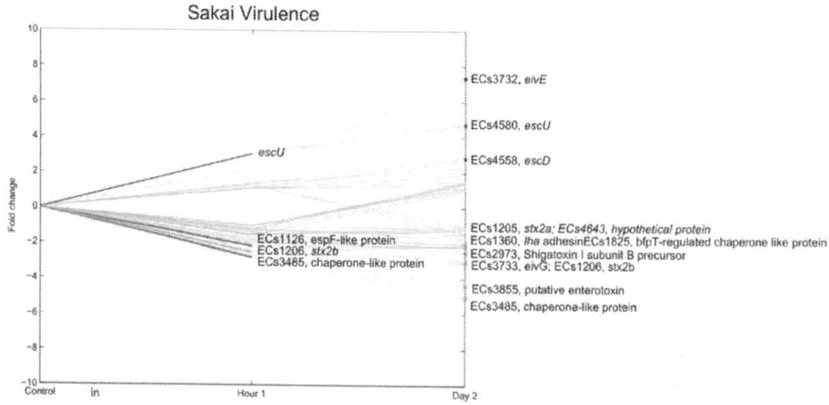

Figure 10. Relative expression of *E. coli* O157:H7 Sakai genes related with virulence (except for fimbrial genes) one hour and two days after inoculation on lettuce plants. A colored line between the control and hour 1 shows that a particular gene was significantly differently expressed one hour after inoculation on the lettuce in comparison with the inoculum. A colored line between hour 1 and day 2 shows that the gene expression of a particular gene was significantly different between these two time points. Colored dots at day 2 show the genes which were significantly differently expressed on day 2 in comparison with the control. Light green: downregulation at $0.01 \le p < 0.05$, dark green: downregulation at $p < 0.01$, light red: upregulation at $0.01 \le p < 0.05$, dark red: upregulation at $p < 0.01$. Grey: no significant difference in expression. Data represent the mean of four biological replicates.

3.8. Unknown Genes

Almost 40% of the genes that were differentially regulated at $p < 0.01$ were poorly characterized or had an unknown function following COG classification. This is likely a reflection of the fact that most research to determine the function of *E. coli* genes is carried out in the context of animal colonization and pathogenicity. The high proportion of uncharacterized genes may indicate that *E. coli* O157:H7 has a specific set of genes to survive in the non-animal environment. Interestingly, several of these genes were also Sakai-specific. Seven Sakai-specific genes were significantly upregulated and 20 downregulated at $p < 0.01$. Alignment analysis of these genes was performed with other *E. coli* strains in an attempt to define a (putative) function (Table 2).

Of note are ECs4115, ECs2526 and ECs0230, three of the most highly expressed Sakai-specific hypothetical genes. ECs4115 has similarity to *yhcR*,

Table 2. Overview of expression levels of genes encoding for hypothetical Sakai-specific proteins at the $p < 0.01$ level at day 0 or day 2 (significant regulation is shown in bold).

GeneName	Foldchange Day 0	Foldchange Day 2	Similarity with Known Gene	Strain with Known Gene with Similar Sequence
ECs0073	−2.3	−2.5	non-LEE-encoded type III secreted effector	*Escherichia* coli O157:H7 str. TW14359
ECs0230	2.0	3.7	VCA0109-like protein (RNA methyltransferase type VI secretion protein)	*Escherichia* coli O157:H7 str. EDL933
ECs0238	−2.4	−4.4	/	/
ECs0281	−1.9	−2.4	Putative tail fiber assembly protein	*Escherichia* coli O157:H7 str. EDL933
ECs0541	−2.1	−2.7	PKD domain protein	*Escherichia coli* O157:H7 str. EC4115
ECs1125	−1.7	−2.7	EspF like protein or Tir-cytoskeleton coupling protein	*Escherichia coli* Xuzhou21/*Escherichia coli* O157:H7 str. TW14359
ECs1204	1.7	2.2	Putative DNA methylase	*Escherichia coli* O157:H7 str. EDL933, complete genome
ECs1238	−2.0	−3.1	/	/
ECs1239	−1.8	−2.3	/	/
ECs1246	−2.2	−3.7	/	/
ECs1445	−1.9	−3.4	/	/
ECs1567	−2.6	−3.3	T3SS effector protein EspO	*Escherichia coli* O145:H28 str. RM12761
ECs1586	−2.9	−4.0	/	/
ECs2291	−1.6	−2.2	Lipoprotein YnfC precursor	*Escherichia coli* O157:H7 str. SS17, complete genome
ECs2473	−2.0	−1.5	Putative lipoprotein	
ECs2526	14.7	8.2	Mg and Co efflux protein CorC	*Escherichia coli* O157:H7 str. EDL933, complete genome
ECs3238	−3.8	−6.2	/	/
ECs3250	−2.1	−2.2	56 bp at 5′ side: hybrid sensory histidine kinase in regulatory system with EvgA 289 bp at 3′ side predicted transporter	a.o. *Escherichia coli* O157:H7 str. EDL933, complete genome
ECs3772	−2.2	−2.9	/	/
ECs3966	2.3	3.5	23S rRNA (guanine−N−2−)−methyltransferase rlmG	a.o. *Escherichia coli* O157:H7 str. EDL933, complete genome
ECs4115	7.6	25.8	similar to *yhcR* which is also known as *aaeX*	*Escherichia coli* O157:H7 str. EC4115
ECs4465	−2.6	−4.4	Fic family protein	*Escherichia coli* O157:H7 str. EC4115, complete genome
ECs4491	3.1	1.8	Periplasmic septal ring factor with murein hydrolase activity EnvC	*Escherichia coli* O157:H7 str. EDL933, complete genome
ECs4537	−2.2	−2.4	transposase ORF B, IS3 family protein	*Escherichia coli* O55:H7 str. RM12579, complete genome
ECs4816	−2.4	−2.4	/	/
ECs5266	2.0	2.6	transposase	*Escherichia coli* O157:H7 str. EDL933, complete genome
ECs5291	−1.9	−2.9	adherence and invasion outermembrane protein (Inv, enhances Peyer's patches colonization)	*Escherichia coli* O157:H7 str. EDL933, complete genome

also known as *aaeX*, which may be related with biofilm formation. The genes in the AaeXAD operon, which are normally not expressed at a significant level [56,57], were some of the most upregulated genes in a study of global gene expression of biofilms of two *E. coli* urinary tract infection strains grown in human urine [58]. Furthermore, Monnappa *et al.* (2013) showed that *aaeXAB* may be involved in the efflux of (even low) concentrations of plant hydrolysate-related phenolics, such as ferulic acid, an abundant phenolic phytochemical found in plant cell wall components and vanillic acids [59]. It was notable here that ECs4115 was induced almost 26-fold. The ECs2526 nucleotide sequence shows some similarity to *corC*, which encodes a Mg and Co efflux protein. ECs0230 potentially encodes an RNA methyltransferase type VI secretion protein. The contributions of the type 6 secretion system (T6SSs) to virulence development are diverse. In cell culture systems, T6SSs have been reported to play crucial roles in cell adhesion and invasion and intracellular growth, but most importantly in inter-bacterial competition [60]. Thus, its expression on lettuce leaves may indicate the presence of other epiphytic bacteria.

3.9. Limitations of the Data

Although efforts were made to mimic a natural contamination event, it should be taken into account that the data presented in this study were generated from artificially inoculated plants and that a high *E. coli* Sakai inoculum level had to be used to be able to extract sufficient levels of bacterial mRNA. As the initial inoculum level was higher than the maximum carrying capacity of the lettuce leaves, which is around 6 log CFU/g [61,62], the pathogen load on the leaves indeed declined. It has, however, already been shown that *E. coli* O157 is able to grow in the lettuce phyllosphere under high humidity conditions [61,63]. Furthermore, we remark that the high inoculum density could also bias other aspects of *E. coli*O157 fitness in the phyllosphere, among which are the transition to a viable but non-culturable state [52] or changes in capacity to attach [64,65].

In addition, the data from this study only describe the gene expression of loosely associated *E. coli* Sakai cells that could be collected by applying the washing protocol as described above. Preliminary tests showed that about 10% of the *E. coli*Sakai population could not be removed from the lettuce leaves with the used washing protocol, and hence were considered either strongly associated or internalized. Attempts to investigate the gene expression profiles of these strongly associated or internalized *E. coli* Sakai cells failed when a wash protocol with glass beads was applied, because not enough high quality mRNA could be extracted due to proportionally high amounts of co-extracted lettuce (chloroplast) RNA.

4. CONCLUSIONS

The main aim of this research was to characterize the gene expression response of *E. coli* O157:H7 in a simulated preharvest contamination event in a greenhouse.

Attention was given to the growth conditions of the plants and bacteria. Plants were grown from seeds under conditions that mimic those in commercial lettuce production greenhouses in Belgium; the bacterial strain used for inoculations, *E. coli*O157:H7 Sakai, previously isolated from a fresh produce outbreak, was cultured at 18 °C and resuspended in a low concentration salt solution to simulate the suboptimal environmental conditions that occur outside the animal host, e.g., in irrigation water. The results support the concept that the plant environment, or, more broadly, the non-animal environment, is part of the ecological niche of human pathogens [66,67,68,69,70]. Firstly, the data showed that genes such as the highly upregulated *azoR* could have been induced in reaction to the plant defense response. Secondly, the fact that almost 40% of the significantly differentially regulated genes were poorly characterized or had an unknown function showed that there are still questions on the function of environment-specific genes, including the plant environment, in comparison with the well-studied animal host environment. Thirdly, there was clearly little need for virulence genes associated with pathogenicity of animal hosts as most virulence genes were downregulated. It was noticed that there was overlap between the results of the present study for growing lettuce leaves and the response to detached lettuce leaves, observed by Fink *et al.* (2012) [4] (for an overview see also supplemental Figure 2), adding strength to the concept of a plant-dependent response. In both studies, stress responsive genes in particular were upregulated, whereas energy metabolism and translation mechanisms were downregulated. Differences, however, could be observed, e.g., for the curli genes which may be due to the different experimental set-up.

The present study provides a basis for further studies using mutants (knock-out, overexpression) and challenge tests such as those performed by Kyle *et al.* (2010) [3] and Oliveira *et al.* (2010) [71]. Such studies would reveal the extent to which the observed gene expression changes influence subsequent survival and antibiotic and sanitation resistance of the bacteria. In order to facilitate choices for experimental set-up, to increase consistency among studies and to enable direct comparison or at least facilitate meta-analysis, the use of standardized protocols would be essential.

In conclusion, we have shown that *E. coli* O157:H7 Sakai actively interacts with the plant environment by adapting its metabolism and responding to oxidative stress. Expression of several other animal-associated virulence genes was decreased in the lettuce phyllosphere. Further research is needed to investigate how these adaptations may affect the pathogen's subsequent survival during processing and consumption.

ACKNOWLEDGMENTS

This study was funded by the Federal Public Service of Health, Food Chain Safety and Environment (contract SALCOSLA RF 6202). The authors would like to thank Louise Birse, Pete Hedley and Jenny Morris for their help and processing the microarrays.

AUTHOR CONTRIBUTIONS

Inge Van der Linden contributed to the design of the study, production, analysis and interpretation of the results, and preparation of the manuscript. Nicola Holden contributed to the design of the study, analysis and interpretation of the results and preparation of the manuscript. Martine Maes, Bart Cottyn, Marc Heyndrickx, Geertrui Vlaemynck and Mieke Uyttendaele contributed to the design of the study and preparation of the manuscript.

CONFLICTS OF INTEREST

The authors declare no conflict of interest.

REFERENCES

1. Klein, S.; Tian, A.; Witmer, J.; DeWaal, C.S. The FDA Top Ten: The Riskiest Foods Regulated by the US Food and Drug Administration; The Center for Science in the Public Interest: Washington, DC, USA, 2009.

2. Delaquis, P.; Bach, S.; Dinu, L.D. Behavior of Escherichia coli O157:H7 in leafy vegetables. J. Food Prot. 2007, 70, 1966–1974.

3. Kyle, J.L.; Parker, C.T.; Goudeau, D.; Brandl, M.T. Transcriptome analysis of Escherichia coli O157:H7 exposed to lysates of lettuce leaves. Appl. Environ. Microbiol. 2010, 76, 1375–1387.

4. Fink, R.C.; Black, E.P.; Hou, Z.; Sugawara, M.; Sadowsky, M.J.; Diez-Gonzalez, F. Transcriptional responses ofEscherichia coli K-12 and O157:H7

associated with lettuce leaves. Appl. Environ. Microbiol. 2012, 78, 1752–1764.

5. Hou, Z.; Fink, R.C.; Black, E.P.; Sugawara, M.; Zhang, Z.; Diez-Gonzalez, F.; Sadowsky, M.J. Gene expression profiling of Escherichia coli in response to interactions with the lettuce rhizosphere. J. Appl. Microbiol. 2012, 113, 1076–1086.

6. Hou, Z.; Fink, R.C.; Sugawara, M.; Diez-Gonzalez, F.; Sadowsky, M.J. Transcriptional and functional responses ofEscherichia coli O157:H7 growing in the lettuce rhizoplane. Food Microbiol. 2013, 35, 136–142.

7. Landstorfer, R.; Simon, S.; Schober, S.; Keim, D.; Scherer, S.; Neuhaus, K. Comparison of strand-specific transcriptomes of enterohemorrhagic Escherichia coli O157:H7 EDL933 (EHEC) under eleven different environmental conditions including radish sprouts and cattle feces. BMC Genomics 2014, 15, 353.

8. Dahan, S.; Knutton, S.; Shaw, R.K.; Crepin, V.F.; Dougan, G.; Frankel, G. Transcriptome of enterohemorrhagicEscherichia coli O157 adhering to eukaryotic plasma membranes. Infect. Immun. 2004, 72, 5452–5459.

9. Michino, H.; Araki, K.; Minami, S.; Takaya, S.; Sakai, N.; Miyazaki, M.; Ono, A.; Yanagawa, H. Massive outbreak ofEscherichia coli O157:H7 infection in schoolchildren in Sakai City, Japan, associated with consumption of white radish sprouts. Am. J. Epidemiol. 1999, 150, 787–796.

10. Watanabe, Y.; Ozasa, K.; Mermin, J.H.; Griffin, P.M.; Masuda, K.; Imashuku, S.; Sawada, T. Factory outbreak ofEscherichia coli O157:H7 infection in Japan. Emerg. Infect. Dis. 1999, 5, 424–428.

11. Neidhardt, F.C.; Bloch, P.L.; Smith, D.F. Culture medium for enterobacteria. J. Bacteriol. 1974, 119, 736–747.

12. Kisluk, G.; Hoover, D.G.; Kneil, K.E.; Yaron, S. Quantification of low and high levels of Salmonella enterica serovar Typhimurium on leaves. LWT Food Sci. Technol. 2012, 45, 36–42.

13. Fislage, R.; Berceanu, M.; Humboldt, Y.; Wendt, M.; Oberender, H. Primer design for a prokaryotic differential display RT-PCR. Nucleic Acids Res. 1997, 25, 1830–1835.

14. Kim, J.; Demeke, T.; Clear, R.M.; Patrick, S.K. Simultaneous detection by PCR of Escherichia coli, Listeria monocytogenesand Salmonella typhimurium in artificially inoculated wheat grain. Int. J. Food Microbiol. 2006, 111, 21–25.

15. Chaudhuri, R.R.; Pallen, M.J. xBASE, a collection of online databases for bacterial comparative genomics. Nucleic Acids Res. 2006, 34, D335–D337.

16. Markowitz, V.M.; Chen, I.-M.A.; Palaniappan, K.; Chu, K.; Szeto, E.; Pillay, M.; Ratner, A.; Huang, J.; Woyke, T.; Huntemann, M. IMG 4 version of the integrated microbial genomes comparative analysis system. Nucleic Acids Res.2013.

17. GenProTec. E. coli genome and proteome database. Available online: http://genprotec.mbl.edu/ (accessed on 31 December 2013).

18. E. coli O157:H7 Sakai genome project. Available online: http://genome.bio.titech.ac.jp/bacteria/o157/search.html(accessed on 10 September 2014).

19. Hayashi, T.; Makino, K.; Ohnishi, M.; Kurokawa, K.; Ishii, K.; Yokoyama, K.; Han, C.-G.; Ohtsubo, E.; Nakayama, K.; Murata, T. Complete genome sequence of enterohemorrhagic Escherichia coli O157:H7 and genomic comparison with a laboratory strain K-12. DNA Res. 2001, 8, 11–22.

20. Rocha, D.J.; Santos, C.S.; Pacheco, L.G. Bacterial reference genes for gene expression studies by RT-qPCR: Survey and analysis. Antonie Van Leeuwenhoek 2015, 108, 685–693.

21. Zhou, K.; Zhou, L.; Lim, Q.E.; Zou, R.; Stephanopoulos, G.; Too, H.-P. Novel reference genes for quantifying transcriptional responses of Escherichia coli to protein overexpression by quantitative PCR. BMC Mol. Biol. 2011.

22. Lemoine, R. Sucrose transporters in plants: Update on function and structure. Biochim. Biophys. Acta 2000, 1465, 246–262.

23. Kang, Y.; Outlaw, W.H.; Andersen, P.C.; Fiore, G.B. Guard-cell apoplastic sucrose concentration—A link between leaf photosynthesis and stomatal aperture size in the apoplastic phloem loader Vicia faba L. Plant Cell Environ. 2007, 30, 551–558.

24. Kroupitski, Y.; Golberg, D.; Belausov, E.; Pinto, R.; Swartzberg, D.; Granot, D.; Sela, S. Internalization of Salmonella enterica in leaves is induced by light and involves chemotaxis and penetration through open stomata. Appl. Environ. Microbiol. 2009, 75, 6076–6086.

25. Leveau, J.H.J.; Lindow, S.E. Appetite of an epiphyte: Quantitative monitoring of bacterial sugar consumption in the phyllosphere. Proc. Natl. Acad. Sci. USA 2001, 98, 3446–3453.

26. Schwartz, C.J.; Djaman, O.; Imlay, J.A.; Kiley, P.J. The cysteine desulfurase, IscS, has a major role in in vivo Fe-S cluster formation in Escherichia coli. Proc. Natl. Acad. Sci. USA 2000, 97, 9009–9014.

27. Hidese, R.; Mihara, H.; Esaki, N. Bacterial cysteine desulfurases: Versatile key players in biosynthetic pathways of sulfur-containing biofactors. Appl. Microbiol. Biotechnol. 2011, 91, 47–61.

28. Brankatschk, K.; Kamber, T.; Pothier, J.F.; Duffy, B.; Smits, T.H. Transcriptional profile of Salmonella enterica subsp.enterica serovar Weltevreden during alfalfa sprout colonization. Microb. Biotechnol. 2013, 7, 528–544.

29. Konstantinidis, K.T.; Tiedje, J.M. Trends between gene content and genome size in prokaryotic species with larger genomes. Proc. Natl. Acad. Sci. USA. 2004, 101, 3160–3165.

30. Matilla, M.A.; Espinosa-Urgel, M.; Rodríguez-Herva, J.J.; Ramos, J.L.; Ramos-González, M.I. Genomic analysis reveals the major driving forces of bacterial life in the rhizosphere. Genome Biol 2007.

31. Liu, G.; Zhou, J.; Fu, Q.S.; Wang, J. The Escherichia coli azoreductase AzoR is involved in resistance to thiol-specific stress caused by electrophilic quinones. J. Bacteriol. 2009, 191, 6394–6400.

32. Goudeau, D.M.; Parker, C.T.; Zhou, Y.; Sela, S.; Kroupitski, Y.; Brandl, M.T. The Salmonella transcriptome in lettuce and cilantro soft rot reveals a niche overlap with the animal host intestine. Appl. Environ. Microbiol. 2013, 79, 250–262.

33. Kortmann, J.; Narberhaus, F. Bacterial RNA thermometers: Molecular zippers and switches. Nat. Rev. Microbiol. 2012,10, 255–265.

34. Duffitt, A.D.; Reber, R.T.; Whipple, A.; Chauret, C. Gene expression during survival of Escherichia coli O157:H7 in soil and water. Int. J. Microbiol. 2011.

35. Lesley, S.A.; Graziano, J.; Cho, C.Y.; Knuth, M.W.; Klock, H.E. Gene expression response to misfolded protein as a screen for soluble recombinant protein. Protein Eng. 2002, 15, 153–160.

36. Xicohtencatl-Cortes, J.; Chacon, E.S.; Saldana, Z.; Freer, E.; Giron, J.A. Interaction of Escherichia coli O157:H7 with leafy green produce. J. Food Prot. 2009, 72, 1531–1537.

37. Berger, C.N.; Shaw, R.K.; Brown, D.J.; Mather, H.; Clare, S.; Dougan, G.; Pallen, M.J.; Frankel, G. Interaction ofSalmonella enterica with basil and other salad leaves. ISME J. 2009, 3, 261–265.

38. Lemuth, K.; Hardiman, T.; Winter, S.; Pfeiffer, D.; Keller, M.; Lange, S.; Reuss, M.; Schmid, R.; Siemann-Herzberg, M. Global transcription and metabolic flux analysis of Escherichia coli in glucose-limited fed-batch cultivations. Appl. Environ. Microbiol. 2008, 74, 7002–7015.

39. Jozefczuk, S.; Klie, S.; Catchpole, G.; Szymanski, J.; Cuadros-Inostroza, A.; Steinhauser, D.; Selbig, J.; Willmitzer, L. Metabolomic and transcriptomic stress response of Escherichia coli. Mol. Syst. Biol. 2010, 6, 1–16.

40. Thilmony, R.; Underwood, W.; He, S.Y. Genome-wide transcriptional analysis of the Arabidopsis thaliana interaction with the plant pathogen Pseudomonas syringae pv. tomato DC3000 and the human pathogen Escherichia coli O157:H7.Plant J. 2006, 46, 34–53.

41. Blomfield, I.C.V.D.M. Regulation of fimbrial expression. In EcoSal; Curtiss, R.I., Kaper, J.B., Squires, C.L., Karp, P.D., Neidhardt, F.C., Slauch, J.M., Eds.; ASM Press: Washington, DC, USA, 2007.

42. Boyer, R.R.; Sumner, S.S.; Williams, R.C.; Pierson, M.D.; Popham, D.L.; Kniel, K.E. Influence of curli expression byEscherichia coli O157:H7 on the cell's overall hydrophobicity, charge, and ability to attach to lettuce. J. Food Prot. 2007,70, 1339–1345.

43. Römling, U.; Bian, Z.; Hammar, M.; Sierralta, W.D.; Normark, S. Curli fibers are highly conserved between Salmonella typhimurium and Escherichia coli with respect to operon structure and regulation. J. Bacteriol. 1998, 180, 722–731.

44. Keseler, I.M.; Collado-Vides, J.; Santos-Zavaleta, A.; Peralta-Gil, M.; Gama-Castro, S.; Muñiz-Rascado, L.; Bonavides-Martinez, C.; Paley, S.; Krummenacker, M.; Altman, T. EcoCyc: A comprehensive database of Escherichia coli biology.Nucleic Acids Res. 2011, 39, D583–D590.

45. Macarisin, D.; Patel, J.; Bauchan, G.; Giron, J.A.; Sharma, V.K. Role of curli and cellulose expression in adherence ofEscherichia coli O157:H7 to spinach leaves. Foodborne Pathog. Dis. 2012, 9, 160–167.

46. Vidal, O.; Longin, R.; Prigent-Combaret, C.; Dorel, C.; Hooreman, M.; Lejeune, P. Isolation of an Escherichia coli K-12 mutant strain able to form biofilms on inert surfaces: Involvement of a new ompR allele that increases curli expression. J. Bacteriol. 1998, 180, 2442–2449.

47. Pomposiello, P.J.; Bennik, M.H.J.; Demple, B. Genome-wide transcriptional profiling of the Escherichia coli responses to superoxide stress and sodium salicylate. J. Bacteriol. 2001, 183, 3890–3902.

48. Zhang, X.S.; Garcia-Contreras, R.; Wood, T.K. YcfR (BhsA) influences Escherichia coli biofilm formation through stress response and surface hydrophobicity. J. Bacteriol. 2007, 189, 3051–3062.

49. Zhu, C.; Feng, S.; Thate, T.E.; Kaper, J.B.; Boedeker, E.C. Towards a vaccine for attaching/effacing Escherichia coli: A LEE encoded regulator (ler) mutant of rabbit enteropathogenic Escherichia coli is attenuated, immunogenic, and protects rabbits from lethal challenge with the wild-type virulent strain. Vaccine 2006, 24, 3845–3855.

50. Carey, C.M.; Kostrzynska, M.; Thompson, S. Escherichia coli O157:H7 stress and virulence gene expression on Romaine lettuce using comparative real-time PCR. J. Microbiol. Methods 2009, 77, 235–242.

51. Sharma, M.; Lakshman, S.; Ferguson, S.; Ingram, D.T.; Luo, Y.G.; Patel, J. Effect of modified atmosphere packaging on the persistence and expression of virulence factors of Escherichia coli O157:H7 on shredded iceberg lettuce. J. Food Prot. 2011, 74, 718–726.

52. Dinu, L.D.; Bach, S. Induction of viable but non-culturable Escherichia coli O157:H7 on the phyllosphere of lettuce: A food safety risk factor. Appl. Environ. Microbiol. 2011, 77, 8295–8302.

53. Wagner, P.L.; Livny, J.; Neely, M.N.; Acheson, D.W.; Friedman, D.I.; Waldor, M.K. Bacteriophage control of Shiga toxin 1 production and release by Escherichia coli. Mol. Microbiol. 2002, 44, 957–970.

54. Grau-Leal, F.; Quiros, P.; Martinez-Castillo, A.; Muniesa, M. Free Shiga toxin 1-encoding bacteriophages are less prevalent than Shiga toxin 2 phages in extraintestinal environments. Environ. Microbiol. 2015, 17, 4790–4801.

55. Noel, J.T.; Arrach, N.; Alagely, A.; McClelland, M.; Teplitski, M. Specific responses of Salmonella enterica to tomato varieties and fruit ripeness identified by in vivo expression technology. PLoS ONE 2010, 5, e12406.

56. Tseng, T.-T.; Gratwick, K.S.; Kollman, J.; Park, D.; Nies, D.H.; Goffeau, A.; Saier, M.H., Jr. The RND permease superfamily: An ancient, ubiquitous and diverse family that includes human disease and development proteins. J. Mol. Microbiol. Biotechnol. 1999, 1, 107–125.

57. Van Dyk, T.K.; Templeton, L.J.; Cantera, K.A.; Sharpe, P.L.; Sariaslani, F.S. Characterization of the Escherichia coliAaeAB efflux pump: A metabolic relief valve? J. Bacteriol. 2004, 186, 7196–7204.

58. Kvist, M.; Hancock, V.; Klemm, P. Inactivation of efflux pumps abolishes bacterial biofilm formation. Appl. Environ. Microbiol. 2008, 74, 7376–7382.

59. Monnappa, A.K.; Lee, S.; Mitchell, R.J. Sensing of plant hydrolysate-related phenolics with an aaeXAB::luxCDABEbioreporter strain of Escherichia coli. Bioresour. Technol. 2013, 127, 429–434.

60. Kapitein, N.; Mogk, A. Deadly syringes: Type VI secretion system activities in pathogenicity and interbacterial competition. Curr. Opin. Microbiol. 2013, 16, 52–58.

61. Van der Linden, I.; Cottyn, B.; Uyttendaele, M.; Vlaemynck, G.; Heyndrickx, M.; Maes, M. Survival of enteric pathogens during butterhead lettuce growth: Crop stage, leaf age, and irrigation. Foodborne Pathog. Dis. 2013, 10, 485–491.

62. Rastogi, G.; Sbodio, A.; Tech, J.J.; Suslow, T.V.; Coaker, G.L.; Leveau, J.H.J. Leaf microbiota in an agroecosystem: Spatiotemporal variation in bacterial community composition on field-grown lettuce. ISME J. 2012, 6, 1812–1822.

63. Brandl, M.T.; Amundson, R. Leaf age as a risk factor in contamination of lettuce with Escherichia coli O157:H7 andSalmonella enterica. Appl. Environ. Microbiol. 2008, 74, 2298–2306.

64. Takeuchi, K.; Frank, J.F. Penetration of Escherichia coli O157:H7 into lettuce tissues as affected by inoculum size and temperature and the effect of chlorine treatment on cell viability. J. Food Prot. 2000, 63, 434–440.

65. Iturriaga, M.H.; Escartin, E.F.; Beuchat, L.R.; Martinez-Peniche, R. Effect of inoculum size, relative humidity, storage temperature, and ripening stage on the attachment of Salmonella Montevideo to tomatoes and tomatillos. J. Food Prot.2003, 66, 1756–1761.

66. Schikora, A.; Garcia, A.V.; Hirt, H. Plants as alternative hosts for Salmonella. Trends Plant Sci. 2012, 17, 245–249.

67. Schikora, A.; Garcia, A.V.; Charrier, A.; Hirt, H. Infection of plants by the human pathogen Salmonella Typhimurium: Challenges and new insights. Biocommun. Plants 2012, 14, 349–360.

68. Schikora, A.; Carreri, A.; Charpentier, E.; Hirt, H. The dark side of the salad: Salmonella typhimurium overcomes the innate immune response of Arabidopsis thaliana and shows an endopathogenic lifestyle. PLoS ONE 2008, 3, e2279.

69. Shirron, N.; Yaron, S. Active suppression of early immune response in tobacco by the human pathogen SalmonellaTyphimurium. PLoS ONE 2011, 6, e18855.

70. Seo, S.; Matthews, K.R. Influence of the plant defense response to Escherichia coli O157:H7 cell surface structures on survival of that enteric pathogen on plant surfaces. Appl. Environ. Microbiol. 2012, 78, 5882–5889.

71. Oliveira, M.; Usall, J.; Solsona, C.; Alegre, I.; Viñas, I.; Abadias, M. Effects of packaging type and storage temperature on the growth of foodborne pathogens on shredded 'Romaine' lettuce. Food Microbiol. 2010, 27, 375–380.

Chapter 11

Antioxidant Capacity and Bioaccessibility of Synergic Mango (cv. Ataulfo) Peel Phenolic Compounds in Edible Coatings Applied to Fresh-Cut Papaya

Gustavo Rubén Velderrain-Rodríguez[1], Maribel Ovando-Martínez[1], Mónica Villegas-Ochoa[1], Jesús Fernando Ayala-Zavala[1], Abraham Wall-Medrano[2], Emilio Álvarez-Parrilla[2], Tomás Jesús Madera-Santana[1], Humberto Astiazarán-García[3], Orlando Tortoledo-Ortiz[3], Gustavo Adolfo González-Aguilar[1*]

[1]Coordinación en Tecnología de Alimentos de Origen Vegetal, Centro de Investigación en Alimentación y Desarrollo, Hermosillo, México

[2]Instituto de Ciencias Biomédicas, Universidad Autónoma de Ciudad Juárez, Chihuahua, México

[3]Departamento de Nutrición Humana, Coordinación de Nutrición, Centro de Investigación en Alimentación y Desarrollo, Hermosillo, México

ABSTRACT

Edible coatings (EC) applied to fresh-cut fruits are used to increase their shelf-life and to deliver antioxidant bioactives such as phenolic compounds (PC) that reduce their oxidative damage while enhance their functional value. However, the combination of different PC may have synergetic, additive or antagonic effects on the final antioxidant capacity (AOXC). The aim of this study was to examine the AOXC of binary combinations of selected PC from mango peel and their bioaccessibility from 6% alginate-based EC applied to fresh-cut papaya, under simulated gastrointestinal conditions. Among equimolar (0.1 mM) combinations, gallic + protocatechuic acids (AB) were synergic in radical scavenging activity (RSA) as assayed by DPPH

(90% RSA) and FRAP (0.39 mg TE/mL) methods; when assayed in 6% alginate-based EC, their RSA increased (117.85% RSA, 0.88 mg TE/mL). The application of EC + AB to papaya cubes and further in vitro digestion decreased their AOXC probably due to interactions between EC and papaya's matrix. Therefore, further studies are needed in order to evaluate the effect of combination of phenolic and EC applied in other fruits matrix on antioxidants bioaccessibility.

Keywords: Edible Coating, Phenolic Acids, Antioxidant Capacity, Bioaccessibility

1. INTRODUCTION

Nowadays, the search for new technologies for food preservation is very important, particularly to inactivate microorganisms and enzymes that affect the nutritional value of foodstuffs. Some of these technologies have the ability to enhance the shelf-life of foods without affecting their quality and safety [1] [2] . Biofilms or edible coatings (EC) are used to reduce decay, weight loss and browning of fresh-cut fruits, prolonging their shelf-life [3] ; when used as carriers to deliver bioactive compounds (e.g. antioxidants (AOX)) they also improve their overall quality and protect them against fungal and bacterial spoilage [4] . EC + AOX also rises the functional value of fresh-cut fruits [4] due to the fact that foods with high antioxidant capacity (AOXC) are suitable for pre- venting and even treating several non-communicable chronic diseases [5] .

Phenolic compounds (PC) are the most abundant AOX in the human diet, and are present in edible vegetables, fruits and their by-products [6] . Tropical fruits such as mango are a good source of antioxidant compounds including phenolic acid and flavonoid compounds. The global production of mangoes exceeds 27 million tons per year [7] , being the variety Ataulfo the most produced in Mexico representing 5% of total mango exportations [8] . This variety stands out among other mango cultivars for its sensorial characteristics and value as functional food due its higher content of Vitamin C, carotenoids and PC [9] - [11] . The use of its pulp to produce juices or minimally processed products, generates vast quantities of peel and seeds (~32%) which may have nutraceutical potential [12] [13] . Particularly, mango peels are good sources of ascorbate, fructose, insoluble fiber (lignin and hemicelluloses) and monomeric (e.g. gallic and protocatehuic acids) and polymeric (e.g. β-PGG) PC with anti-obesigenic, anti-inflammatory, anti-carcinogenic and anti-diabetic potential [11] .

However, PC present in mango pulp or peel interact with its food matrix, other complex carbohydrates (like those present in GI tract) and with other PC. These interactions can either increase or decrease the AOXC due a synergic or antagonic effect [14] and its potential delivering within the GI tract. Such interactions have repercussions on the functional value of mangoes for human health, since strong interactions significantly limit their bioaccesibility. Hence, the application of novel technologies for controlling an efficient delivery of bioactive compounds might be the next big step toward the functional market consolidation. In a previous work we have reported that the in vitro molecular interaction of four of the most important phenolic acids present in mango pulp (chlorogenic, gallic, protocatechuic and vanillic acids) has a synergic effect on the AOXC [15] [16] . How- ever, there is a lack of reports on the use of these PC as additives in edible coating to preserve fresh-cut fruits and its potential bioaccesibility. The aim of this study was to examine the AOXC of binary combinations of se- lected PC from mango peel and their bioaccessibility from 6% alginate-based EC applied to fresh-cut papaya, under simulated gastrointestinal conditions.

2. MATERIAL AND METHODS

2.1. Materials

Papaya (Carica Papaya L. cv Maradol) and mango (Mangifera indica L. cv. Ataulfo) fruits were purchased at a local marked in Hermosillo, Sonora, Mexico. Alginate, ACS phenolic standards (gallic, ferulic, syringic, protocatechuic and 2-hydroxycinnamic acids), DPPH (2,2-diphenyl-1-picrylhydrazyl), TPTZ (2,4,6-Tris(2-pyridyl)-s- triazine) and Trolox (6-hydroxy-2,5,7,8-tetramethylchromane-2-carboxylic) were purchased form Sigma Aldrich, Toluca, Mexico. All chemicals were of analytical or HPLC grade.

2.2. Methods

2.2.1. Phenolic Compounds Aqueous Extraction

Mango peel PC was extracted as described by Wong, Leong [17] . Mango peels were grounded using a domestic blender, and PC were extracted using distilled water. The mixture was refrigerated at −20˚C and kept in the dark until further analysis.

2.2.2. HPLC-DAD PC Identification

PC profile was done using high-performance liquid chromatography (HPLC; Varian 920-LC) equipped with a diode array detector (DAD). Prior injection (10 μL), 2 mL of the aqueous extract was hydrolyzed by mixing it with 1 mL of 6 N HCl during 3 h at 80˚C. The hydrolyzate was passed through a filter (0.22 μm), transferred to sealed vials and injected into the HPLC system. The UV detector was set at 280 nm. Each PC was identified by comparing its retention time and absorption spectra with those obtained with authentic standards under the same HPLC operating conditions. The mobile phase was composed of 1% formic acid (A) and acetonitrile (B). Initially, a pre-run with 90% A and 10% B with a flow rate of 1 mL/min was conducted. Afterwards, the gradient program was as follows (A:B): a) hold at 90:10 for 1 min; b) 95:5 at 12 min; c) 90:10 at 14 min; d) 80:20 at 20 min and hold for 8 min; e) 74:26 at 30 min and hold for 20 min; f) 50:50 at 60.1 min and hold for 1.9 min; g) 90:10 at 70 min and hold for 5 min. The flow rate was 1 mL/min for the first 30 min and 0.75 mL/min for the last 45 min.

2.2.3. Antioxidant Capacity (AOXC)

The AOXC of individual and combinations of selected mango peel PC (gallic acid, protocatechuic acid, syringic acid, ferulic acid and 2-hydroxycinnamic acid) in methanol was assayed by two radical scavenging capacity (RSC) methods. The DPPH (2,2-diphenyl-1-picrylhydrazyl) assay was performed according to Brand-Williams [18] with some modifications. The DPPH radical solution was prepared by dissolving 2.5 mg of DPPH radical in 100 mL of methanol. The absorbance of solution was adjusted to 0.7 ± 0.02 at 515 nm. Aliquots of 20 μL of sample were placed in a microplate and 280 μL of DPPH radical solution was added to each sample. Samples were incubated for 30 min in the dark and the absorbance was measured with a microplate reader (FLUOstar Omega, BMG LABTECH). Results were expressed as the percentage of remaining RSC. The FRAP assay was carried out following the method reported by Benzie and Strain [19] . Twenty μL of sample were placed into each microplate well and mixed with 280 μL of FRAP solution. The samples were incubated at room temperature in the dark for 30 min and the absorbance was measured at 630 nm. The results were reported as mg of Trolox equivalents per mL of solution (mg TE/mL) using a standard curve of Trolox. Based on AOXC results, the higher synergic combination was selected and evaluated in the next experiments.

2.2.4. Preparation of Alginate-Based EC

Alginate solutions (6% w/v) were prepared adding 0.1 M of the synergic PC combination (gallic acid + protocatechuic acid; combination AB). The AOXC was then assayed as previously described and compared to the corresponding aqueous solution. EC were prepared using the casting method where the alginate solutions with or without the synergic PC combination were placed into plates and dry overnight at 40°C. The alginate solution was used as a control in both EC and biofilms in further analysis.

2.2.5. EC on Fresh Cut Papaya

Papaya was cut into pieces of 1 cm^3. The EC were applied to the fresh-cut papaya by immersion and dried by air at room temperature for 5 min. Samples were used for AOXC and in vitro bioaccessibility of PC.

2.2.6. Bioaccessibility of PC

To analyze the effect of alginate on the release of PC under simulated gastrointestinal conditions, the bioaccessibility of PC in fresh cut papaya with EC and biofilms was determined according to the methodology of Saura-Calixto [20] , which mimics the physiological conditions of the digestive tract. Samples (300 mg) with 10 mL of HCl-KCl buffer (pH 5.0) were incubated with a pepsin solution (300 mg/mL) for 60 min at 40°C in a water bath with constant shaking, simulating gastric conditions. Then 4.5 mL of phosphate buffer (pH 7.5) were added to the samples and pH was adjusted to 7.5. Additionally, to simulate intestinal digestion conditions, 1 mL of pancreatin solution (5 mg/mL) was added. Samples were incubated at 37°C during 6 h with constant shaking. Finally, the samples were used for AOXC analysis.

2.2.7. Statistical Analysis

All tests were performed in triplicate, one-way ANOVA and the Tukey-Kramer multiple comparison test was used to determine significant differences between groups ($p < 0.05$), using the commercial software JMP Software, Version 7.0.

3. RESULTS AND DISCUSSIONS

3.1. Mango Peel PC Profile

Figure 1 shows the phenolic profile of the aqueous extract from mango peel. The numbered peaks in the chromatogram correspond to gallic (1), protocatechuic (2), ferulic (3), syringic (4) and 2-hydroxycinnamic acid (5). These results are similar to those found by Palafox-Carlos, Yahia [16] in mango pulp, where they asserts that the main PC found in mango Ataulfo pulp were phenolics acids, especially gallic and protocatechuic acids. Likewise, Sáyago-Ayerdi, Moreno-Hernández [7] found that Ataulfo peel is rich in hydrolysable tannins, which are polymers of phenolic acids [21] . A research was conducted using standard PC, in order to investigate the synergism or antagonism of these compounds when combined and added to an alginate EC. The information generated can provide new knowledge to understand the behavior of single PC and its contribution to the total AOXC. The individual contribution of these PC has been reported previously by Palafox-Carlos, Yahia [16] .

3.2. AOXC of the Identified PC Standards

Table 1 shows the AOXC of individual and combined solutions of PC in methanol. Beker, Nissen [14] , suggested that synergism is a phenomenon that can be more evident in certain compounds which, when combined, can present a more pronounced effect compared to the sum of their individual activity. In this study, most of the combined solutions presented an antagonic effect when DPPH assay was used, except for the combination of gallic-protocatechuic acid (AB) which had a synergic effect. These results are similar to those reported by Palafox-Carlos [15] [16] , who found an increase in the inhibition percentage of the DPPH radical with the combination of gallic-protocatechuic acid compared with the theoretical sum of its individual AOXC. Regarding to FRAP assay, the combinations AB and protocatechuic-ferulic acid (BD) showed a synergic effect, whilst the rest combinations had mostly an additive effect. These results indicate that the combination AB enhance both, the transference of electron and hydrogen atoms [DPPH is a method for hydrogen atom transference (HAT), and FRAP is a specific method for electron transference (SET)] [22] . Among the binary PC combinations studied, the AB combination was selected for its application in the 6% alginate EC.

3.3. AOXC of PC in Alginate Solution

Alginate molecules are unbranched polysaccharides consisting of 1 → 4 linked β-D-mannuronic acid (M) and its C-5 epimer α-L-guluronic acid (G) [23] . Falkeborg, Cheong [24] suggested that the antioxidant contribution of alginate molecules is due to hydrogen abstraction from the carbon-bonded hydrogens in glucuronic acid. Hernandez-Marin and Martinez [25] , after theoretical studies, have concluded that the antioxidant mechanism of carbohydrates is mainly by HAT from carbon-bonded hydrogens and lees likely to be SET. However, it was observed that the PC (gallic and protocatechuic acids) added into alginate solutions resulted in a slight increase of the AOXC of the protocatechuic acid measured by the DPPH assay (Figure 2(B)), whilst the AOXC by FRAP assay showed 2 fold increase (Figure 2(D)). These results, according to Peshev and Vergauwen [26] , suggest that the antioxidant mechanism followed by the combination of alginate, gallic and protocathechuic acid are not entirely by hydrogen donation. As Peshev, Vergauwen [26] described, if double bonds are present in the antioxidant molecules they follow the radical addition mechanism, which is not considered as a possible mechanism for carbohydrates, but it is for PC [24] . It means that the capacity of the synergic combination to donate electrons is not restricted, but enhanced by the presence of alginate, following diverse possible antioxidant mechanisms.

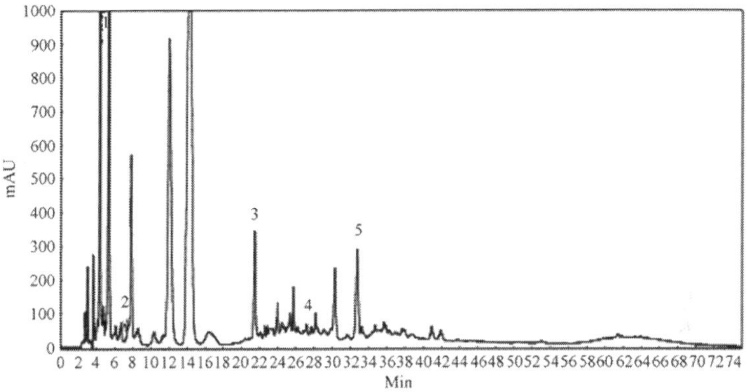

Figure 1. Identified phenolic compounds on mango cv Ataulfo peel aqueous extract by HPLC- DAD. (1) Gallic acid, (2) Protocatechuic acid, (3) Syringic acid, (4) Ferulic acid, (5) 2-hydro- xycinnamic acid.

Table 1. Antioxidant capacity of individual and mixtures of phenolic acids representative of those found in mango peel. Gallic acid (A), protocatechuic acid (B), siringic acid (C), ferulic acid (D), 2-hydroxycinamic acid (E).

Individual (0.1 mM)	DPPH			FRAP		
	% RSA Real	% RSA Theoretical (Sum)	Type of Interaction	mgTE/mL	mgTE/mL Theoretical (Sum)	Type of Interaction
A	49.26a			0.14a		
B	27.26b			0.07b		
C	28.90b			0.09b		
D	19.61c			0.07b		
E	9.819d			0.01c		
Combination	**% RSA Real**	**% RSA Theoretical (Sum)**	**Type of Interaction**	**mgTE/mL**	**mgTE/mL Theoretical (Sum)**	**Type of Interaction**
AB	90.04a	76.53b	Synergic	0.36a	0.22b	Synergic
AC	61.47b	78.16a	Antagonist	0.24a	0.24a	Additive
AD	52.56b	68.87a	Antagonist	0.23a	0.22a	Additive
AE	48.65b	59.08a	Antagonist	0.10b	0.16a	Antagonist
BC	45.76b	56.17a	Antagonist	0.17a	0.17a	Additive
BD	35.47b	46.87a	Antagonist	0.18a	0.15b	Synergic
BE	25.99b	37.08a	Antagonist	0.11a	0.09a	Additive
CD	31.34b	48.51a	Antagonist	0.18a	0.17a	Additive
CE	32.31b	38.72a	Antagonist	0.11a	0.11a	Additive

Different letters in each column of individual standards indicates significant differences ($p \le 0.05$) between standards for each method; whereas in combination, different letters indicates differences ($p \le 0.05$) between real and theoretical values for each method.

3.4. AOXC and Bioaccessibility of PC Added in Alginate Edible Films

Figure 3 shows the results of the AOXC by FRAP method in the EC with 6% alginate and PC. These results indicates that there is an antioxidant synergism related with the transference of electrons, which can be determined by the FRAP assay in the simulated physiological conditions. Nonetheless, these values are lower than those determined in the pure standards in methanol and standards with alginate under the same method (Figure 2). Such differences in the AOXC could be due to the low stability of the PC at basic pH. These results agreed with those reported by Friedman and Jürgens [27] , that states that simple PC with low molecular weight such as phenolic acids are susceptible to oxidation under high pH conditions. Under this condition, PC are oxidized and a chemical structure change takes place, being irreversible and directly affecting the AOXC in both transference of hydrogens (HAT) and electrons (SET) mechanisms.

On the other hand, from the nutritional point of view, some carbohydrates polymers like alginate have gelation properties, which can affect the bioaccessibility of bioactive compounds being trapped within the cross- linked chains of the polymers. Those cross-linked chains are the result of the carbohydrates gelification proper-

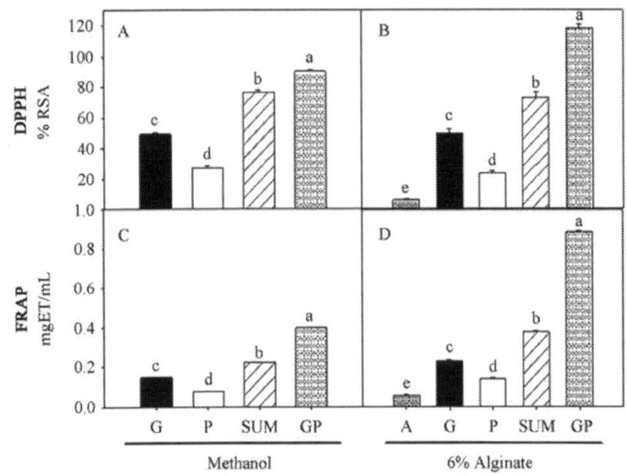

Figure 2. Comparison of antioxidant activity by FRAP and DPPH methods of gallic and protocatechuic acids standards in methanol and 6% alginate (A) solutions. Each bar represents the standard solutions with gallic acid (G), protocatechuic acid (P), gallic and protocatechuic acid theoretical sum (SUM) and the real values of gallic-protocatechuic combination (GP). Values are the mean ± Std error (n = 3). Different letters indicate statistical differences ($p \leq 0.05$).

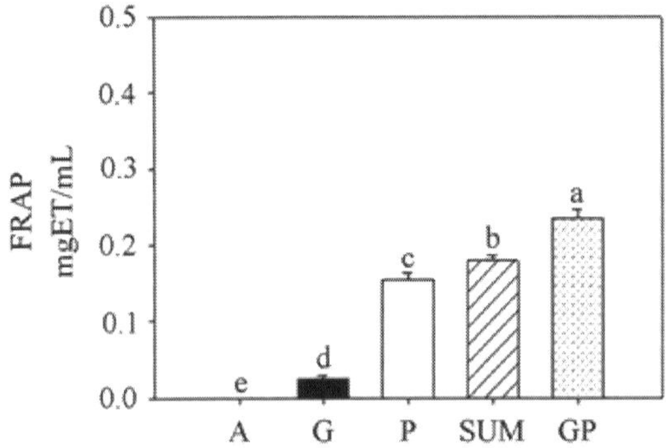

Figure 3. Antioxidant activity by FRAP assay of 6% Alginate (A) edible films added with phenolic compounds gallic-protocatechuic synergic combination. Each bar represents the edible films with gallic acid (G), protocatechuic acid (P), gallic and protocatechuic acid theoretical sum (SUM) and the real values of gallic-protocatechuic combination (GP). Values are the mean ± Std error (n = 3). Different letters indicate statistical differences ($p \leq 0.05$).

ties, i.e. alginate undergoes the sol-to-gel transition upon lowering pH even in the absence of cations [28] . Therefore, this indicates that PC added to the EC have low bioaccessibility due to the changes mentioned previously, affecting the rheological properties of alginate. This suggests that during the digestion in the upper gastrointestinal tract, the acid conditions enhance the interaction of alginate with PC, decreasing their bioaccessibility in the lower gastrointestinal tract.

3.5. AOXC and Bioaccessibility of PC Added to EC Applied in Papaya

The study of the AOXC in FRAP assay of PC released from the in vitro digestion of fresh-cut papaya with 6% alginate EC is presented in Figure 4. These results indicated that the AOXC in FRAP assay decreased, compared to theoretical values, which suggests a lower electron transfer. Nonetheless, when the EC mixed with gallic acid was applied to the papaya fruit a strong antagonism was observed, resulting in lower FRAP values than those found in papaya without EC. This could be due to the chemical-enzymatic conditions involved in the in vitro digestion [28] [29] . During the study of the bioaccessibility of PC with the in vitro digestion model dif-

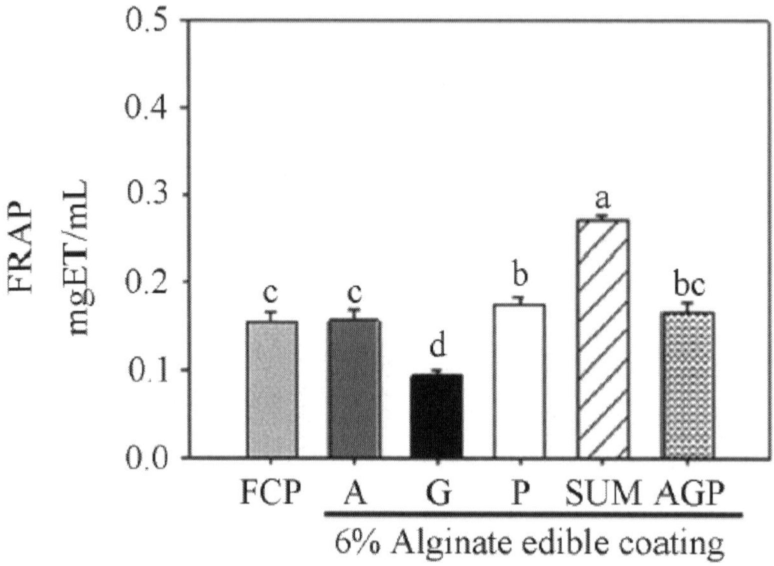

Figure 4. Comparison of antioxidant activity by FRAP assay of fresh cut papaya (FCP) with different 6% Alginate (A) edible coatings. Each bar represents the edible coatings with gallic acid (G), protocatechuic acid (P), gallic and protocatechuic acid theoretical sum (SUM) and the real values of gallic-protocatechuic combination (GP). Values are the mean ± Std error (n = 3). Different letters indicate statistical differences ($p \leq 0.05$).

ferent digestion phases were emulated. In the gastric phase the environment was acid (pH 1.5) while in the intestinal phase the acid environment decreased (pH 7.5). The AOXC measurement was performed at the intestinal phase, which has a less protonated environment, as compared to that of standards in methanol solution or in alginate, where the pH was between 6 - 7.

On the other hand, papaya is a fruit rich in soluble dietary fiber, especially pectin [30] . It has been reported that pectin produces an antioxidant antagonism in AOXC methods due to an interaction between pectin and gallic acid forming hydrogen bonds [31] . In the same way, the alginate chains contain electronegative D-mannuric and D-guluronic groups, so, the hydrogen atoms present in gallic acid could interact with both the pectin from papaya and the alginate chains in the EC [32] . As pectin and alginate have similar gelation properties, both molecules are able to form gel at low pH in presence of hydrophilic molecules. In this case, as gallic and protocatechuic acids are hydrophilic molecules, their presence may induce hydrophobic interactions or hydrogen bonds between pectin and alginate at low pH. These hydrogen bonds reduce the ability of gallic acid molecules to donate either electrons or hydrogen atoms, either due to their inaccessible position within the pectin/alginate branches or to the changes in the molecules reactivity. This can help us to explain the antagonistic effect found in the AOXC of the EC with gallic acid. However, this interaction between PC and alginate was not observed in the absence of papaya fruit, which supports the idea that antagonistic effect is given due to the cross-linked chains formed between papaya's pectin and alginate molecules under acid conditions. In the same way, it suggest that the bioaccessibility of the PC added in the alginate EC is decreasing by the interactions of them with the alginate and dietary fiber present in papaya fruit.

4. CONCLUSION

According to these results, we concluded that alginate EC kept the antioxidant properties, maintaining the synergic effect reported for gallic-protocatechuic acids combination. However, the presence of dietary fiber, such as pectin in the food matrix of the papaya fruit, appears to be a limiting factor in the synergic effect of PC that negatively affect its bioaccessibility. Further studies using UV-VIS, fluorescence, ITC, and FT-IR techniques can be performed to elucidate the possible interactions among alginate coatings, edible films and gallic acid.

ACKNOWLEDGEMENTS

Gustavo Rubén Velderrain-Rodríguez, Maribel Ovando-Martínez, Mónica Villegas-Ochoa, Jesús Fernando Ayala-Zavala, Abraham Wall-Medrano, Emilio Álvarez-Parrilla, Tomás Jesús Madera-Santana, Humberto Astiazarán-García, Orlando Tortoledo-Ortiz, Gustavo Adolfo González-Aguilar (Mangifera indica, L.) en un sistema murino" Project 179574CB2012-01 and Red AlFaNutra.

REFERENCES

1. Kester, J. and Fennema, O. (1986) Edible Films and Coatings: A Review. Food Technology, 40, 47-59.

2. Alikhani, M. (2014) Enhancing Safety and Shelf Life of Fresh-Cut Mango by Application of Edible Coatings and Microencapsulation Technique. Food Science & Nutrition, 2, 210-217.

3. Rojas-Graü, M.A., Soliva-Fortuny, R. and Martín-Belloso, O. (2009) Edible Coatings to Incorporate Active Ingredients to Fresh-Cut Fruits: A Review. Trends in Food Science & Technology, 20, 438-447.

4. Ponce, A.G., et al. (2008) Antimicrobial and Antioxidant Activities of Edible Coatings Enriched with Natural Plant Extracts: In Vitro and in Vivo Studies. Postharvest Biology and Technology, 49, 294-300.

5. Valko, M., et al. (2007) Free Radicals and Antioxidants in Normal Physiological Functions and Human Disease. The International Journal of Biochemistry & Cell Biology, 39, 44-84.

6. Balasundram, N., Sundram, K. and Samman, S. (2006) Phenolic Compounds in Plants and Agri-Industrial By-Products: Antioxidant Activity, Occurrence, and Potential Uses. Food Chemistry, 99, 191-203.

7. Sáyago-Ayerdi, S.G., et al. (2013) Mexican "Ataulfo" Mango (Mangifera indica L.) as a Source of Hydrolyzable Tannins. Analysis by MALDI-TOF/TOF MS. Food Research International, 51, 188-194.

8. Villa-Corrales, L., et al. (2010) Numerical and Experimental Analysis of Heat and Moisture Transfer during Drying of Ataulfo Mango. Journal of Food Engineering, 98, 198-206.

9. Manthey, J.A. and Perkins-Veazie, P. (2009) Influences of Harvest Date and Location on the Levels of β-Carotene, Ascorbic Acid, Total Phenols, the in Vitro Antioxidant Capacity, and Phenolic Profiles of Five Commercial Varieties of Mango (Mangifera indica L.). Journal of Agricultural and Food Chemistry, 57, 10825-10830.

10. García-Solís, P., Yahia, E.M. and Aceves, C. (2008) Study of the Effect of "Ataulfo" Mango (Mangifera indica L.) Intake on Mammary Carcinogenesis and Antioxidant Capacity in Plasma of N-Methyl-N-Nitrosourea (MNU)-Treated Rats. Food Chemistry, 111, 309-315.

11. Wall-Medrano, A., Olivas-Aguirre, F.J., Velderrain-Rodriguez, G.R., González-Aguilar, G.A., De la Rosa, L.A., López- Díaz, J.A. and Álvarez-Parrilla, E. (2015) El mango: Aspectos agroindustriales, valor nutricional/funcional y efectos en la salud. Nutricion Hospitalaria, 31, 67-75.

12. Schieber, A., Ullrich, W. and Carle, R. (2000) Characterization of Polyphenols in Mango Puree Concentrate by HPLC with Diode Array and Mass Spectrometric Detection. Innovative Food Science & Emerging Technologies, 1, 161-166.

13. Ayala-Zavala, J., Rosas-Domínguez, C., Vega-Vega, V. and González-Aguilar, G.A. (2010) Antioxidant Enrichment and Antimicrobial Protection of Fresh-Cut Fruits Using Their Own Byproducts: Looking for Integral Exploitation. Journal of Food Science, 75, R175-R181.

14. Becker, E., Nissen, L. and Skibsted, L. (2004) Antioxidant Evaluation Protocols: Food Quality or Health Effects. European Food Research and Technology, 219, 561-571.

15. Palafox-Carlos, H., Gil-Chávez, J., Sotelo-Mundo, R.R., Namiesnik, J., Gorinstein, S. and González-Aguilar, G.A. (2012) Antioxidant Interactions between Major Phenolic Compounds Found in "Ataulfo" Mango Pulp: Chlorogenic, Gallic, Protocatechuic and Vanillic Acids. Molecules, 17, 12657-12664.

16. Palafox-Carlos, H., Yahia, E. and González-Aguilar, G. (2012) Identification and Quantification of Major Phenolic Compounds from Mango (Mangifera indica, cv. Ataulfo) Fruit by HPLC-DAD-MS/MS-ESI and Their Individual Contribution to the Antioxidant Activity during Ripening. Food Chemistry, 135, 105-111.

17. Wong, S.P., Leong, L.P. and William Koh, J.H. (2006) Antioxidant Activities of Aqueous Extracts of Selected Plants. Food Chemistry, 99, 775-783.

18. Brand-Williams, W., Cuvelier, M. and Berset, C. (1995) Use of a Free Radical Method to Evaluate Antioxidant Activity. LWT-Food Science and Technology, 28, 25-30.

19. Benzie, I.F. and Strain, J. (1996) The Ferric Reducing Ability of Plasma (FRAP) as a Measure of "Antioxidant Power": The FRAP Assay. Analytical Biochemistry, 239, 70-76.

20. Saura-Calixto, F., García-Alonso, A., Goñi, I. and Bravo, L. (2000) In Vitro Determination of the Indigestible Fraction in Foods: An Alternative to Dietary Fiber Analysis. Journal of Agricultural and Food Chemistry, 48, 3342-3347.

21. Okuda, T., Yoshida, T. and Hatano, T. (1995) Hydrolyzable Tannins and Related Polyphenols. In: Fortschritte der Che- mie organischer Naturstoffe/Progress in the Chemistry of Organic Natural Products, Springer, Berlin, 1-117.

22. Rojano, B., Saez, J., Schinella, G., Quijano, J., Vélez, E., Gil, A. and Notario, R. (2008) Experimental and Theoretical Determination of the Antioxidant Properties of Isoespintanol (2-Isopropyl-3,6-Dimethoxy-5-Methylphenol). Journal of Molecular Structure, 877, 1-6.

23. Pawar, S.N. and Edgar, K.J. (2012) Alginate Derivatization: A Review of Chemistry, Properties and Applications. Biomaterials, 33, 3279-3305.

24. Falkeborg, M., Cheong, L.Z., Gianfico, C., Magdalena Sztukiel, K., Kristensen, K., Glasius, M., et al. (2014) Alginate Oligosaccharides:

Enzymatic Preparation and Antioxidant Property Evaluation. Food Chemistry, 164, 185-194.

25. Hernandez-Marin, E. and Martínez, A. (2012) Carbohydrates and Their Free Radical Scavenging Capability: A Theoretical Study. The Journal of Physical Chemistry B, 116, 9668-9675.

26. Peshev, D., Vergauwen, R., Moglia, A., Hideg, É. and Van den Ende, W. (2013) Towards Understanding Vacuolar Antioxidant Mechanisms: A Role for Fructans? Journal of Experimental Botany, 64, 1025-1038.

27. Friedman, M. and Jürgens, H.S. (2000) Effect of pH on the Stability of Plant Phenolic Compounds. Journal of Agricultural and Food Chemistry, 48, 2101-2110.

28. Saura-Calixto, F., Serrano, J. and Goñi, I. (2007) Intake and Bioaccessibility of Total Polyphenols in a Whole Diet. Food Chemistry, 101, 492-501.

29. Palafox-Carlos, H., Ayala-Zavala, J.F. and González-Aguilar, G.A. (2011) The Role of Dietary Fiber in the Bioaccessibility and Bioavailability of Fruit and Vegetable Antioxidants. Journal of Food Science, 76, R6-R15.

30. Manrique, G.D. and Lajolo, F.M. (2002) FT-IR Spectroscopy as a Tool for Measuring Degree of Methyl Esterification in Pectins Isolated from Ripening Papaya Fruit. Postharvest Biology and Technology, 25, 99-107.

31. Martínez-de-Santiago, M. (2014) Interacciones entre pectnia y ácidos fenólicos presentes en pulpa de mango (Man- gifera indica L.) cv. "Ataulfo" y su efecto en la capacidad antioxidante. Departamento de Ingeniería Agroindustrial, Universidad Autónoma Chapingo, Chapingo.

32. Fundueanu, G., Nastruzzi, C., Carpov, A., Desbrieres, J. and Rinaudo, M. (1999) Physico-Chemical Characterization of Ca- Alginate Microparticles Produced with Different Methods. Biomaterials, 20, 1427-1435.

Chapter 12

Effect of Storage Conditions on the Sensory Quality, Colour and Texture of Fresh-Cut Minimally Processed Cabbage with the Addition of Ascorbic Acid, Citric Acid and Calcium Chloride

Eleni Manolopoulou[1], Theodoros Varzakas[2*]

[1]Department of Plant Production, School of Agricultural Technology, Higher Institute of Kalamata, Kalamata, Greece; [2]Department of Food Technology, School of Agricultural Technology, Higher Institute of Kalamata, Kalamata, Greece.

ABSTRACT

The aim of this study was to investigate the effect of storage conditions on the sensory quality, colour and texture of fresh-cut cabbage during the addition of ascorbic acid, citric acid and calcium chloride. Ascorbic acid maintained the overall quality for 14 days at 0°C and 7 days at 5°C; no difference, however, was observed regarding browning of cut surface compared to the control sample at both storage temperatures. Calcium chloride maintained the overall quality and cut surface browning for 14 days at both storage temperatures. It was also found that citric acid 1% can be used for minimally processed cabbage. Soaking with citric acid helped retain the color and increased the overall acceptance and organoleptic quality of fresh cut cabbage; it reduced browning of the cut surface and protected against formation of black specks. Citric acid treatment combined with low temperature storage (0°C) prolonged the shelf life of minimally processed cabbage for 22 days, time sufficient for acceptable marketing of the product. The lightness of minimally processed cabbage decreased linearly from 70.94 ± 6 to 63.8 ± 8.5 - 61.3 ± 8 units for the chemical treatments during 22 days of storage at 0°C. Hue angle values during storage time were also significantly influenced by chemical treatments mainly at 0°C.

Keywords: storage, sensory quality, colour, texture, fresh-cut minimally processed cabbage, ascorbic acid, citric acid, calcium chloride

1. INTRODUCTION

Minimally processed vegetables have become widely accepted in restaurants, catering services, salad bars and consumer packs for home use.

The primary quality attributes of a fresh-cut vegetable include colour, texture, flavour and odour [1].

Minimally processing operations such as peeling, coring and cutting damage the tissue integrity resulting in cellular decompartmentation. Contact of enzymes and substrates lead to biochemical injuries such as browning, off-flavours and texture breakdown. Enzymatic browning and textural defects reduce product quality and shelf-life [2,3].

Degree of browning depends on phenol content, polyphenol oxidase (PPO) activity [4] and oxygen concentrations.

Brown discolorations in foods are complex because of the large number of secondary reactions that may occur [5].

Various approaches are now applied to limit enzymatic browning such as modified atmosphere [6], chemicals inhibitors and gaseous chlorine dioxide [7,8] or novel decontamination methods such as neutral electrolysed oxidising water [9]. Chemicals exhibit different modes of action on preventing enzymatic browning; the following groups may be used: reducing agents, chelating agents, inorganic salts.

Ascorbic acid is a moderate reducing component, acidic in nature. It forms neutral salts with bases and is water soluble. Ascorbic acid is generally recognized as safe (GRAS) for use in fruits and vegetables [10] and is probably the most widely used.

Citric acid is the best known chelating agent for use in fruits and vegetables [11,12] and is a phenolase Cu-chelating agent. The inhibition of polyphenol oxides (PPO) was attributed to the chelating action [13]. Citric acid is usually used in combination with other anti-browning agents [14].

Salts of calcium have been tested as anti-browning agents. Van Rensburg and Engelbrecht [15] studied the effect of calcium salts on susceptibility to browning of avocados. Varella et al. [16] proposed the use of minimally processed apples "Fuji" in a treatment with 1% $CaCl_2$ for 16 days storage at 8°C - 10°C. Kim and Klieber [17] reported that calcium chloride reduce browning in minimally processed Chinese cabbage.

Cabbage is an important member of cruciferous vegetables being also an important dietary vegetable consumed as fresh cut salad or cooked. Recently it was widely marketed as a minimally processed product. Brassica vegetables (cabbage, cauliflower, broccoli, Brussels sprouts, Kale) possess antioxidant and anticarcinogenic properties [18] and a large group of glycosinolates. Their hydrolysis products can protect against cancer [19].

Finally Rinaldi et al. [20] studied the shelf life of minimally processed cabbage stored under controlled atmospehere (CA) and found that CA did not prolong the shelf life of 10 days at 5°C. Rinaldi et al. [21] also reported that LDPE packaging was more adequate for fresh-cut cabbage storage.

The objective of this study was to investigate the potential of ascorbic acid, citric acid and calcium chloride, individually, to extend the shelf life and maintain the quality of fresh-cut cabbage.

2. MATERIALS AND METHODS

2.1. Raw Material and Treatments

White Cabbage (Brassica oleracea L. var. capitata) var. Bunner harvested in March and April from a farm located in Messinia (Peloponnese, Greece) was minimally processed using the best preparation techniques. The leaves were cut into quarters and then sliced into 3 cm wide strips with a sharp stainless steel knife sterilized in ethanol. The shredded cabbage was washed for 4 min in chlorinated (100 ppm NaOCl) water at 5°C (at a solid/ liquid ratio of 1:3) to remove released nutrients. Then the cabbage was rinsed with tap water for 4 min and centrifuged for 1 min at a speed of 500 rpm.

The treatments were conducted in parallel and prepared for the same batch of product. To minimize product heterogeneity, processed cabbage was pooled, mixed and subsequently divided in four equal parts. The first part was imbibed in distilled water and was used as a control. The second, third and fourth parts were imbibed in ascorbic acid, citric acid and calcium chloride solution, respectively.

Imbibition treatments with solutions of ascorbic acid (Serva, Germany), citric acid (Merck, Germany) and calcium chloride (Carlo Erba, Italy) were applied separately for 3 min, after draining. All dipping solutions were made with distilled water at 5°C and the concentrations were 1% (w/v) at all cases.

100 ± 2 g cabbage slices treated with ascorbic acid, citric acid, calcium chloride or distilled water (control) were placed in clear polystyrene trays (15.5 × 22.5 × 3 cm) covered with PVC film (13 μm) and stored at 0°C and 5°C and 90% E.R.H. in darkness.

Mass loss, instrumental texture, instrumental colour, total soluble solids (TSS), visual quality were evaluated at days 0, 7, 14 and 22. Six packs (replicates) were prepared for each sampling date and treatment. For mass loss ten packs per treatment were prepared.

Mass loss (%) was calculated as percentage of the initial mass of each sample using an electronic scale with an accuracy of ±0.01 g. Weighing was carried out in air conditioned area to avoid package condensation.

2.2. Texture Analysis

Firmness was evaluated using a Texture Analyser TAXT2i (SMS, UK) with a 100 N load cell attached. A Kramer shear cell with a five probe attached to the instrument was used. The speed of the probe was set to 1 $mm \cdot s^{-1}$. Firmness was evaluated in terms of maximum force (rupture force in N) recorded in a typical forcedistance diagram.

2.3. Colour

Colour was measured with a Minolta CR-300 chromameter (Minolta Corp. Japan) on CIE L*, a*, b* chromatic space. The degree of browning was analysed in terms of changes in L*, and hue angle values (h°).

Total soluble solids (TSS) was measured using a digital refractometer (SR-400) in a 10 g homogeneous and filtered cabbage pulp sample.

Quality changes (cut surface browning, black speck, microbial spoilage and overall visual acceptance) were evaluated in 6 packs per treatment, by sensory evaluation performed by 6 semi trained panellists using a score system from 1 to 5 (**Table 1**) with the score of 3 as the cut off for marketing.

2.4. Statistical Analysis

The experiment was performed according to a full factorial design (treatments × storage time) and subjected to an analysis of variance (ANOVA) using the Statgraphics

Table 1. Estimation of visual quality characteristics of minimally processed cabbage (Adapted from Kim and Klieber, 1997).

Characteristics	Score				
	5	4	3	2	1
Cut surface browning	None	Traces in <5 pieces	Very light in <10 pieces	Light in >10 pieces	Dark in >10 pieces
Black speck	None	Few small specks on <5 pieces	Some small specks on <10 pieces	Some large specks on <20 pieces	Many large specks on >20 pieces
Microbial spoilage	None	Traces in 1 piece	Traces in 2 - 5 pieces	Spoilage on 6 - 10 pieces	Spoilage on >10 pieces
Overall visual quality	Very fresh	Fresh	Marketable	Edible non marketable	Non edible.

Plus 5.1 (Statpoint Technologies, Inc, VA, USA) statistical software. Mean values were subjected to Fisher's Least Significant Difference test (LSD) at $p = 0.05$. This statistical test is liberal with respect to the comparison wise error rate, however, it is powerful for detecting true difference [22]. Data shown in the figures and tables are the mean values of two experimental series since no significant difference between the series was found.

3. RESULTS AND DISCUSSION

3.1. Weight Loss

The multifactor ANOVA indicated that storage temperature, chemical treatments and storage time affected ($p = 0.05$) weight loss. After 22 days at 0°C in air, weight loss (**Table 2**) was similar for the control and the treatments with ascorbic acid and calcium chloride (2.28% - 2.81%). The treatment with citric acid showed significantly higher weight loss (3.17%). Similar results were presented at 5°C. A linear correlation was found between weight loss and storage time ($R^2 = 0.985 - 0.995$).

Leafy vegetables are particularly vulnerable to rapid water loss [23]. The loss in moisture results in a reduction of fresh weight accompanied by the loss of freshness, appearance and texture. The maximum acceptable weight loss of cabbage is 7% [23,24]. In the present study, weight loss was below 7% at both storage temperatures (0°C, 5°C).

3.2. Texture

Firmness changes of minimal processed cabbage during storage at 0˚C and 5˚C depend upon the soaking solution and the time of storage (**Figure 1**).

The rupture values for minimally processed cabbage treated with different chemicals as well as these of control increased over the 22 days storage period. This increase varied between 2.2% - 5% at 0˚C for the control and the treatments with ascorbic and citric acid. At 5˚C the increase varied between 6% - 8% for the control and treatment with citric acid and 13% for the treatment with ascorbic acid. Calcium chloride had a greater firming effect ($p = 0.05$) at both storage temperatures compared to other treatments. This was due to cross-linking of both cell wall and middle-lamella pectin by calcium ions [25]. The texture of a plant food is affected by the genetic background, morphology, cell turgor, water content, biochemical components and growing conditions [26,27]. Treatments with ascorbic and citric acid did not result in tissue softening, results not agreeable to those reported by Gil et al. [28]. Cocci et al. [29] showed a structural breakdown with consequent softening in apple tissue.

Table 2. Weight loss of minimal processing cabbage treated with different chemicals after 22 days of storage at 0˚C and 5˚C.

Temperature	Ascorbic acid	Citric acid	Calcium chloride	control
0˚C	2.45^{Aa}	3.17^{Ab}	2.28^{Aa}	2.81^{Aa}
5˚C	3.63^{Ba}	4.2^{Bb}	3.35^{Ba}	3.22^{Ba}

Values followed by different superscript **lowercase** letters in the same row show statistically significant differences. $p = 0.05$; Different superscript **uppercase** letters show statistically significant differences in the same column (among temperatures). $p = 0.05$; Mean values. N = 2 repetitions × 10 samples per repetition = 20.

3.3. Colour

The L* colour value and hue angle (h˚) were selected as the most suitable parameters to measure the colour change of cut surface of minimally processed cabbage. When the colour became darker, a decrease in L* and hue angle (h˚) was observed. The hue value represents true colour which is effective for visualizing the colour appearance of food products [30]. The storage time and the interaction between storage time and temperature had significant effect on L*. Among the parameters which were studied (temperature, chemical treatments, storage time and their interactions) only temperature did not have a significant effect ($p = 0.05$) on hue values.

The lightness of minimally processed cabbage decreased linearly (**Figure 2**) from 70.94 ± 6 to 63.8 ± 8.5 −61.3 ± 8 for the chemical treatments during 22 days of storage at 0°C. The control sample presented a decrease of L* values to 59.4 ± 5.4.

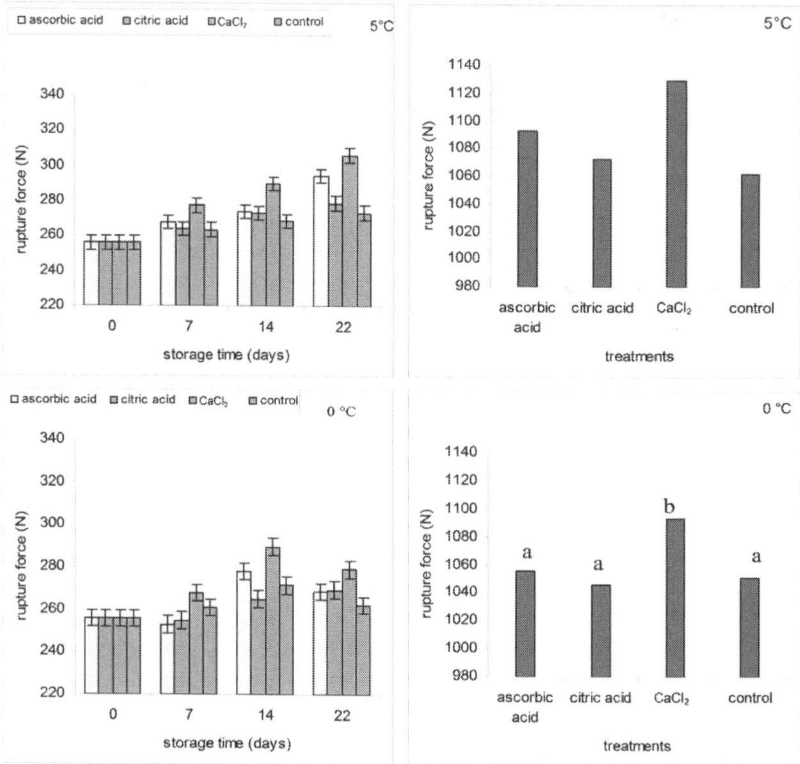

Figure 1. Effects of treatment with chemicals on the firmness of minimal processed cabbage stored at 0°C and 5°C. Data observed on the left set of histograms show the changes during the 22 days of storage (Mean values. N = 2 repetitions × 6 samples = 12). Vertical bars represent LSD at the 5% level. The data observed in the right set of histograms are the mean of all times.

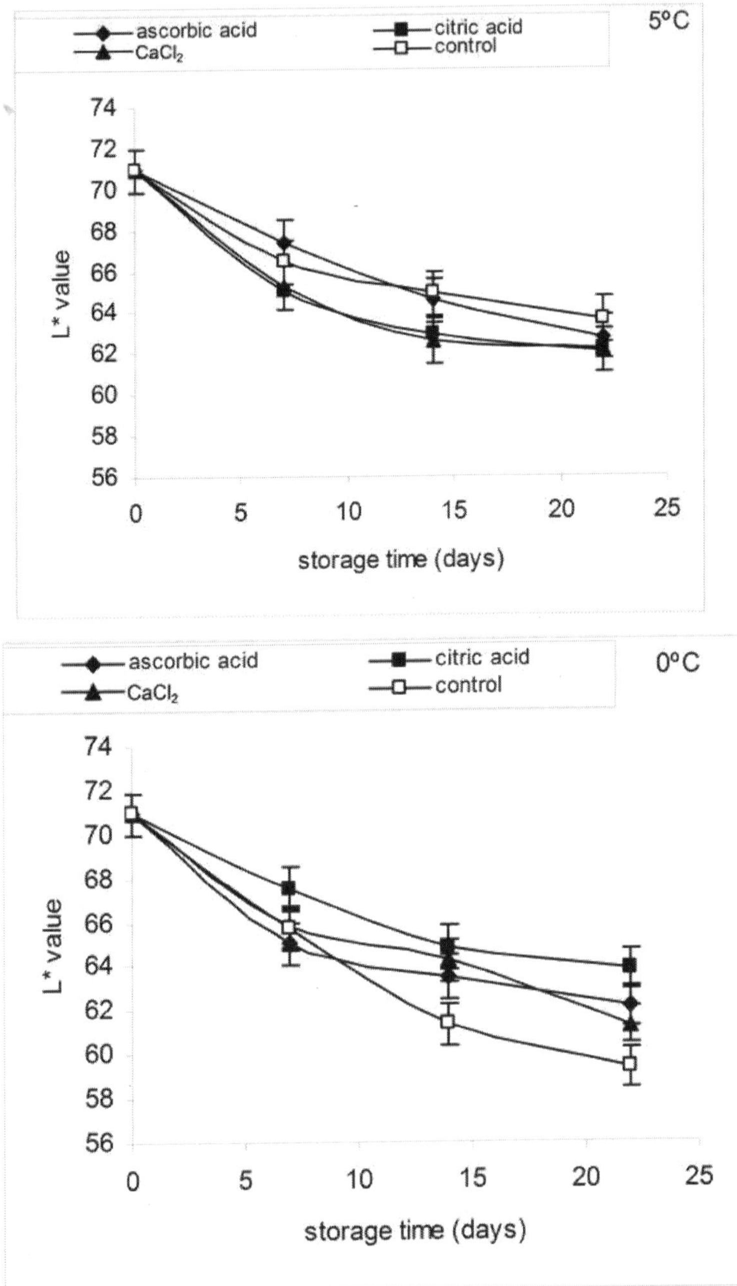

Figure 2. Effects of chemical treatments on changes in L* value of minimal processed cabbage stored at 0˚C and 5˚C for 22 days. Vertical bars represent LSD at the 5% level (Mean values. N = 2 repetitions × 10 samples = 20).

At 5˚C the lightness decreased from 70.94 ± 6 to 63.6 ± 7.9 - 62.07 ± 8.2 at all cases during 22 days of storage. By day 22, at 0˚C the L* value was lower in non treated cabbage (control) than in treated cabbage with chemicals. Among the chemical treatments citric acid was the best inhibitor showing the least change in lightness. At 5˚C, no significant difference in L* values between the control and treated samples was observed after 22 days of storage.

Hue angle values (**Figure 3**) during storage time were significantly affected by chemical treatments mainly at 0˚C.

Hue angle values of control decreased rapidly at 0˚C. Among the chemical treatments, citric acid retained better the color of fresh-cut cabbage. Citric acid has been reported extensively for its inhibitory activity on polyphenol oxidase, PPO [31] and its antibrowning activity in minimally processed fruits and vegetables [32]. Our results come in agreement with Kim and Klieber [17] who worked on minimally processed Chinese cabbage and Kaur and Kapoor [33] working on minimally processed Indian cabbage. Ascorbic acid and calcium chloride are intermediate inhibitors as seen in their hue angle (h˚) values (**Figure 3**). This result was similar to results of other researchers [16,34] working on banana slices and minimally processed apples. Ascorbic acid is a reducing agent widely used as an antibrowning agent for processing of fruits and vegetables. Reductants are irreversibly oxidized during the reaction, which means that the protection they confer is only temporary since they are consumed in the reaction [34]. The action of calcium chloride might be due to the PPO inhibition by chloride ions [35].

3.4. Soluble Solids Contents (SSC)

The soluble solids content in minimally processed cabbage during storage at 0˚C and 5˚C are presented in **Table 3**.

At 0˚C the control exhibited higher SSC and the cabbage treated with calcium chloride showed the smallest. At 5˚C the treatments with ascorbic acid and citric acid exhibited the highest SSC. Similar results were reported by Jiang et al. [36] in Chinese water chestnut. Gonzalez-Aguilar et al. [37] reported an appreciable increase in fructose and glucose in fresh-cut mangoes treated with antibrowning agents and stored at 10˚C.

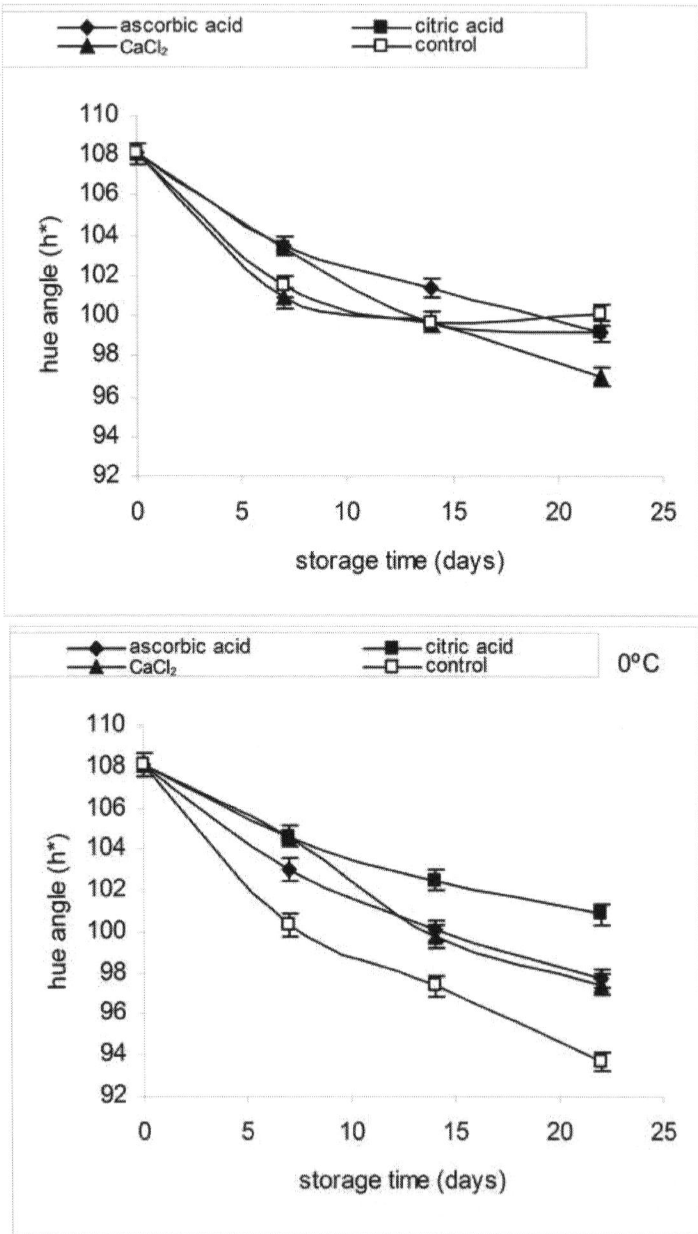

Figure 3. Hue angle changes of minimal processing cabbage stored at 0˚C and 5˚C for 22 days. Vertical bars represent LSD at the 5% level (Mean values. N = 2 repetitions × 10 samples = 20).

Table 3. Soluble solids content of minimal processed cabbage treated with different chemicals after 22 days of storage at 0˚C and 5˚C.

Temperature (˚C)	Time of storage (days)				
	0 day	22 days			
		Ascorbic acid	Citric acid	Calcium chloride	control
0	5.88	4.5^{aA}	4.4^{aA}	4.0^{bA}	4.9^{cA}
5	5.88	4.8^{aB}	4.4^{aA}	3.8^{bA}	3.9^{bB}

Values followed by different superscript **lowercase** letters in the same row show statistically significant differences. $p = 0.05$; Different superscript **uppercase** letters show statistically significant differences in the same column (among temperatures). $p = 0.05$; Mean values. N = 2 repetitions × 6 samples per repetition = 12.

3.5. Overall Organoleptic Quality

Organoleptic quality evaluation was carried out by human volunteers in conjunction with quantifiable physical and chemical measurements to determine if a shelf-life extension treatment has merit.

Overall acceptance together with cut surface browning determined the shelf life of control samples which was 7 days at both storage temperatures (**Table 4**) using a score of 3 as the limit. Treatment with citric acid extended the shelf life of shredded cabbage stored at 0˚C up to 22 days. During marketing the predominant temperature is 5˚C and at this temperature treatment with citric acid extended the shelf life from 7 days (control) to 14 days. Ascorbic acid maintained the overall quality for 14 days at 0˚C and 7 days at 5˚C, however, no difference was observed regarding cut surface browning compared to the control sample at both storage temperatures. Similar results were reported by Rosen and Kader [38] working with pear slices. Calcium chloride maintained the overall quality and cut surface browning for 14 days at both storage temperatures. Similar results in apples were reported by Drake and Spayed [39]. However, Bolin and Huxoll [40] reported that calcium chloride is not effecttive in preventing browning in peaches. Therefore, 1% citric acid can be used for minimally processed cabbage.

Citric acid is the most commonly used acidifier in combination with other types of antibrowning agents to control browning in fruits and vegetables.

During storage, the quality of minimally processed cabbage was affected by black speck. Black speck is a non-pathogenic disorder appearing during storage. It consists of very small to moderate size discolored lesions on the midrib and veins of the leaves [41]. At 0˚C citric acid reduced the effects of black speck at the end of shelf-life (22 days) resulting in a better quality. Citric acid may reduce black speck through its inhibitory effect on polyphenoloxidase by lowering the pH and resulting in the chelation of copper at the active site of the enzyme [42]. At 5˚C the chemical treatments affected black speck and reduced the incidence at the end of shelf-life (14 days).

Table 4. Changes in organoleptic quality properties of minimally processed cabbage with different chemical treatments during storage at 0˚C and 5˚C.

Treatments	Temperature 0˚C				Temperature 5˚C			
	Storage time (days)				Storage time (days)			
	0	7	14	22	0	7	14	22
Overall acceptance								
Ascorbic acid	5^{aA}	4.75^{aA}	3.8^{bA}	2.1^{cA}	5^{aA}	4^{bA}	2^{cA}	1.1^{dA}
Citric acid	5^{aA}	4.9^{aA}	4^{bA}	3.3^{cB}	5^{aA}	4.8^{aB}	3.3^{bB}	2^{cB}
Calcium chloride	5^{aA}	4.9^{aA}	3.4^{bA}	2.1^{cA}	5^{aA}	3.4^{bC}	3^{bB}	1.13^{cA}
control	5^{aA}	3.9^{bB}	2.8^{cB}	2.1^{dA}	5^{aA}	3.3^{bC}	2.4^{cC}	1.8^{dB}
Cut surface browning								
Ascorbic acid	5^{aA}	4.6^{aA}	2.5^{bA}	1.6^{cA}	5^{aA}	3.9^{bA}	2.5^{cA}	1.4^{dA}
Citric acid	5^{aA}	4.9^{aB}	3.5^{bB}	3^{cB}	5^{aA}	4.6^{aB}	3.1^{bB}	1.8^{cB}
Calcium chloride	5^{aA}	4.5^{aA}	3.5^{bB}	2.4^{cC}	5^{aA}	2.9^{bC}	3^{bB}	1^{cC}
control	5^{aA}	4.4^{bA}	2.6^{bA}	2.1^{cC}	5^{aA}	3.8^{bA}	2.5^{cA}	1.8^{dB}
Black speck								
Ascorbic acid	5^{aA}	5^{aA}	2.1^{bA}	1.9^{cA}	5^{aA}	4.8^{aA}	3.3^{bA}	1.6^{cA}
Citric acid	5^{aA}	5^{aA}	3.8^{bB}	3.1^{cB}	5^{aA}	5^{aA}	3.9^{bB}	2.8^{cB}
Calcium chloride	5^{aA}	5^{aA}	3.3^{bC}	2.6^{cC}	5^{aA}	5^{aA}	3.1^{bA}	1^{cC}
control	5^{aA}	5^{aA}	3^{bC}	2^{cA}	5^{aA}	5^{aA}	2.8^{bC}	2.3^{cD}
Microbial spoilage								
Ascorbic acid	5^{aA}	5^{aA}	3.6^{bA}	1.9^{cA}	5^{aA}	5^{aA}	3.3^{bA}	2^{cA}
Citric acid	5^{aA}	5^{aA}	4.6^{bB}	3.4^{cB}	5^{aA}	5^{aA}	3.8^{bB}	3.3^{cB}
Calcium chloride	5^{aA}	5^{aA}	4.1^{bA}	1^{cC}	5^{aA}	5^{aA}	3.3^{bA}	2.9^{cC}
control	5^{aA}	5^{aA}	3.9^{bA}	2.8^{cD}	5^{aA}	5^{aA}	2.9^{bA}	2^{cA}

Values followed by different superscript **lowercase** letters in the same row show statistically significant differences. $p = 0.05$; Different superscript **uppercase** letters show statistically significant differences in the same column (among treatments). $p = 0.05$; Mean values. N = 2 repetitions × 6 samples per repetition × 6 panellists = 72.

Citric acid at 0˚C and all chemical treatments at 5˚C prevented microbial spoilage at the end of shelf-life (22 and 14 days, respectively).

4. CONCLUSIONS

The results presented indicate that soaking with citric acid better retains the color and the overall acceptance and organoleptic quality of fresh cut cabbage. It can reduce the cut surface browning and protect from black speck. Citric acid treatment combined with low temperature storage (0˚C) prolonged the shelf life of minimally processed cabbage for 22 days, time sufficient for acceptable marketing.

REFERENCES

1. E. Watada and L. Qi, "Quality of Fresh-Cut Produce," Postharvest Biology and Technology, Vol. 15, No. 3, 1999, pp. 201-205.

2. D. King and H. R. Bolin, "Physiological and Microbiological Storage Stability of Minimally Processed Fruits and Vegetables," Food Technology, Vol. 43, No. 2, 1989, pp. 132-139.

3. J. K. Brecht, "Physiology of Lightly Processed Fruits and Vegetables," Horticultural Science, Vol. 30, No. 1, 1995, pp. 18-22.

4. J. Zawistowshy, C. G. Biliaderis and N. A. M. Eskin, "Polyphenol Oxidase," In: D. S. Robinson and N. A. M. Eskin, Eds., Oxidative Enzymes in Foods, Elsevier, London, 1991, pp. 217-273.

5. G. M. Sapers, "Browning of Foods: Control by Sulfites. Antioxidants and Other Means," Food Technology, Vol. 47, No. 10, 1993, pp. 75-84.

6. M. E. Pirovani, D. R. Güemes, A. M. Piagentini and J. H Di Pentina, "Storage Quality of Minimally Processed Cabbage Packaged in Plastic Films," Journal of Food Quality, Vol. 20, No. 5, 1997, pp. 381-389.

7. M. Friedman, "Food Browning and Its Prevention—A Review," Journal of Agricultural and Food Chemistry, Vol. 44, No. 3, 1996, pp. 631-653.

8. V. M. Gomez-Lopez, P. Ragaert, V. Jeyachchandran, J. Debevere and F. Devlieghere, "Shelf-Life of Minimally Processed Lettuce and Cabbage Treated with Gaseous Chlorine Dioxide and Cysteine," International Journal of Food Microbiology, Vol. 121, No. 1, 2008, pp. 74-83.

9. V. M. Gomez-Lopez, P. Ragaert, J. Ryckeboer, V. Jeyachchandran, J. Debevere and F. Devlieghere, "Shelf-Life of Minimally Processed Cabbage Treated with Neutral Electrolysed Oxidizing Water and Stored under Equilibrium Modified Atmosphere," International Journal of Food Microbiology Vol. 117, No. 1, 2007, pp. 91-98.

10. L. D. Alvarez and A. Chiralt, "Color of Minimally Processed Fruits and Vegetables as Affected by Some Chemical and Biochemical Changes," In: S. M. Alzamora, M. S. Tapia and A. Lopez-Malo Eds., Minimally Processed Fruits and Vegetables, Aspen Publishers, Frederick, 2000, pp. 111-126.

11. F. Pizzocaro, D. Torreggiani and G. Gilardi, "Inhibition of Apple Polyphenol Oxidase by Ascorbic Acid, Citric Acid and Sodium Chloride," Journal of Food Processing and Preservation Vol. 17, No. 1, 1993, pp. 21-30.

12. M. A. Rojas-Graü, Solina-Fortuny and O. Martin-Belloso, "Effect of Natural Antibrowning Agents on Color and Related Enzymes in Fresh-Cut 'Fuji' Apples as an Alternative to the Use Ascorbic Acid," Journal of Food Science, Vol. 73, 2008, pp. 267-272.

13. Y. M. Fu, J. R. Jiang, G. Zanberman and Y. Fuchs, "Purification of Polyphenol Oxidase and the Browning Control of Litchi Fruit by Glutathiose and Citric Acid," Journal of the Science of Food and Agriculture, Vol. 79, No. 7, 1999, pp. 950-954.

14. H. E. Moline, J. G. Buta and I. M. Newman, "Prevention of Browning of Banana Slices Using Natural Produce and Their Derivatives," Journal of Food Quality, Vol. 22, No. 5, 1999, pp. 499-511.

15. E. Van Rensburg and A. H. P. Engelbrecht, "Effect of Calcium Salts on Susceptibility to Browning of Avocados Fruit," Journal of Food Science, Vol. 51, No. 4, 1986, pp. 1067-1068.

16. P. Varela, A. Salvador and S. M. Fiszman, "The Use of Calcium Chloride in Minimally Processed Apples," European Food Research and Technology, Vol. 224, No. 4, 2007, pp. 461-467.

17. S. Kim and A. Klieber, "Quality Maintenance of Minimally Processed Chinese Cabbage with Low Temperature and Citric Acid Dip," Journal of the Science of Food and Agriculture, Vol. 75, No. 1, 1997, pp. 31-36.

18. Y. F. Chu, J. Sun, X. Wu and R. H. Liu, "Antioxidant and Anti Proliferative Activities of Common Vegetables," Journal of Agricultural and Food Chemistry, Vol. 50, No. 23, 2002, pp. 6910-6916.

19. Y. S. Keum, W. S. Jeong and A. N. T. Kong, "Chemoprevention by Isothiocyanates and Their Underlying Molecular Signalling Mechanisms," Mutation Research, Vol. 555, No. 1-2, 2004, pp. 191-202.

20. M. M. Rinaldi, B. C. Benedetti and C. L. Moretti, "Respiration Rate, Ethylene Production and Shelf Life of Minimally Processed Cabbage under Controlled Atmosphere," Acta Horticulturae, Vol. 876, 2010a, pp. 279- 286.

21. M. M. Rinaldi, C. I. G. L. Sarantopoulos, B. C. Benedetti and C. L. Moretti, "Storage of Minimally Processed Cabbage in Different Packaging Systems," Acta Horticulturae, Vol. 877, 2010b, pp. 597-602.

22. M. C. Gacula and J. Singh, "Statistical Methods in Food and Consumer Research," Academic Press Inc., New York, 1984.

23. S. J. Kays, "Postharvest Physiology of Perishable Plant Products," AVI Publisher, New York, 1991, pp. 356-357.

24. H. M. Kang, K. W. Park and M. E. Saltveit, "Elevated Growing Temperatures during the Day Improve the Postharvest Chilling Tolerance of Greenhouse—Grown Cucumber," Postharvest Biology and Technology Vol. 24, 2002, pp. 49-57.

25. W. Poovaich, "Role of Calcium in Prolonging Storage Life of Fruits and Vegetables," Food Technology, Vol. 40, No. 5, 1986, pp. 86-89.

26. F. R. Harker, R. J. Redgwell, I. C Hallet and S. H. Murray, "Texture of Fresh Fruit," Horticultural Reviews, Vol. 20, 1997, pp. 121-224.

27. E. Sams, "Preharvest Factors Affecting Postharvest Texture," Postharvest Biology and Technology, Vol. 15, No. 3, 1999, pp. 249-254.

28. M. I. Gil, J. R. Gorny and A. A. Kader, "Responses of 'Fuji' Apple Slices to Ascorbic Acid Treatments and Low Oxygen Atmospheres," Horticultural Science, Vol. 33, No. 2, 1998, pp. 305-309.

29. Cocci, P. Rocculi, S. Romani and M. Dalla Rosa, "Changes in Nutritional Properties of Minimally Processed Apples during Storage," Postharvest Biology and Technology, Vol. 39, No. 3, 2006, pp. 265-271.

30. R. G. Mcguire, "Reporting of Objective Colour Measurements," Horticultural Science, Vol. 27, No. 12, 1992, pp. 1254-1255.

31. T. T. Langdon, "Preventing of Browning in Fresh Prepared Potatoes without the Use of Sulfiting Agents," Food Technology, Vol. 41, No. 64, 1987, pp. 66-67.

32. R. Ahvenainen, "New Approaches in Improving the Shelf Life of Minimally Processed Fruit and Vegetables,". Trends in Food Science and Technology, Vol. 7, No. 6, 1996, pp. 179-187.

33. C. Kaur and H. C. Kapoor, "Antioxidant Activity and Quality of Minimally Processed Indian Cabbage (Brassica Oleracea var. Capitata)," Journal of Food Processing and Preservation, Vol. 25, No. 5, 2001, pp. 367-380.

34. M. Apintanapong, K. Cheachumluang, P. Suansawan and N. Thongpraset, "Effect of Antibrowning Agents on Banana Slices and Vacuum-Fried Slices," Journal of Food Agriculture and Environment, Vol. 5, No. 3-4, 2007, pp. 151-157.

35. Garcia and D. M. Barette, "Preservation Treatments for Fresh Cut Fruits and Vegetables," In: O. Lamikanra, Ed., Fresh Cut Fruits and Vegetables: Science, Technology and Market, CRC Press, Boca Raton, 2002, pp. 267-303.

36. Y. M. Jiang, L. T. Peng and J. R. Li, "Use of Citric Acid for Shelf Life and Quality Maintenance of Fresh-Cut Chinese Water Chestnut," Journal of Food Engineering, Vol. 63, No. 3, 2004, pp. 325-328.

37. A. Gonzalez-Aguilar, C. Y. Wang and J. C. Buta, "Maintaining Quality of Fresh-Cut Mangoes Using Antibrowning Agents and Modified Atmosphere Packaging," Journal of Agricultural and Food Chemistry, Vol. 48, No. 9, 2000, pp. 4204-4208.

38. J. C. Rosen and A. A. Kader, "Postharvest Physiology and Quality Maintenance of Sliced Pear and Strawberry Fruits," Journal of Food Science, Vol. 54, No. 3, 1989, pp. 656-659.

39. S. R. Drake and S. E. Spayed, "Influence of Calcium Treatment on 'Golden Delicious' Apple Quality," Journal of Food Science, Vol. 48, No. 2, 1983, pp. 403-405.

40. R. Bolin and C. C Huxoll, "Storage Stability of Minimally Processed Fruits," Journal of Food Processing and Preservation, Vol. 31, No. 4, 1989, pp. 281-292.

41. R. Walsh, E. C. Lougheed and P. M. A. Toivenon, "The Effect of Benomyl. Sodium Hypochlorite and Controlled Atmospheres upon the Incidence of Black Speck of Stored Cabbage," Journal of American Society of Horticultural Sciences, Vol. 108, No. 4, 1983, pp. 533-536.

42. R. Lyengar and A. J. Mcevily, "Anti-Browning Agents: Alternatives to the Use of Sulfites in Foods," Trends in Food Science and Technology, Vol. 3, 1992, pp. 60-64.

Chapter 13

Antioxidant Enrichment and Antimicrobial Protection of Fresh-Cut Mango Applying Bioactive Extracts from Their Seeds By-Products

Violeta Vega-Vega, Brenda A. Silva-Espinoza, Manuel Reynaldo Cruz-Valenzuela, Ariadna Thalía Bernal-Mercado, Gustavo A. González-Aguilar, Irasema Vargas-Arispuro, Consuelo G. Corrales-Maldonado, J. Fernando Ayala-Zavala*

Centro de Investigacion en Alimentacion y Desarrollo, Carretera a la Victoria Km 0.6, La Victoria, Hermosillo, Mexico.

ABSTRACT

The effect of the application of ethanolic extract of "Haden" mango seeds to increase the antioxidant and antimicrobial capacity of fresh-cut mango flesh was evaluated. Phenolic HPLC analysis revealed that gallic acid was the major identified compound in the ethanolic extract (586.68 mg/g). Fresh-cut mango was treated for immersion within the ethanolic extract, packed and stored at 5°C during 15 days. The content of phenolic compounds, antioxidant capacity and microbial growth of the treated mango were evaluated during the storage. Compared with the controls the treated fruit presented the highest content of the total phenolic (7.4 times), flavonoids (3.1 times), and antioxidant capacity DPPH$^{\bullet}$, TEAC and ORAC (2.9, 2.3 and 2.8 times, respectively), and showed a microbial reduction of 80% of mesophilic plate count and 97% of total molds. This study demonstrates the potential of phenolic compounds derived from mango seed as antimicrobials and antioxidants.

Keywords: Mangifera indica; Seed; Ethanolic Extract; Phenolic Compounds; Antioxidant Capacity; Antimicrobial Properties

1. INTRODUCTION

Mangoes are a good source of natural antioxidants. In addition to the usual nutrients, such as vitamins and minerals, mangoes are also rich in flavonoids and phenolic acids. Interest in the role of antioxidants in human health has promoted research in the field of food science to evaluate fruit and vegetable antioxidants and to determine whether their content and activity can be maintained or even improved through crop breeding, cultural practices, postharvest storage, fresh-cut processing, UV-C irradiation and the addition of plant extracts [1].

The antimicrobial and antioxidant activities of a variety of naturally occurring phenolics from different plant sources have been studied in detail [2-4]. The proposed mechanisms to explain the antimicrobial activity of these compounds include inhibition of extracellular microbial enzymes, deprivation of the substrates required for microbial growth, or direct action on microbial metabolism through inhibition of oxidative phosphorylation or iron deprivation [5,6]. The ability of phenolic compounds to quench free radicals arises because of both their ability to donate protons and to transfer electrons while remaining relatively stable [7]. Therefore, plant phenolic compounds present remarkable antimicrobial and antioxidant potential [6,8-10].

The phenolic content of byproducts obtained during the minimal processing of mango fruits has been analyzed. Fresh-cut mango is a demanded product, however, during its processing the generation of byproducts represents important economic and environmental issues [11]. Previously it has been observed that the byproducts from the varieties Haden and Ataulfo had the highest phenolic content, antioxidant and antimicrobial capacity [12]. Among tissues, seed showed the highest bioactive content and activity. The non polar fraccion of the ethanolic-extraction showed the highest phenolics, flavonoids, antioxidant and antimicrobial activity, followed by the methanolic-non-polar, methanolic-polar, ethanolic-polar fractions and water infusion, respectively. Combining factors, the ethanolicnon-polar extract obtained from seed and peel of mango “Ataulfo” and “Haden” had the highest bioactive properties. The quantities of mango seed byproducts yielded by its minimal processing are very high, which could represent an economic and environmental problem [13]. As an example in India, the mango processing industry annually generates 3×10^5 tons of seed approximately [13], representing the tissue from 9% to 40% by weight of the fruit. This demonstrates the potential that byproducts of fresh mango could have as a good source of functional extracts [12].

In this context, the objective of this study was to show the antimicrobial protection and the increment in the antioxidant capacity of fresh-cut mango (var. Haden) treated with ethanolic extract of their seed by-products.

2. MATERIALS AND METHODS

2.1. Sample Preparation

Mangoes variety "Haden" were obtained from a local market in the city of Hermosillo, Sonora, Mexico. Mangoes were selected according to homogenous color and free of physical defects in a commercial maturity stage. Fruits were washed with chlorinated water (250 ppm) for 3 min and air dried at 25°C, and the peel and seed were removed, the pulp was cut into cubes of 2 × 2 cm. Seeds were used for the extraction of phenolic compounds.

2.2. Preparation of Ethanolic Extract of Mango Seeds

Ethanolic extracts of mango seeds was performed in accordance to a previously reported technique that showed to generate a potent in vitro antimicrobial and antioxidant extract [12]. The seed was weighted (10 g) and placed in containers with 100 mL of ethanol: water (7:3). The sample was left to macerate in darkness for 10 days at 25°C. After that time, the extract was filtered and the solvent removed using a rotary evaporator at reduced pressure and temperature of 45°C. The aqueous fraction was lyophilized, forming the dry extract, which was subject to alkaline hydrolysis (10 mL of NaOH 4 M) for 4 h in the absence of light. Subsequently, an acid hydrolysis was performed with HCl 4 M taken every sample to pH 2. In the next step, the hydrolyzed extract was subject to liquid-liquid separation by 2 washes with 20 mL of ethyl acetate [14]. The ethyl acetate phase was used and evaporated; the obtained extract was re-suspended in de-ionized water to a concentration of 25 mg of extract per mL.

2.3. Identification and Quantification of the Major Phenolic Compound in the Extract of Mango Seeds

Identification and quantification of the major phenolic compound was completed by high performance liquid chromatography. We used a HP (1100) equipment with a quaternary pump and connected to a HP diode array detector at wavelengths of 280 and 320 nm was used. The equipment was controlled by Chem Station program for the acquisition and data analysis. The runs were performed with an analytical column Nucleosil 120 C-18 with dimensions of 250 × 4.6 mm, particle size of 5 micras. The used mobile phases were: 1% formic acid (A) and 100% acetonitrile (B), with a step gradient started with 98% formic acid and 2% acetonitrile, reaching 100% acetonitrile in 60 min at a flow of 0.5 mL/min [15]. The running time was 75 min and the injection volume of 20 μL of sample (0.3 mg). The identification was made by comparison of UV spectra, using a database previously made with reference substances. Quantification was performed using standard curves of the corresponding compounds and reported as mg of the compound/g dry weight (dw).

2.4. Application of Ethanolic Extract of Mango Seeds to Fresh-Cut Mango Flesh

The fresh-cut mango was divided into two groups: control and treated (25 mg/mL of ethanolic extract of mango seeds). The application of the extract was performed by immersion of 100 g of mango cubes for 2 min in the ethanolic extract and dried at room temperature for 30 s. The control and treated fresh-cut fruit were packed in polypropylene trays and stored at 5°C during 15 days. Sampling was performed every 5 days to assess the effect of the extract applied on the content of phenolic compounds, antioxidant capacity and microbial populations (mesophilic aerobic microorganisms, total fungi and yeasts).

2.5. Total Phenolic and Flavonoid Contents of Treated Mango Flesh

Total phenol and flavonoids compounds were extracted from fresh-cut mango tissue (10 g), which was homogenized in a 15 mL of methanol (80%) and sodium bisulfate (0.5%). The homogenate was sonicated 30 min at 1°C, centrifuged at 1200 g for 15 min at room temperature (Beckman Coulter Centrifuge, Allegra 64R) and filtering. This was repeated two times with 10 mL of methanol (80%), and the volume of pooled supernatant was brought up to 40 mL. Total phenolic content was measured by the method described by Singleton and Rossi [16] with some modifications. The samples of the previous extraction (50 μL) were mixed with 3 mL of H_2O and 250 μL of Folin-Ciocalteu's phenol reagent 1 N. After 8 min of equilibration time, 750 μL of Na_2CO_3 (20%) and 950 μL of H_2O were added to the mixture, and incubated for 30 min at room temperature. After incubation, the absorbance was read at 765 nm with an UV-Vis spectrophotometer (Cary, model 50 Bio, Varian, Italy). Concentration of total phenolic compounds was calculated using a standard curve of gallic acid equivalents (GAE) and expressed as milligrams per 100 g of fresh weight. Flavonoid content was determined based on the method described by Zhishen et al. [17] with slight modifications. One mL of each sample was mixed with 5% $NaNO_2$, 10% $AlCl_3$ and 1 mol/L NaOH and measured spectrophotometrically at 415 nm using quercetin as standard. The results were expressed as mg of quercetin equivalents (QE) per 100 g of fresh weight.

2.6. DPPH$^{\bullet}$ Radical Scavenging Activity

This assay is based on the measurement of the scavenging ability of antioxidants towards the stable radical DPPH$^{\bullet}$ (2,2-diphenyl-1-picrylhydrazyl) [18]. A 3.9 mL aliquot of a 0.0634 mM of DPPH$^{\bullet}$solution, in methanol was added to 0.1 mL of the methanolic extraction of mango flesh. The mixture was shaken in a vortex and kept 30 min in the dark. Absorbance was then read it in an UVVis spectrophotometer (Cary, model 50 Bio, Varian, Italy), at a wavelength of 515 nm. Results were expressed in DPPH$^{\bullet}$ inhibition %. Analyses were performed in triplicate.

2.7. Trolox Equivalent Antioxidant Capacity (TEAC)

This assay is based on the ability of the antioxidants to scavenge the blue-green ABTS$^{\bullet+}$ radical (2,2'-azino-bis(3- ethy-lbenzothiazoline-6-sulphonic acid) compared to the scavenging ability of the water-soluble vitamin E analogue Trolox [19]. The ABTS$^{\bullet+}$ radical cation was generated by the interaction of 5 mL of 7mM ABTS solution and 88 µL of 0.139 mM $K_2S_2O_8$ solution. After the addition of 2970 µL of ABTS solution to 30 µL of the methanolic extraction of mango flesh or trolox standards (0 to 20 µM range), the absorbance was monitored 1 and 6 min after the initial mixing. The percentage of absorbance inhibition at 734 nm was calculated and plotted as a function of that obtained for the extracts and the standard reference (trolox). The final TEAC value was calculated by using a regression equation between the trolox concentration and the inhibition percentage and expressed as trolox equivalents (µmol TE) per 100 g of fresh weight.

2.8. Oxygen Radical Absorbance Capacity Assay (ORAC)

This assay measures the effect of antioxidant components in the sample on the decline in fluorescein fluorescence induced by a peroxyl radical generator, 2',2'-azobis (2- amidino propane) dihydrochloride (AAPH) [18]. Different dilutions of Trolox (200 µM - 12.5 µM) and methanolic mango samples were tested. All solutions were prepared fresh daily. In every working well the following was pipetted in triplicate: Fluorescein, 150 µl of a 10 nM solution; for standard, 25 µl of Trolox dilution; for sample, 25 µl of sample dilution; for blank, 25 µl of phosphate buffer. The microplates were incubated at 37°C and fluorescence (Ex. 485 nm, Em. 520 nm) were taken every 90 s using a BMG Labtech's multimode plate reader FLUOstar Omega. After 3 cycles, 25 µl (240 mM) of AAPH was added manually with a multi-channel-pipette. The test was resumed and fluorescent measurements were taken up to 90 minutes. The final ORAC value of cut mango was calculated by using a regression equation between the trolox concentrations and the fluorescence intensity and expressed as µmol TE per 100 g of fresh weight.

2.9. Microbial Growth in the Treated Mango

Microbial growth was determined initially and during the storage period. Mangoes samples were diluted 1:9 in a solution containing 0.1 g/100 mL of peptone and 0.5 g/100 mL of sodium chloride and homogenized for 1 min, subsequently 10-fold dilutions were also made in this diluent. Each dilution was plated in duplicate for the different tested microorganisms. Aerobic plate count was incubated for 3 days at 30°C in plate count agar. Yeast and mold counts were incubated for 5 days at 25°C in potato dextrose agar. Total coliforms were measured by the solid medium method incubating samples for 24 h at 35°C using violet red bile agar. All analyses were performed according to the US Food and Drug Administration Bacteriological Analytical Manual methodology [20].

2.10. Statistical Analysis

The experimental design was completely randomized blocks, where the factors were the treatment (0 and 25 mg/mL of ethanolic seed extract) and the time, which was blocked to see the effect of the treatment on the phenolic compounds, antioxidant capacity and microbial growth in mango cubes. Analysis of variance ANOVA was performed ($p \leq 0.05$) to estimate significant differences between treatments, Tukey-Kramer was the mean test used for comparison ($p \leq 0.05$) performed with the Number Cruncher Statistical Systems [21].

3. RESULTS AND DISCUSSION

3.1. Identification and Quantification of the Major Phenolic Compound in the Ethanolic Extract of Mango Seeds

Figure 1 shows the chromatogram corresponding for the ethanolic extract of mango seed. Three compounds were found, of which one was identified, corresponding to gallic acid as the main component of the extract, in a concentration of 586.68 mg/g dw, and with a retention time of 15.07 min.

Previous studies by Abdalla et al. [22], reported that methanolic extracts obtained from the Egyptian mango seeds contain different phenolic compounds such as tannins (20.7 mg/100 g) and vainillin (20.2 mg/100 g). Also, was found gallic acid (6.0 mg/100 g), cumaric acid (12.6 mg/100 g), caffeic acid (7.7 mg/100 g), mangiferin (4.2 mg/100 g), ferulic acid (10.4 mg/100 g), cinnamic acid (11.2 mg/100 g) and unidentified compounds (7.1 mg/ 100 g). Puravankara et al. [23], identified six phenolic compounds in mango seed extracts, especially gallic acid, ellagic and gallates. While Arogba [24], reported the presence of gallotannins and condensed tannins.

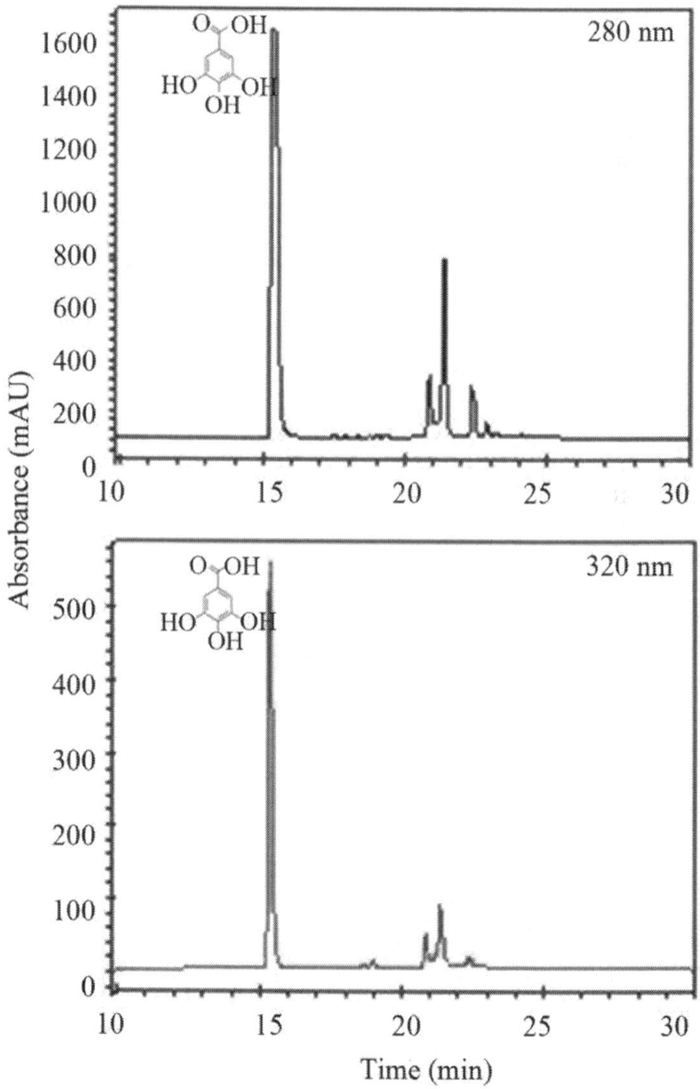

Figure 1. Phenolic compounds of the ethanolic extract of mango seed. Reverse phase chromatogram at 280 and 320 nm.

The differences in composition presented between the extracts reported in the literature and in the present study may be due variety, degree of ripeness of the fruit used and the method of extraction. An important factor is the solvent used for the extraction of phenolic compounds, because solvents with different polarity extracted different compounds in varying quantities [25]. Another factor to consider is the hydrolysis with acids and bases [26], a step that was performed to development of the ethanolic extract, in where the tannins present in the mango seed could be hydrolyzed to simpler compounds such as gallic acid.

3.2. Phenolic Compounds and Antioxidant Capacity of Fresh-Cut Mango

Figures 2 and 3 show the phenolic content and antioxidant capacity of fresh-cut mango untreated and treated with the ethanolic extract of seeds. A significant effect ($p < 0.05$) of the treatment on the phenolic content and antioxidant capacity of fresh-cut mango was found. Treated mango showed higher values of total phenolic and flavonoid compounds than the controls, being statistical differences between them. At the end of the storage, the the treated fresh-cut mango presented a highest content of the total phenolic (7.4 times), flavonoids (3.1 times), and antioxidant capacity evaluated for methods $DPPH^{\bullet}$, TEAC and ORAC (2.9, 2.3 and 2.8 times, respectively), compared to control. The ethanolic extract prepared from the mango seeds increased the antioxidant capacity of fresh-cut mango. In general, it was a positive correlation between phenolic content and antioxidant capacity, apparently such compounds are responsible for the antioxidant capacity in the mango fruit [27].

The usage of bioactive extracts applied to fruit presservation is an alternative to chemical preservatives and helps to achieve consumer demand for fresh, nutritious and safe fruits, and vegetables that are free of synthetic additives [4]. Presently, there are very few studies that provide information about the effect of bioactive compounds that are extracted from plant extracts and applied to fresh-cut fruits with the purpose of increasing the bioactive content [2,3,6,9,28]. However, the effect of antioxidant extracts obtained from fresh-cut mango byproducts as food preservatives has not been reported.

Some bioactive extracts have been proven to be effective antioxidants [4]. For example green tea extract has been evaluated as being able to act in the preservative treatment of fresh-cut lettuce. Different quality markers, such as respiration, browning, ascorbic acid, and carotenoid content were evaluated. Several tea extract concentrations (0.25, 0.5, and 1 g/100 mL) at different temperatures (20°C and 50°C) were tested. Optimal tea extract treatments (0.25 g/100 mL at 20°C) were compared with chlorine (120 ppm at 20°C). High tea extract concentrations (0.5 g/100 mL and 1.0 g/100 mL) prevented ascorbic acid and carotenoid losses. No significant differences were observed between chlorine and the optimal tea extract concentrations (0.25 g/100 mL at 20°C) in the browning appearance and sensory properties. Tea extract kept the antioxidant activity of the samples better than chlorine did [9].

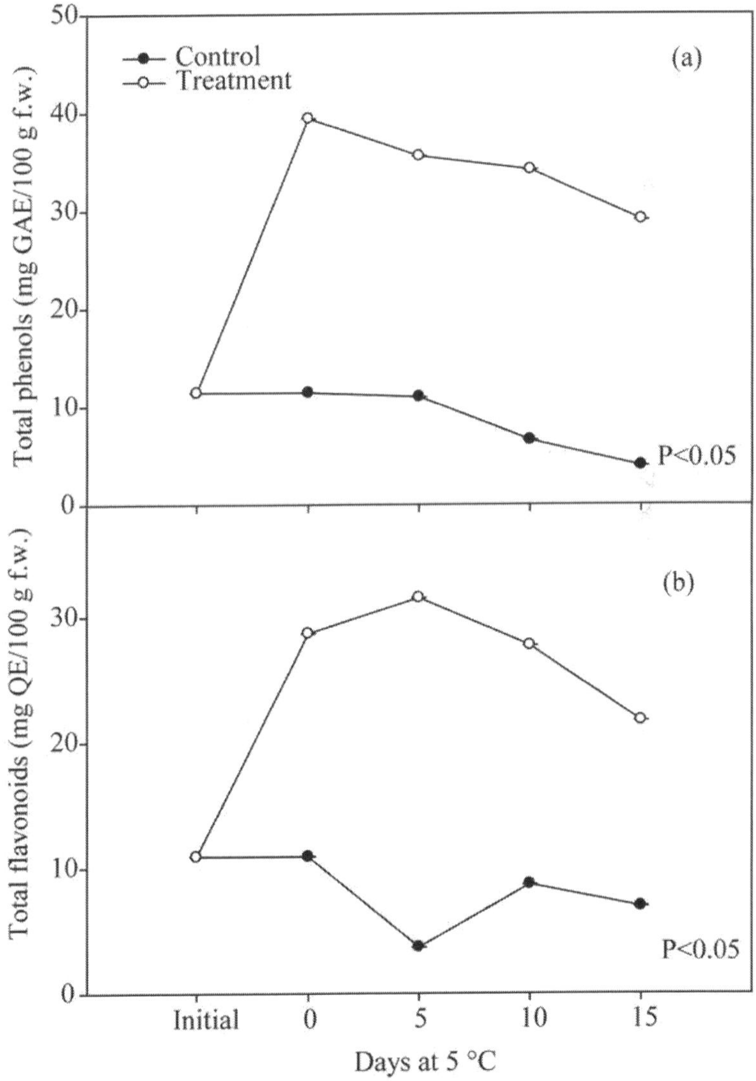

Figure 2. Changes on the total phenolic (a) and flavonoid (b) compounds of fresh-cut mango untreated and treated with the ethanolic extract of mango seeds, stored at 5°C during 15 days.

3.3. Microbial Growth on the Treated Mangoes

Table 1 shows the antimicrobial activity of fresh-cut mango untreated and treated with the ethanolic extract of seeds, stored at 5°C during 15 days. A significant effect ($p \leq 0.05$) of the treatment on the growth of the mesophilic aerobic microorganisms and total fungi and yeast was found. On days 0 and 5 microbial counts were under the limit of enumeration, however, on days 10 and 15 the control had a higher population of these microorganisms, being statistical differences amongst them. At the end of the storage, the treatment inhibited the growth of mesophilic aerobic microorganisms, total fungi and yeast causing an 80% and 97% of microbial reduction compared to control fruit.

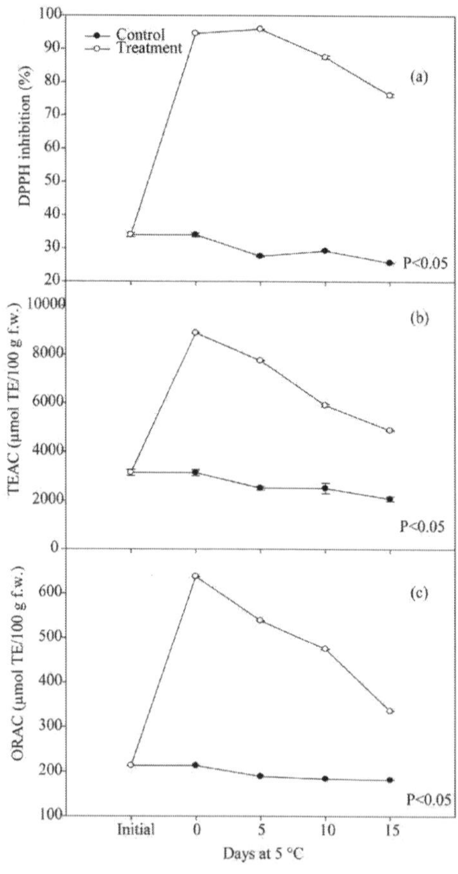

Figure 3. DPPH radical scavenging (a), TEAC (b) and ORAC (c) values of fresh-cut mango untreated and treated with the ethanolic extract of mango seeds, stored at 5°C during 15 days.

Table 1. Antimicrobial activity of fresh-cut mango untreated and treated with the ethanolic extract of mango seeds, stored at 5°C during 15 days.

Days	Mesophilic aerobic microorganisms		Total fungi and yeast	
	Control	Treatment	Control	Treatment
0	<250 ± 0*	<250 ± 0	<250 ± 0	<250 ± 0
5	<250 ± 0	<250 ± 0	<250 ± 0	<250 ± 0
10	19,000 ± 605	2000 ± 48	26,000 ± 774	<250 ± 0
15	121,000 ± 7315	23,000 ± 800	472,000 ± 341	14,000 ± 5567

*Average of three determinations ± standard error ($P < 0.05$).

No reports of the antioxidant capacity or the antimicrobial activity of extracts obtained from fruit byproducts as mango preservatives were found. But it has been reported that plant derived compounds like eugenol, thymol and carvacrol applied to grapes reduced growth of yeast, mold and mesophilic aerobics but the odor was detected after opening the package [10]. In addition, in a study with grapes wrapped in 2 distinct films having different permeabilities, and treated with or without the addition of a mixture of eugenol, thymol, and carvacrol [28], the microbial counts (of molds, yeasts, and mesophilic aerobics) drastically decreased, and consequently diminished berry decay.

Ethanolic extract of cinnamon bark (1% w/v) reduced the aerobic growth of bacteria inoculated fresh-cut apples significantly during storage at 6°C up to 12 d. Catechin, chlorogenic acid, and phloridzin, 3 phenolic compounds that are abundant in apple processing byproducts, exhibited varying degree of inhibitory action toward the growth of tested food pathogenic and spoilage bacteria, fungi, and yeasts [2]. However, it is important to note that these phenolics (except 25 mM phloridzin) did not inhibit the probiotic bacterium Lac. rhamnosus suggesting no or minimal threat to the beneficial colon micro-flora, if the phenolics are used as food additives at the desirable concentrations. Also these authors suggest that the major phenolic compounds of apple byproducts could find use as food additives. The use of natural, non-synthetic, additives is of great interest to the food industry due to consumer demand for products free of toxic residues. Therefore, there is a need to explore the possibility of using natural products with antioxidant and antimicrobial activity [29].

4. CONCLUSION

The application of ethanolic extract of mango seed byproducts from minimal processed mango enhance the increment of total phenolic content, antioxidant capacity, while inhibited the microbial growth of the treated mango flesh. This demonstrated the potential of phenolic compounds derived from mango seed as antimicrobials and antioxidants in the fresh-cut mango industry.

5. ACKNOWLEDGEMENTS

Authors thank the technical assistance of Monica Alejandra Villegas Ochoa and the financial support of the national council for science and technology CONACYT and the national program for the improvement of the professorate PROMEP.

REFERENCES

1. J. F. Ayala-Zavala, S. Y. Wang, C. Y. Wang and G. A. González-Aguilar, "Methyl Jasmonate in Conjunction with Ethanol Treatment Increases Antioxidant Capacity, Volatile Compounds and Postharvest Life of Strawberry Fruit," European Food Research and Technology, Vol. 221, No. 6, 2005, pp. 731-738.

2. S. Muthuswamy and H. P. V. Rupasinghe, "Fruit Phenolics as Natural Antimicrobial Agents: Selective Antimicrobial Activity of Catechin, Chlorogenic Acid and Phloridzin," International Journal of Food, Agriculture and Environment, Vol. 5, No. 3-4, 2007, pp. 81-85.

3. S. Muthuswamy, H. P. V. Rupasinghe and G. W. Stratton, "Antimicrobial Effect of Cinnamon Bark Extract on Escherichia coli O157:H7, Listeria innocua and Fresh-Cut Apple Slices," Journal of Food Safety, Vol. 28, No. 4, 2008, pp. 534-549.

4. J. F. Ayala-Zavala, C. Rosas-Domínguez, V. Vega-Vega and G. A. González Aguilar, "Antioxidant Enrichment and Antimicrobial Protection of Fresh Cut Fruits Using Their Own Byproducts: Looking for Integral Exploitation," Journal of Food Science, Vol. 75, No. 8, 2010, pp. 175- 181.

5. T. P. Cushnie and A. J. Lamb, "Antimicrobial Activity of Flavonoids," International Journal of Antimicrobial Agents, Vol. 26, No. 5, 2005, pp. 343-356.

6. R. M. Raybaudi-Massilia, J. Mosqueda-Melgar, R. Soliva-Fortuny and O. Martin-Belloso, "Control of Pathogenic and Spoilage Microorganisms in Fresh-Cut Fruits and Fruit Juices by Traditional and Alternative Natural Antimicrobials," Comprehensive Reviews in Food Science and Food Safety, Vol. 8, No. 3, 2009, pp. 157-180.

7. P. G. Pietta, "Flavonoids as Antioxidants," Journal of Natural Products, Vol. 63, No. 7, 2000, pp. 1035-1042.

8. S. Burt, "Essential Oils: Their Antibacterial Properties and Potential Applications in Foods-a Review," International Journal of Food Microbiology, Vol. 94, No. 3, 2004, pp. 223-253.

9. A. B. Martín-Diana, D. Rico and C. Barry-Ryan, "Green Tea Extract as a Natural Antioxidant to Extend the ShelfLife of Fresh-Cut Lettuce," Innovative Food Science and Emerging Technologies, Vol. 9, No. 4, 2008, pp. 593-603.

10. M. A. Rojas-Graü, R. M. Raybaudi-Massilia, R. C. Soliva-Fortuny, R. J. Avena-Bustillos, T. H. McHugh and O. Martín-Belloso, "Apple Puree-Alginate Edible Coating as Carrier of Antimicrobial Agents to Prolong Shelf-Life of Fresh-Cut Apples," Postharvest Biology and Technology, Vol. 45, No. 2, 2007, pp. 254-264.

11. Y. Kim, J. K. Brecht and S. T. Talcott, "Antioxidant Phytochemical and Fruit Quality Changes in Mango (Mangifera Indica L.) Following Hot Water Immersion and Controlled Atmosphere Storage," Food Chemistry, Vol. 105, No. 4, 2007, pp. 1327-1334.

12. V. Vega-Vega, "Enriquecimiento de la Capacidad Antioxidante y Protección Antimicrobiana del Mango Fresco Cortado Aplicando Compuestos Fenólicos de sus Subproductos," Master Thesis, Centro de Investigación en Alimentación y Desarrollo, A.C, Hermosillo, Sonora, Mexico, 2011. [Paper reference:3]

13. Y. Y. Soong and P. J. Barlow, "Antioxidant Activity and Phenolic Content of Selected Fruit Seeds," Food Chemistry, Vol. 88, No. 3, 2004, pp. 411-417.

14. G. Oboh and J. B. T. Rocha, "Distribution and Antioxidant Activity of Polyphenols in Ripe and Unripe Tree Pepper (Capsicum pubescens)," Journal of Food Biochemistry, Vol. 31, No. 4, 2007, pp. 456-473.

15. M. J. Simirgiotis, P. D. S. Caligari and G. Schmeda-Hirschmann, "Identification of Phenolic Compounds from the Fruits of the Mountain Papaya Vasconcellea pubescens A. Dc. Grown in Chile by Liquid ChromatographyUv Detection-Mass Spectrometry," Food Chemistry, Vol. 115, No. 2, 2009, pp. 775-784.

16. V. L. Singleton and J. A. Rossi, "Colorimetry of Total Phenolics with Phosphomolybdic-Phosphotungstic Acid Reagents," American Journal of Enology and Vitiviniculture, Vol. 16, No. 3, 1965, pp. 144-158.

17. J. Zhishen, T. Mengcheng and W. Jianming, "The Determination of Flavonoid Contents in Mulberry and Their Scavenging Effects on Superoxide Radicals," Food Chemistry, Vol. 64, No. 4, 1999, pp. 555-559.

18. J. F. Ayala-Zavala, S. Y. Wang, C. Y. Wang and G. A. González-Aguilar, "High Oxygen Treatment Increases Antioxidant Capacity and Postharvest Life of Strawberry Fruit," Food Technology and Biotechnology, Vol. 45, No. 2, 2007, pp. 166-173.

19. R. Re, N. Pellegrini, A. Proteggente, A. Pannala, M. Yang and C. Rice-Evans, "Antioxidant Activity Applying an Improved ABTS Radical Cation Decolorization Assay," Free Radical Biology & Medicine, Vol. 26, No. 9-10, 1999, pp. 1231-1237.

20. CFSAN/FDA, "Bacteriological Analytical Manual Online," 2001. http://www.fda.gov/Food/FoodScienceResearch/LaboratoryMethods/ucm2006949.htm

21. NCSS—Number Cruncher Statistical Systems, "Statistical Program for Windows," Kaysville, Utah, 2007.

22. A. E. M. Abdalla, S. M. Darwish, E. H. E. Ayad and R. M. El-Hamahmy, "Egyptian Mango By-Product 2: Antioxidant and Antimicrobial Activities of Extract and Oil from Mango Seed Kernel," Food Chemistry, Vol. 103, No. 4, 2007, pp. 1141-1152.

23. D. Puravankara, V. Boghra and R. S. Sharma, "Effect of Antioxidant Principles Isolated from Mango (Mangifera indica L.) Seed Kernels on Oxidative Stability of Buffalo Ghee (Butter Fat)," Journal of the Science of Food and Agriculture, Vol. 80, No. 4, 2000, pp. 522-526.

24. S. S. Arogba, "Mango (Mangifera indica) Kernel: Chromatographic Analysis of the Tannin, and Stability Study of the Associated Polyphenol Oxidase Activity," Journal of Food Composition and Analysis, Vol. 13, No. 2, 2000, pp. 149-156.

25. Y. C. Chen, J. T. Lin, S. C. Liu, P. S. Lu and D. J. Yang, "Composition of Flavonoids and Phenolic Acids in Lychee (Litchi chinensis Sonn.) Flower Extracts and Their Antioxidant Capacities Estimated with Human Ldl, Erythrocyte, and Blood Models," Journal of Food Science, Vol. 76, No. 5, 2011, pp. 724-728.

26. K. T. Chung, C. I. Wei and M. G. Johnson, "Are Tannins a Double-Edged Sword in Biology and Health?" Trends in Food Science & Technology, Vol. 9, No. 4, 1998, pp. 168-175.

27. S. M. R. Ribeiro, L. C. A. Barbosa, J. H. Queiroz, M. Knödler and A. Schieber, "Phenolic Compounds and Antioxidant Capacity of Brazilian Mango (Mangifera indica L.) Varieties," Food Chemistry, Vol. 110, No. 3, 2008, pp. 620-626.

28. F. Guillen, P. J. Zapata, D. Martinez-Romero, S. Castillo, M. Serrano and D. Valero, "Improvement of the Overall Quality of Table Grapes Stored under Modified Atmosphere Packaging in Combination with Natural Antimicrobial Compounds," Journal of Food Science, Vol. 72, No. 3, 2007, p. 185.

29. K. N. Seneviratne and R. T. Kotuwegedara, "Antioxidant Activities of the Phenolic Extracts of Seed Oils and Seed Hulls of Five Plant Species," Food Science and Technology International, Vol. 15, No. 5, 2009, pp. 419-425.

Index